FIVE SUNS

STEPHEN J. PYNE

FIVE SUNS

A FIRE HISTORY OF MEXICO

THE UNIVERSITY OF
ARIZONA PRESS
TUCSON

The University of Arizona Press
www.uapress.arizona.edu

We respectfully acknowledge the University of Arizona is on the land and territories of Indigenous peoples. Today, Arizona is home to twenty-two federally recognized tribes, with Tucson being home to the O'odham and the Yaqui. Committed to diversity and inclusion, the University strives to build sustainable relationships with sovereign Native Nations and Indigenous communities through education offerings, partnerships, and community service.

ISBN-13: 978-0-8165-5339-6 (hardcover)
ISBN-13: 978-0-8165-5340-2 (ebook)

Cover design by Leigh McDonald
Cover art by Angel Martínez Martinez
Typeset by Leigh McDonald in Warnock Pro 10.5/14, Brandon Grotesque and Kilroy WF (display)

Library of Congress Cataloging-in-Publication Data
Names: Pyne, Stephen J., 1949– author.
Title: Five suns : a fire history of Mexico / Stephen J. Pyne.
Description: Tucson : University of Arizona Press, 2024. | Includes bibliographical references and index.
Identifiers: LCCN 2023031861 (print) | LCCN 2023031862 (ebook) | ISBN 9780816553396 (hardcover) | ISBN 9780816553402 (ebook)
Subjects: LCSH: Fires—Mexico—History. | Wildfires—Mexico—History. | Fire ecology—Mexico.
Classification: LCC SD421.34.M48 P96 2024 (print) | LCC SD421.34.M48 (ebook) | DDC 577.2/40972—dc23/eng/20231212
LC record available at https://lccn.loc.gov/2023031861
LC ebook record available at https://lccn.loc.gov/2023031862

Printed in the United States of America
♾ This paper meets the requirements of ANSI/NISO Z39.48-1992 (Permanence of Paper).

For Sonja

a new fire that never went out

Behold what was done when the years were bound—
when was reached the time when they were to draw the new fire . . .

—BERNARDINO DE SAHAGÚN,
GENERAL HISTORY OF THE THINGS OF NEW SPAIN
(FLORENTINE CODEX), 1545–90

CONTENTS

Prologue: New Fire 1
First Sun: Fire and Earth 7
Second Sun: Fire and Water 21
Third Sun: Fire and Sword 53
Fourth Sun: Fire and Iron 117
Fifth Sun: Fire and Air 223
Epilogue: Between Three Fires 343

Author's Note *347*
Abbreviations *351*
Notes *353*
Credits *405*
Index *407*
Illustrations follow page *107*

FIVE SUNS

PROLOGUE

NEW FIRE

IT BEGINS *in darkness, closing one world, and ends in light, opening another. The focal point between those worlds, old and new, is fire.*

Every 52 years the two calendars that guided the Aztec universe met when the Pleiades reached their zenith. One calendar, of 365 days, recorded secular time; the other, of 260 days, ritual time. The years between their merging were bundled like firewood. The overlay of calendars marks the potential ending and starting of worlds; four times the world had ended and then rekindled, each ruin and rise coinciding with the crossing of the two calendars. The current world (or "sun") originated when, amid the death of the fourth world, Nanahuatzin and Tecuciztecatl threw themselves onto a pyre, followed by an eagle and a jaguar. By their sacrifice they reappear as sun and moon to brighten the fifth world.

The New Fire Ceremony occurred at those moments when the two calendars overlapped, when the world tottered between wreckage and renewal. Like the greater cosmos of which it was a microcosm, the ritual commenced with total darkness as five days before its onset every fire throughout the dominion was extinguished. Temple fires, royal fires, fires in shops, fires in households—all went out. Wood and stone statues of the gods were tossed into water; cooking utensils from pestles to hearthstones were dumped into rubbish pits; ashes, along with everything else loose, were swept away. "Everywhere there was much sweeping." The old world was cleansed.

As the last day of the great cycle approached and "it came to pass that night fell," Bernardino de Sahagún recorded, "all were frightened and filled with dread . . . it was claimed that if fire could not be drawn, then the sun would be destroyed forever. . . . Night would prevail forever and the demons of darkness would descend, to eat men." The climax came on the Cerro de la Estrella (or Huixachtlan), a volcanic dome overlooking the Valley of Mexico: at midnight, when "the night divided in half," the fire priest of Copulco loomed over a sacrificial victim, a captured well-born warrior, laid out on a stone altar. Upon his chest, using a fire drill, he kindled a blaze, then with an obsidian knife he cut out the victim's heart and tossed it into the flames; the body was immolated. The rite thus repeated the origin myth of the fifth sun. The drawing of the fire assured the continuance of the world.[1]

The source fire (tlepili) *was carried first to the Templo Mayor and placed in the fire holder of Huitzilopochtli, then was brought in a downward hierarchy to the lesser temples and the schools of the Aztec nobility; carried by runners specially selected for their speed and endurance, it sped outward to the fire priests of every village and from there to every household. At noon the next day the sacrifice was repeated over and over, perhaps as many as 400 times, as the sun rose to its height. The midnight fire met the noonday sun. A new fire, a new sun, a renewed world.*

The symbolism is so perfect in its connections it almost seems a parody. The ceremony joins the three realms of existence—heaven, earth, and underworld. Its incense rises to the sky, its fire sparks the sun. Its altar occupies the axis of the four cardinal directions, linking all the lands. Its ritual rekindling binds all the orders of society, from lowly household hearths to lordly temples. It penetrates into the *inframundo* to prompt new growth from the soil, releasing the dead to life. It re-creates the slashing and burning that was the essence of agriculture. It bundles years like firewood, such that its flames start and end worlds the way hearth fires start and end days. It confirms the order of the universe, and the legitimacy of those who rule. Five times a new sun had been created to preside over a new world. It is, in all its symbolism, a cameo of the cosmos.

That cosmology was chiseled in stone in the form of the Aztec solar calendar created in 1479, when the Fifth Sun emerged at Teotihuacan.[2]

At the center of its concentric rings sits Tonatiuh, the sun god, the Fifth Sun, as he passes through the day. Around that core reside the four previous suns, representing the passage of earlier epochs. Circling them are the 20 days that make a Mesoamerican month. They are, in turn, encircled by the five elements, populated by glyphs that "resemble the symbol of the old Fire God, Xiuhtecuhtli-Huehuetéotl." Around that ring, completing the cycle-circle, are two fire serpents, Xiuhcoatls, that, emulating the sun, begin their journey in the east and end in the west, thus sealing off the cosmogony of sun and time.[3]

Fire is present throughout: it features in almost every figure, event, and symbol. The god of the sun is also a god of fire. Each new sun is birthed from fire. The Old God encompasses the days, months, years, and epochs as they wheel through space and time. The fire serpent is the weapon of Xiuhtecuhtli, who appears to carry the Old God. In that fusion he is also the Lord of the Year. From solar core to the circumference of the universe there is some feature of fire.[4]

The immense sun stone itself rested in the temple of Huitzilopochtli at the Templo Mayor. It thus resided at the center of the political world whose cosmology it expresses.[5]

A new fire ceremony is not unique to the Valley of Mexico. Versions were common throughout Mesoamerica, and for that matter throughout the world. Europe knew several versions, most familiarly the needfire, kindled during times of crisis. In Mexico a succession of ruling peoples added to and reorganized what existed. For households, probably not much changed, perhaps over millennia. For empires, the ruling peoples relocated and rewrote the rite to suit their tastes, theology, and seat of power.

The new fire rite originated, it seems, "in time immemorial." Aztec lore describes their emergence as one of the seven tribes of Chichimec from a cave in a hill on whose summit the First Fire appeared. Fire and

Aztecs thus shared a common creation story. The New Fire Ceremony itself apparently appeared at Teotihuacan, and was long celebrated at the Cerro de la Estrella, perhaps since Early Classic times. It then disseminated throughout the cultural and political imperium of the Toltecs, notably those sites that echoed Teotihuacan; at Tula, at Becán, at Zaculeu, at Xochicalco, at Chichén Itzá, where the 52-year cycle merged with the Mayan Long Count into a doubled 110-year Old Age. The triumphant Aztecs merged it with their own legends, as they did with their fire gods, thus affirming their claim to be the Toltecs' rightful successors. With the founding of Tenochtitlan, the New Fire Ceremony was held at Chapultepec, then at the Templo Mayor, and finally in 1507 CE at the Cerro de la Estrella. As the empire expanded, the rite was relocated to each new imperial center.[6]

The New Fire was preserved with perpetual fires maintained in sacred *braseros* (some 600 may have existed in the capital alone), which likely doubled as a public utility where household fires could be rekindled. Households replicated that arrangement with small, permanent fires, the practical and spiritual center of the home. The household fire was the center of daily life, as the temple fire was the center of the life of the people, and the New Fire the center of the world.[7]

So also the ceremony spawned secondary rituals that absorbed and emulated it and carried fire into other aspects of life; these rituals joined nature and society. The festival of Xocotl Huetzi celebrated fire in the cycle of agriculture, ensuring the sun would follow the rains. The festival of Huey Tozoztli, principally dedicated to the gods of rain, incorporated aspects of fire in the context of a sacred tree (*tota*) and post (*xocotl*). The annual festival of Izcalli included two fire ceremonies that, on a smaller scale, replayed the New Fire. One offered plants (maize), the other animals. The role of fire in helping provide for and then cook food united nature with people. The month of Izcalli, mid-January to mid-February, especially celebrated fire. The fire rites took on particular significance in a 4-year cycle, then became cosmically important with the 52-year cycle that led to a New Fire.[8]

Yet there was hardly any ceremony that did not include fire in some way, if only to illuminate and assist the transformation of raw to cooked, or profane to sacred. Fire cooked food; it cooked landscapes; it could, in a sense, cook the world. (The first portion of cooked food was tossed into the hearth in appreciation of what fire made possible.) Fire was always there, as it was in Mexican life and nature. Sometimes it might be the special concern of a rite, but inevitably it was present as an enabler, a catalyst, ready to accept whatever meaning priest or peasant wished to invest it with. The New Fire Ceremony was unique in focusing so exclusively on fire, though even here it is not fire as fire that is worshipped but fire as an indispensable catalyst in a ritual renewal of the world.[9]

As with ritual, so with religion: there were many fire gods, or many versions of an ancestral god, each slightly different, each preserving a history of contact among peoples. They were layered, they overlapped. They were, as with the New Fire rite, a bundling of peoples and years. Much as the New Fire was distributed through the land, so the ancestral god found himself parceled out to a pantheon of deities, great and small.

The core rite built on a core deity, Huehueteotl, the Old God. His effigy shows an ancient man, as old as time, bearded and often toothless, with a brazier on his shoulder, which he bears like a Mexican Atlas. Early effigies appear in Cuicuilco and Monte Albán: he is indeed the ancient god. As the fire he carries spreads, so his hybrid forms acquire new identities. Like the hero of a thousand faces, he became the god of a thousand names. Like fire he was a shapeshifter, taking his identity from his context. He was associated with time, the North Star, the Lords of the Night, the Sun. Turquoise was his stone. His symbols were flint and sticks that could be rubbed together to re-create him. Interestingly, he had many associations with water—the other vital ingredient for agriculture.

The Aztecs interbred him with affairs both routine and profound. He appears as Huitzilopochtli, their patron god, born with a burning weapon. He is Xiuhcoatl, a god of fire and war; he was Xiuhtecuhtli, god of fire and time. He was Teteo Innan Teteo Inta (mother and father of gods); Tocenta (the unitary father); Huehue (elder); Ixcozauhqui (he who has a yellow face); Tlamacazqui (yellow priest); Ocopilli (noble of the pine); Tlahuizcalpantecuhtli (lord of the place of the house of twilight); Chantico (a domestic fire goddess); and a score of others. He was a god of volcanoes. He linked the world of the dead with that of the living. He was the emblem of transformation. He was cause, catalyst, and companion.[10]

He was truly the *dios viejo*, the Old God, the patriarchal god, the father of humanity, the god whose ceremony marked the intersection of the three realms and the four cardinal directions, the god who bound time, the god who joined hearth to sun, who tied the cycle of quotidian life to the cycles of worlds broken and reborn. He was both common and cosmic. He occupied the central axis of the Earth. He was a unifying presence for a complex, syncretic cosmogony, integrating beliefs and rituals as he did natural processes. Mostly he was always, in some form, just there.

To those accustomed to Judeo-Christian tradition, the Old God makes an odd deity. He is not a Yahweh or his Christian successor—all-powerful, all-knowing, singular, with no other god before him. Huehueteotl is a god who glows in the background and appears when needed, who connects and acts in concert with others, open to many hybrid identities depending on circumstances, yet seemingly everywhere and through all time. He is omnipresent, but not omnipotent. He expresses himself through how his flames interact with the world around him. He is not one Form, immutable, but many, as conditions warrant.

He is a god of fire who resembles the way real fire works in the lands that became Mexico and among the people who became Mexicans. As a hearth, fire organizes the house; as a tool, it makes agriculture possible, sparking from the cycle of drying and wetting the new growth that will feed nature and people; as ritual, it invests the world with meaning. It bundles the pieces of nature and the firewood of time to renew lands and epochs. With fire, the rain, sun, wind, grasses, trees, and humans can renew and continue. Without fire, Mexico would face the endless dark and its demons.

Just as there is no reigning fire god, so there is no single fire, or fire regime, or fire-catalyzed landscape. In Mexico a nominally simple story fragments across terrain and time. It is customary to characterize landscapes, particularly those shaped by fire, in terms of mosaics. Pieces join with pieces to make larger assemblages that join with others. Fire puts those pieces in motion. But in Mexico, there is a terrain of time as well as of space. The mosaic becomes a kaleidoscope. Like peoples, their gods, and their temples, the pieces may come and go, reconfigured through history. What remains—patient, ancient, persistent—is fire.

FIRST SUN

FIRE AND EARTH

NEW FIRE

CHICXULUB

IT IS a common practice, in many cultures, to give the name of fire to anything that emanates heat and light. The sun, the stars, and comets are celestial fire; lightning, electrical fire; volcanoes, terrestrial fire. Some 65 million years ago those definitions converged when an asteroid 7.5 miles in diameter, slightly smaller than San Francisco, slammed into the Yucatán Peninsula, left a crater 125 miles wide and 25 to 30 miles deep, and spewed smoke along with 325 gigatons of sulfur into the atmosphere.[1]

The Chicxulub event ripped the heart out of one world and kindled another. The collision sparked tsunamis and fires, and plunged the planet into a small glacial epoch. Most dinosaurs died out; enough mammals survived to slowly replace them. After the initial flash ignition, fires feeding on disturbed or dying vegetation spread year after year, to all corners. A scrim of iridium identifies the asteroid. A layer of charcoal marks the pivot between worlds. Together they trace the planet's greatest recorded example of smash and burn.

It was a world-ending, world-renewing event, the first of the New Fires that have shaped Mexico. Its legacy is still visible regionally in rocks and landforms. It's a reminder that geologically Mexico is chronically disturbed and episodically unsettled by massive eruptions, a condition that is also true for its biota and its human societies, and it's a reminder that fire has likely been present throughout. In Mexico even the stones tell a story of fire.

LANDED MEXICO

In its beginning, as today, Mexico's saga is about land, always about land, for land is the matrix and arena for everything else. In the lands that have become the political entity of Mexico, informing principles compete and antitheses merge into mash-ups; even a dominant process will fracture and fuse as it expresses itself across space and time. Earth, air, water, and life—each element undergoes its own peculiar *mestizaje*, the mixing, sorting, and layering that seems inescapable in Mexican geography and history. Ideas and institutions, no less than biotas, fracture and hybridize as each encounters the splintering effects that result from Mexico's rocky substrate. The Mexican mural fragments into a mosaic.

Over eons Mexico's rocky substrate has been pulled, pushed, squeezed, and occasionally blasted. A swarm of tectonic plates, large and small, converges on Mesoamerica from all directions and leaves Mexico as the largest in a series of steplike anglings that track the long chain of mountains beginning in Panama and ending in Southern California. The cordillera bends and narrows at Tehuantepec, then spreads and splits into mountain ranges that, running northward, inscribe the eastern and western flanks of the country, leaving a high, broad plateau between them. The congestion of stresses and strains makes for a complicated history, an active landscape, and a largely elevated country, with a great swath of volcanoes, the Trans-Mexican Volcanic Belt, that dapples the center and seems to hold the contrasting geologies of north and south like a belt of rivets.[2]

Some odd symmetries result. North and south: the mountainous country is wide in the north and narrow in the south. East and west: above the volcanic belt the cordillera divides into two long-trending sierras, east and west, until farther north the ranges splinter into isolated sky islands. Northwest and southeast: two oppositely structured peninsulas, like charged poles of a magnet, frame the country diagonally. In the northwest, Baja California offers a narrow miniature of Mexico, with a central range of mountains. In the southeast, the Yucatán Peninsula features a vast limestone shelf, alternately flooded and exposed, joined to and sundered from Cuba by changing sea levels.

Most of Mexico formed during the Mesozoic, some 245 to 65 million years ago. But the underlying dynamics remain active, and they concentrate in the east-west-trending Trans-Mexican Volcanic Belt, which traverses the country from the north of Veracruz on the Gulf to Puerto Vallarta on the Pacific. In the east, the belt consists largely of high mountains with shallow basins, notably the Valley of Mexico. In the west, it displays deep basins with sentinel mountains, most famously Lake Chapala. Depending on perspective, a lot either converges into or diverges out of this exceptionally active zone.

However imagined, it's a zone of tension, of strains and stresses that can literally move mountains. Earthquakes are common, and volcanoes vigorous. Since 1910 Mexico has suffered five quakes comparable to the 1906 earthquake that left San Francisco, California, in ruins (Richter 8.0). Mount Parícutin rose out of a cornfield in 1943. El Chichón erupted in Chiapas in 1982, the second-largest eruption on twentieth-century Earth. There is a lot of violence latent below the surface that frequently rumbles and erupts onto the surface. *México profundo* is a place where the present and deep time brush against each other and occasionally interpenetrate.

Mexico the country is an accident of human history. It has been both less and more than it is today as a result of migration and conquest. But it has always had a core where people and power have concentrated, each successor overlying the other, like layers of flood silt, every new arrival marked by the charcoal of fire and sword. A geologic center corresponds to the Trans-Mexican Volcanic Belt. A geographic center resides near Zacatecas. A historic center lies between the Valley of Mexico and Guadalajara, dividing the country between the Gran Chichimeca of the north and the sedentary civilizations of the south—the largest pairing of the many Mexicos and the basis for the dual "personality" of Mexico famously identified by Carl Sauer. Mexico's striking geomorphic polarities emphasize a fundamental paradox of Mexican history: the power of the Mexican earth to reconcile antitheses not so much by a melting-pot synthesis as by a mash-up amid often unpredictable, if ceaseless, upheavals.[3]

Mexico the landscape has its pulls and pushes. Its textured terrain of rock, mountains, tablelands high and low, faults, and subterranean magma fractures the larger movements of air, water, flora, fauna, and of course people. Its geologic activism created the mineral wealth that would, upon European contact, inspire invasion and fund a Spanish empire, and later, with deposits of coal and especially petroleum, power Mexico's push to industrialize. Yet the absence of navigable rivers, the frost zones in the mountainous tropics, a climate of extremes rather than averages, the subterranean hydrology of the Yucatán, the intricate carving of mountain mass into nooks and niches—all frustrated any simple process that humans might wish to impose. That deep tension accounts for the country's fabled megadiversity even as it endlessly challenges the center to hold.

WEATHERED MEXICO

By latitude Mexico lies between a temperate north and a tropical south—the exact border shifting with Earth's seasonal orientations to the sun. Roughly half of the country's landmass lies north of the Tropic of Cancer, a line that runs from Ciudad Victoria to Mazatlán; Tijuana occupies the same latitude as Dallas, and Monterrey is a hair below the southern tips of Texas and Florida. In winter, the temperate realm moves south, blustering with rains and *nortes*, sometimes reaching as far as central Mexico. In summer, the tropical realm moves north, pushing the Mexican monsoon into the U.S. Southwest along with stubborn dry spells. Those seasonal shiftings of the temperate-tropical border provide the winds to move moisture over land.[4]

As always in Mexico, the general splinters into particulars, in this case as the atmospheric circulations encounter a deeply textured terrain. Broad belts of climate are transformed into elevations of climate. Below 3,000 feet lies the *tierra caliente* (hot land); between 3,000 and 6,000 feet lies the *tierra templada* (temperate land); and above 6,000 feet lies the *tierra fría* (cold land). Yet temperature is a less useful index for fire than moisture. The source wetness comes from two oceans that lap against Mexico's coasts, and little of Mexico (other than the far north) is distant from their moderating influence. One is warm, the other cool. The Gulf of Mexico is a cauldron of heat and moisture; water and wind circulate clockwise and spin inland, seasonally drenching Yucatán and the coasts of Veracruz and Tamaulipas, striking the eastern Sierra Madre, and carrying moist air masses inland and northward. The Pacific sends a cool current and westerlies to the southern and western Sierra Madres, the western coast of Baja California, and the Chiapas Highlands. The Gulf of

California acts as a source of minor moderation for Sonora, Sinaloa, and Baja California. The rains are heaviest in the south, lightest in the north.

Seasons are thus defined by wet and dry rather than cold and hot. June through October are the principal rainy months; January through May, the dry. But while it is common to characterize Mexico as possessing seven climates—two tropical, two dry, and three temperate—only a few places like Tabasco are chronically wet, or like the Chihuahuan desert chronically dry. Compounding the annual cycle are episodic events and eccentric rhythms of drought and deluge. The episodic events are largely hurricanes that can slam both coasts; Quintana Roo, Tamaulipas, and Sinaloa are storm magnets. The Atlantic hurricanes strike mostly in July through October, the Pacific ones in September and October. Larger climatic cadences, notably the El Niño–Southern Oscillation (ENSO), boost winter precipitation in both north and central Mexico and shrink summer rainfall in the heavily populated center.

The telling details follow from how that moisture interacts with terrain. Clouds pile up along the flanks of the sierras; they rise like bubbles in a vat around the peaks of the volcanic belt and the sky islands of the northwest; they stall on the lee sides, leaving rain shadows. In the Yucatán Peninsula a karstified limestone shelf latticed with sinkholes and caverns carries rain below ground. On the northern plateau and central volcanic belt, the rain ends in lakes without outflows, which rise and fall with the rainy seasons. Everywhere it flows in rivers that are often intermittent, and always unnavigable.

As with years, so with epochs. Sedimentary records speak of a Late Pleistocene climate much different from that apparent today, or rather a time when traits that characterize today were exaggerated and relocated. With the advent of the Holocene, southern Mexico (especially the Yucatán Peninsula but also the volcanic belt) was dry, and the north wet. Only after 9,000 years BP did the temperate system retire northward and the summer monsoon establish itself. By 6,000 years BP Mexico was mostly wet. Then northern Mexico seems to have dried to something like its current state around 4,000 years ago, while southern Mexico remained wet, save for occasional (if massive) droughts. Probably the driest moment of the Holocene for Yucatán and central Mexico occurred 1,000 years BP.[5]

It isn't just that climate is variable: it's that norms are not very useful in describing that variability. There are few average storms or seasons; climatic graphs resemble scatter diagrams. What matters is how the rains come, whether gently or in convulsions, whether consistently, year after year, or potholed with droughts and floods, decade upon decade. Life in Mexico exists on a knife-edge, precariously balanced on the right amounts and timing of the rains.

Almost everywhere in Mexico fire is possible, and most everywhere inevitable. Wet-dry seasons are perfect for combustion, and what is true annually also holds between years. The ideal rhythms are wet years followed by dry—the first growing more combustibles than normal grazing and decay can remove before the second makes them available for fire. Thunderstorms hurl lightning that can kindle the ready fuels—dry lightning in the early season is especially effective. (Mexico has a higher density of lightning strikes than the United States: 25 per km^2 versus 18 per km^2. The Sierra Madre Oriental has the highest lightning density in North America.) Hurricanes are nature's great practitioners of slash and burn. What the winds throw down and the surge of rain grows is succeeded by dry spells that allow the mangled landscape to catch fire. The erratic pulsing of ENSO extends the same dynamic across years and decades. Even the enormous upheavals of the Pleistocene tended to accentuate and shift the geography of wet and dry: Mexico was not buried under ice or lakes or swept by periglacial sands and silt. Instead, four weather patterns converged, and they adjusted as each swelled or shrank.[6]

With or without people, Mexico will burn. It will burn one way in the Mediterranean climate of northern Baja California, another in the wet-dry tropics of Quintana Roo. It will burn with one regime on Chihuahuan plains and another in the gorges of the Sierra Madre Occidental, and with yet another in the rumpled swells of the Sierra Madre del Sur. It burned through the Pleistocene, and now it burns in the Pyrocene. It burns because it occupies an axis along which four cardinal directions of climate converge in ways that link ignition with fuels.

In some respects fire is just another disturbance in a land adapted to droughts, floods, earthquakes, volcanic eruptions, and hurricanes, a regimen of spasms and cataclysms for which averages mean little and which the megadiverse biota of Mexico not only has to adapt to but also emerged out of. Yet in one vital trait fire differs. The great and common disturbances of earth, wind, and water can occur without a particle of life present; fire can't. While not alive, fire is a creation of the living world, not unlike a virus, and its chemistry is a *bio*chemistry. It thrives by combusting plants. It inhabits the living landscape.

LIVING MEXICO

Geographically Mexico resembles a cornucopia, coiled tightly in the Yucatán Peninsula and spiraling outward in a broad swath to the northwest. Located at the crossroads of two continents and an island arc, on a mountainous isthmus between two seas, in a land where two climate systems meet and mix and where the cultivars and weeds of Eurasia and the Americas mingle—it should surprise no one that Mexico is renowned for its megadiversity.

In broad strokes, its biogeography divides into two floral realms. The Nearctic Region lies to the north and the Neotropical Region to the south. But the mountainous geography scrambles a simple division: Mexico is megadiverse because it overflows with mega-niches. The Neotropical realm extends northward along the coastal plains, and the Nearctic plunges south along the elevated plateau and over the central volcano belt. The numbers are impressive. Mexico holds over 10 percent of all Earth's species. It has 67 percent more floral species than the United States and 150 percent more than Europe. It ranks first in the world for its diversity of reptiles, second for mammals, and fourth for amphibians, and it holds the world's richest collection of cacti. It boasts over 1,100 species of birds. With 157 species of oak and 48 of pine (half of the world's species), it is a center of diversification for both genera.[7]

But if this cornucopia is a global biodiversity hot spot, that is also because it is prone to burn. Its myriad microbiomes support micropyromes.

The case for fire's natural character is simply put. Because so much of Mexico is amenable to burning, it burns, and because it burns, its biota has adapted to that burning. Fire is not some alien visitation: it emerges out of the conditions of the living landscape. Little in Mexico is not touched by fire, and little does not, in turn, shape fire's manifestations.

Fire is no more unique to Mexico than are its oaks and pines, yet just as Mexico displays more species of oak and pine than anywhere else, so it shows an astonishing menagerie of fires. Like Mexico's other disturbances, fire defies norms. In 1998 fires flashed through highland Chiapas; in 2011 they swept over large swaths of northern Coahuila. In 1981 Mexico recorded 2,500 fires; in 1998, 14,445. In 1992 some 35,000 hectares burned; in 2011, 956,505.[8]

Lightning is abundant, wet-dry rhythms affect nearly all regions, and Mexico's biomes have plenty of adaptations to assorted disturbances. Even today lightning accounts for 3 to 12 percent of fires in official records, and probably a higher percentage in actual fact. Official records, moreover, are notorious for the magnitude of starts from "unknown" origins, a sensible fraction of which are surely lightning, although that label can also incorporate some exotic ignitions. On April 26, 1910, the *New York Times* reported that a "gigantic" meteor in Chihuahua "burst, and some of the fragments started a forest fire." Such events happen, but far too infrequently to account for the abundant adaptations encoded in the biota. Similarly, an earthquake was reported at Bavispe in northern Mexico, "so strong it did not leave a single habitable house standing"—an event alternatively identified as "a volcano of fire" that set "fire to the forest."[9]

Yet even amid a land prone to agitations of all kinds, fire does seem special. Some 50 percent of Mexico's species are fire-adapted, and another 40 percent fire-influenced. All the usual evidence for fire's ecological longevity is present: thick bark; self-pruning of low limbs; lignotubers; sprouting from trunks, branches, and even crowns; rapid seeding in ash; smoke-stimulated germination; the liberation of scarce nutrients otherwise locked up in wood; a tendency toward higher combustibility that drives off less-adapted competitors; serotiny (closed cones that release seeds when heated). Different species show various combinations of such traits, adapted to particular patterns (or regimes) of fire, of which Mexican researchers have identified 14. Mexico's arid grasslands are susceptible to fire and expect it; its woodland savannas veer either to grass or forest depending on the character of fire; its tropical grasslands can only survive with fire. Some landscapes recover from a burn within days; some require decades; and a few, centuries. Of Mexico's burned area, 42 percent occurs in temperate biomes, 38 percent in croplands, 18 percent in tropical forests, and 12 percent in shrublands, which tend toward fewer fires but greater area burned by each fire.[10]

Such breadth of traits speaks to a suite of adaptations to a suite of stresses that yields a hardy biota capable of coping across a spectrum of cataclysms and the spasmodic upheavals of history. Much of Mexico is a landscape episodically prone to violent outbreaks—nature's *tumultos*. Floods, droughts, frosts, landslides, storms, hurricanes, earthquakes, eruptions—it's a cauldron of cataclysms, and any landscape may experience all or many of them, perhaps as they interact. Between 1871 and 1975, some 31 hurricanes struck the Yucatán Peninsula. On September 14, 1988, Hurricane Gilbert, then a Category 5 storm with winds at 185 miles per hour, slammed into Cozumel before sweeping across the peninsula, regrouping in the Gulf of Mexico, and again making landfall, this time in Tamaulipas. In the months that followed, an estimated 39 percent of Quintana Roo's forest burned. The fires had their ecological role: They decomposed the tough, larger debris left in Gilbert's wake, woody boles that would not decay naturally, at least not in a time frame needed to keep the nutrient cycles working. They rejuvenated vast patches of woods. They invited, for a while, new species to claim a habitat. They did on a vast scale what slash-and-burn cultivation does on a tiny one. The forests regenerated seamlessly, one way with wind, another with fire, both leaving the landscape dappled rather than scraped clean.[11]

What Mexico doesn't yet have are celebrity fire species and habitats—not like the United States' sequoia and longleaf pine, tallgrass prairie and chamise chaparral, South Africa's fynbos and sourveld, Australia's tropical savannas and jarrah forests, Canada's black spruce and jack pine. That's likely a function of where research gets done and how it is publicized, not the intrinsic character of a land and its resident pyrophytes. And recognition can change. The mountain forests of the Sierra San Pedro Mártir in Baja California, the pines of El Ajusco in the central volcano belt, the pine-oak woodlands of the Sierra de Manantlán, and the pine savannas of the Chiapas Highlands are all becoming staples of the international literature on fire.[12]

FIERY MEXICO

Mexico did not require people in order to have fire. Nor did it need to rely on eruptions and meteors; its characteristic lightning made fire inevitable. When people first arrived, the Mexican biota was already a tangled trellis across a complex landscape of mountains, gorges, volcanoes, limestone lowlands, and upland plateaus. Even in recent times there remain places that have burned routinely without any human contribution, like the Sierra San Luis and Sierra de los Ajos in northern Chihuahua and Sonora, where fires have swept through pine steppes every two to five years (it was enough that people did not put fires out). And there were places where people rarely entered, yet which burned on long rhythms of millennia, like the cloud forests of Los Chimalapas in Oaxaca. Yet all those adaptations to fire made a fire-receptive environment especially attractive to a fire-wielding creature.[13]

Humans proved a force multiplier for fire. Lightning could not slash woods and *matorral*, could not grow plants, could not extinguish a fire it started, could not kindle a fire amid snow or downpour, could not time ignition to align with wind, slope, and fuels. People could. With their arrival Mexico experienced one of the most profound of the many influences that have shaped its living landscapes and history, for here was a creature that could start and stop fires, terrace hillsides, dam streams, relocate and rearrange plants and animals, fell and plant forests, and ultimately unhinge even the climate. Wherever people went, fire was a companion, and whatever they did, fire was present as an enabler. Humans co-opted nature's fire, tamed it, and, rather like a software virus, used it to recode and redirect ecosystem processes in ways they preferred. So integral has anthropogenic fire been in Mexico, and for so long, that removing it can itself be a profound ecological disturbance.

The power and the glory of Mexican fire is its bonding with people. The synthesis of fire and humans became a symbiosis so long-standing that its intertwining threads are almost impossible to disentangle—a Gordian knot of Mexican ecology. Humans have, with increasing heft, influenced every aspect of fire's environment. They are the keystone species for fire—in reality, a species monopolist. By controlling ignition, they complete the cycle of fire for the circle of life. They loom over the scene like the priests around the altar on Cerro de la Estrella. While they couldn't extinguish all of nature's fires, they could coax a new fire out of their circumstances and then distribute it to the four corners of their world.

✸

Mexico.

Its geology is a triple junction where North and South America and the Caribbean converge. Its climate is a fluid axis where four weather systems—north, south, east, and west; temperate and tropical, Atlantic and Pacific—meet and mix. Its biology is a cornucopia, celebrated for overflowing with megadiversity. For humans, it is a kaleidoscope, where history constantly reorganizes the natural mosaic of rock, water, plants, and fire, ever overwriting the past without fully erasing it.

For North America, Mexico's natural history proposes a contrast of narratives. The Canadian Shield is a long-stable craton for the continent, only active along its western cordillera. The Laurentide Ice Sheet wiped the past away and left, at the end of the Pleistocene, a blank page. Mexico is a landscape more recent and more active, never quite scrubbed of its heritage. Canada offers a tabula rasa: it's easy to identify a beginning and a narrative arc as ice retreated and greenery and fire advanced. Mexico presents a palimpsest, cluttered with inscriptions of the past that are only partially scrubbed away, proposing no clear origin or plot.

More than an ecological disturbance or a convenient tool, anthropogenic fire is a relationship. It is a pact between humans and fire: it mediates between a people and their place. Like the Old God, fire is a foundational presence that is dispersed across almost everything people do on the land. It crosses social orders, cultures, and history. For Mexico it is the axis where the three geologic worlds join, where the four forces of climate converge, where the hierarchies of society mingle, and where, amid ceaseless cycling, worlds end and begin.

SECOND SUN

FIRE AND WATER

NEW FIRE
TOLUCA

IN 1478, *the Aztecs, assisted by their allies from Tenantzinco, went to war against the Matlatzincas. In Aztec theology fire and war were fused: Huitzilopochtli has attributes of Xiuhtecuhtli, a fire god; he is born "shooting out burning rays" and comes fully armed with Xiuhcoatl, a fire serpent. These are emblems of fire and war—war being the primary basis for Aztec power. (The Aztec symbol for war, a feathered coyote, spat fire and water.) As Karl Taube notes, veneration of the gods involved "burning of blood-spattered paper, burning of perfumed copal incense, and immolation of hundreds or thousands of human captives yearly."*[1]

In exhorting his troops arraigned against Cuetlaxtla, Moctezuma had likened war to fire: "War is like a field of dry grass set on fire, which while burning is swept away by the wind, and all is consumed." The battlefield is "where your hearts will be set on fire," and one must "also set on fire, consume, sweep away, those barbarians who you see before you." Small wonder that it was "well known," as Diego Durán records, "that the Aztecs were like fire that burned the fields of grain and the fruit trees, that consumed with its heat the maguey plants and the houses." They and their fires were, it seemed, interchangeable.[2]

The battle (and war) against the Matlatzincas, now led by Axayacatl, was framed by fire imagery. The assault began with the fire-themed "usual address." When the speech ended, Axayacatl "ordered the flaming torches raised high on the hilltop," and when they saw the flames the troops "shrieked, whistled, and yelled so loudly that the clamor reached the sky."

The battle ended when the Aztecs entered Toluca, "removed the god of that place," and then "burned the sanctuary." The immolation of a temple was "customary in war," the signal that a city had fallen.[3]

The fires real and figurative continued, helping add new vassals and devising a political alliance that ruled the Valley of Mexico. Out of the flame-galvanized wreckage emerged a new social order, a new empire, and in the Aztec calendar a new sun that was carried throughout central Mexico and beyond.

Yet the episode, so symbolically rich, even allegorical, can mislead. The new fire destined to remake Mexico was the fire in the field that Axayacatl urged his troops to emulate. The transfiguring flames were those that burned fallow, woods, and maguey, that further reforged Mexico's raw nature into a second nature. It's easy to focus on the epic and the monumental, and overlook the humble and mundane. The burning of temples and capital cities gets written. The fires that made life work are buried beneath the smoke and ash of empires. Yet the enduring makeover of Mexico by fire came through millions of milpas, *hogares*, and other human-kindled fires over thousands of years.

On February 20, 1943, outside a Purépecha village, the earth split, ash spit out, lava spilled over fields, and a cinder cone grew. Even amid World War II the Parícutin volcano was a sensation; within a month geologists William F. Foshag and Jenaro González Reyna were documenting its biography. Dionisio Pulido, on whose field it emerged, became, for a brief spell, a celebrity.

It's a famous event, recorded in the still-active grumblings of a full-spectrum volcano, and a predictable event, occurring within the central volcanic belt. What makes it instructive in Mexico's fire history is not that the Parícutin volcano was a fire, which it wasn't, or that it started fires, which it did, but that it broke out on ground that Dionisio Pulido was preparing to burn for his crops. All that morning Dionisio, his wife Paula, and their son, along with a neighbor, Demeterio, had cut, gathered, and burned to ready their plot for plowing and sowing. Dionisio had left one pile to ignite another when he smelled an acrid smoke and heard a crack like thunder, and a volcano sprang up before him.[4]

✸

Mexico's volcanoes laid down an earthy matrix for its living landscapes. Its anthropogenic fires would remake those landscapes; they could penetrate into every nook and niche. Temples came and went, but the brasero and the milpa endured. They were the essentials of a household and of cultivated life. Fetching water and gathering firewood were among the core duties of daily routine. Fire and water made agriculture possible, which rooted society, fed armies, and allowed monuments to rise or be razed.

Theology was an abstraction of everyday activity. Routine became ritual, symbols were set in stone. "Always and everywhere in Mesoamerica, the hearth comprises three stones," note Michael Coe and Rex Koontz, "and being the conceptual center of the world, is semi-sacred." The principle of duality that informed Mesoamerican mythology reflected the dualities of existence. The major centers of civilization, Toltec, Aztec, and Mayan, celebrated the role of fire and water in temples and ceremony. Even at the Sacred Cenote at Chichén Itzá, the putative eye of the world (*ophthalmos mundi*), most offerings were burned or sacrificed in flame before being tossed into the well. The Pyramid of Tenayuca, begun in Chichimec times and concluded in Aztec, an antecedent to the Templo Mayor that dominated and was the axis of the Aztec capital Tenochtitlan, was deftly cleaved in two. Its north part was dedicated to Tlaloc, the god of rain and lightning; its south, to Huitzilopochtli, the cult god of the Aztecs and a fire chimera. (Paradoxically, the Aztec avatar of the Old God, Xiuhtecuhtli, was associated with many properties of water; he was even the day glyph for water.) The annual fire ceremonies of Xocotl Huetzi and Izcalli divided the calendar in two, corresponding to the seasons of wet and dry. In the New Fire Ceremony household goods were tossed into water when the old fire was extinguished.[5]

The splendor of fire sacred—if not on the altar, then burning the altar—could obscure the universal presence of fire profane. Yet Mexico's second sun arose out of the second nature its peoples fashioned from fire and water, milpa and *quema*, tile by tile, over millennia.

THE COMING OF ANTHROPOGENIC FIRE

There is no definitive date or place for human entry into Mexico. But if the when and the where are unsettled, it's certain the firstcomers carried fire among their possessions. The domestication of the dog likely occurred between 12,000 and 40,000 years BP; the domestication of fire dates at least to *Homo erectus.*

The traditional origin narrative tracks hunters and foragers spilling into the interior of North America as the receding ice sheets opened up corridors. More recently, evidence suggests that ancestral Americans moved down the coasts as well, with the bulk of that evidence long lost to rising seas. By 12,000 BP they had reached Tierra del Fuego, and elsewhere in Chile possibly by 12,500 BP. But research continues to unearth still earlier dates. In the Valley of Mexico artifacts suggest possible occupation by 21,000 to 24,000 BP, and recent discoveries in Zacatecas located cultural artifacts that date to the Last Glacial Maximum (26,500–19,000 BP), arguing for human migration through the region as early as 33,000 to 31,000 years ago.[6]

The record sharpens with the advent of Clovis spearpoints, and soon afterward with Folsom modifications. The classic sites for both lie just north of the present-day U.S.-Mexico border (11,000 BP and 10,500 BP, respectively). Hunting sites in the Valley of Mexico date soon afterward. Spears boosted by atlatls made possible the hunting of the Pleistocene's fabled megafauna; along the coast the hunt included marine mammals. Small groups, bands not yet even clans, no doubt followed the game into new territory.[7]

But while big game are dramatic, most subsistence came from small fauna, fish, waterfowl, mussels, and above all plants. If deer, rabbits, and peccary marked a downsizing from the exalted megafauna of the high Pleistocene, they were also more abundant. Almost coincident with the bones of mammoths are the pollen of teosinte (the ancestor of maize) and cacao (the predecessor of chocolate). Other plants crowded the scene: fruits, grains, vegetables, squash, beans, berries, millet—whatever might be found among the landscape larder. The emerging megabiodiversity of Mexico had people amid the mix, omnivores able to forage widely and harvest whatever the seasons, fast-morphing climate, and niche-thickened landscapes made available. They began to tame species and landscapes as they had fire, which meant fire was also changed.[8]

The cultural diversity of Mexico matches that of its flora and fauna. Genetic tracking substantiates both the general migration of people north to south and the astonishing differentiation that resulted as, once again, Mexico's fracturing terrain broke up a general movement into a mosaic of intricate pieces. Groups split off like lithic flakes into smaller population clusters. Once settled and separated, the isolated populations showed a strong rootedness.[9]

Genes and geography—what was true for plants and animals proved true for people, and, equally to the point, those initial patterns have persisted. Even after the shock of European encounter and conquest, amid the dominantly mestizo population that followed, the initial matrix of human settlement continues with a "strong correlation between Native American ancestry and geography within Mexico." This observation applies to cultures as well as to genes, which also means fire practices. Despite adoptions, hybridizing, and a centuries-long top-down push to reform, ancient fire customs have tended to abide.[10]

Biodiversity, cultural diversity, pyrodiversity—thus did the great fire triangle of Mexico emerge.

There is an old quarrel—actually there are many—about which factor drives fire. It's a discourse that trends toward scholasticism, because fire behaves like a driverless car, speeding or crawling down a road integrating everything around it. The real question is not whether natural processes such as climate and terrain dominate over human activities, but how people and nature interact. Where a place favors fire, humans can expand its range and season; where fire is rare, people can, within limits, introduce it. Like the Old God, fire rarely stands alone, but almost never does it stand apart.

Reconstructing Mexico's ancient fire history is tricky because there are no written records, no narratives chiseled into basalt, few glyphs in murals to document everyday practices, only a carbonized chronicle here and there. While fire pervades and endures on the Mexican landscape, it is dispersed in Mexico's mythology and overwhelmingly absent in its formal history. Yet there are consistencies among fire economies around the world. Aboriginal use of fire, adapted to local circumstances, looks very similar in mediterranean Australia and California, in southern Africa and Great Plains North America; agricultural practices of slash and burn, tweaked to local species and circumstances, look familiar in upland Thailand, the Coastal Plain of North America, and boreal Scandinavia. Given a few preserved relics it is possible to construct a plausible account of how fire functioned in pre-Columbian Mesoamerica. These fire sherds testify to the anthropogenic landscape as the fundamental unit of Mexican land and life.

FORAGING

Fire went where people went; it interacted with whatever they did. What we might call aboriginal fire practices rely on the capacity to control ignition, the ability to determine where and when a fire starts—"aboriginal" here referring to a state of technology, a suite of fire practices associated with foraging and hunting rather than with a people.

The general formula for use appeals to lines of fire and fields of fire. Lines of fire are routes of travel, typically part of a seasonal walkabout that exploits resources of different places as they become available across a calendar year. Where corridors become overgrown, they are burned open, and where not deliberately lit, they are subject to inevitable fire littering—fire is humanity's oldest camp follower. Fields of fire are places that are routinely burned to stimulate grass or browse attractive to game, to help gather acorns and nuts, to promote twigs suitable for baskets, to prune berries, or to encourage edible tubers; they are often sites of regular hunting. Because aboriginal economies follow circuits (or songlines) around the countryside, fire propagates twice—once by means of people as they move, once on its own since it must spread to survive. The nearly universal wet-dry climates of Mexico made almost every place, with some tinkering, amenable to fire.[11]

It is surely no accident that the world's hearths for domestication are all places that undergo routine rhythms of wetting and drying, which makes them all amenable not only to finagling by human hands but to tended fires—a kind of pyric horticulture that Australian anthropologist Rhys Jones famously termed "fire-stick farming." In effect, domestication brings flora to the hearth. When Father Junípero Serra entered a region south of present-day San Diego, California, in 1796, he described a valley "more than a league in width, and in parts so green that, if I did not know in what country I was, I would have taken it, without any hesitation, for land under cultivation." It was, in fact, under cultivation, but with a torch rather than a plow. Such practices surely extended southward.[12]

The abundance of accessible foodstuffs astonished early Spanish observers, who noted about the Chichimecs and Otomís that their diet, in the phrasing of Sherburne Cook, "included every plant that grew and every creature that walked, flew, or crawled," as well as extensively relying on maguey and nopal. The usual interpretation has been that the Indigenous people, pressed by necessity, had found uses for what their environment held. A better notion might be that by patient selection, not least with fire, they had promoted those plants they found helpful and removed those they didn't.[13]

The best-recorded practices concern hunting. Fire hunting involved both drives and draws—the first propelling animals and the second enticing them with fresh fodder into traps and favored hunting grounds that fire could renew annually. Cabeza de Vaca noted how the indigenes "go about with a firebrand, setting fire to the plains and timber so as to drive off the mosquitoes, and also to get lizards and similar things which they eat, to come out of the soil. In the same manner they kill deer, encircling them with fires, and they do it also to deprive the animals of pasture, compelling them to go for food where the Indians want." Gonzalo Fernández de Oviedo y Valdés quotes another survivor who added mice to the mix. Bartolomé de Las Casas describes fire drives for rabbits in the savannas of Hispaniola. Early Spaniards reported similar practices in California, pitting hunting fires for rabbits against pastoral burns for cattle, and they found such usage common around the Caribbean coasts. Pascal de Andagoya recorded how Cuevans hunted deer and peccary: "The chiefs had hunting tracts (*cotos*) to which they went in summer [the dry season] to hunt deer. Fire was set on the windward side and since the plants (*yerba*) grew tall the fire was great. Indians were placed in file at a position where the fire would come to a stop. The deer, massed in their flight and blinded by the smoke, were thus driven by fire to the place where the Indians were waiting with their dart throwers and stone points so that few creatures escaped." Apaches in the U.S. Southwest used smudge fires to entice deer to the smoke in order to escape flies. Reportedly, they also set fire "to the grass of large areas" in the mountains and drove "the game against some narrow mountain-pass or abyss, where they could easily be killed." With mild fires Sudanese nudge elephants into marshes, where they can be more easily hunted; early Mesoamericans may have done the same with mammoths. In Southern California, fire drives entangled megafauna in the La Brea tar pits (and resulted in a "state change" in the landscape). Tribes along the Colorado River burned riverine rushes to flush out rabbits; around the Great Basin, people set fires to encircle and harvest grasshoppers; and across the Americas, they used torches at night to "spotlight" deer, hunt waterfowl, and draw fish to spears.[14]

Lt. Col. Don Antonio Cordero recounted a similar practice among the Apaches in 1796.

> The present object of description is the hunt which is made usually for deer, burro, antelope, javelina, porcupine, mountain lion, bear, wolves, coyotes, hare and rabbits. Having reconnoitered the valleys, mountain ranges, plains, and brush which they inhabit, for the traces of these animals, and having decided on the day, the leader of the undertaking determines at dawn the places, where the different groups who are to start the hunt should be, the points which are to be occupied by archers on horseback and on foot, and those who are to serve at a distance as lookouts in order to guard against attacks of enemies, and in these places those appointed take up their posts. In this way at dawn a piece of terrain is encircled, which frequently is five or six leagues in circumference. The sign to commence the chase, and consequently to close the circle, is given by smoke signals. There are men on horseback assigned to this project, which consists in setting fire to the grass and herbage of the whole circumference; and since for this purpose they are already placed ahead of time in their posts with torches ready which they make from dried bark or dried palmella, it takes only a moment to see the whole circle flare up. At the same instant, the shouts and noise commence, the animals flee, they find no exit, and finally they fall into the hands of their astute adversaries.

Cordero also notes that "smoke is the most efficient means by which they communicate." Size and location shaped its grammar. And "there is no man or woman who does not carry with him the implements necessary to make fire." Even children could start fire with ease.[15]

José de Acosta observed a fire drive, embedded in ceremony, in the province of Tlaxcala, a place "very suitable for hunting" whose residents were "much given to it" and offered their hunting god "great festivities."

> They represent the idol in a certain form, and I will not waste time in describing it; but the festivity that they offered to him is very colorful. And it was that at daybreak they blew upon a horn and everyone assembled with their bows and arrows, nets, and other hunting implements and went in procession with their idol, a large crowd following, to a high hill, on the top of which they had built a bower with a richly adorned altar in it, where they placed the idol. And marching along with a great noise of trumpets, conch shells, flutes, and drums they reached the place and surrounded the whole slope of the hill, and when they had set fire to it everywhere many and very diverse animals came out, such as deer, rabbits, hares, foxes, wolves, and so on, which ran toward the summit fleeing from the fire. Then the hunters went after them with much shouting and yelling, playing different instruments; they steered them to the summit before the idol, where the press of game was so great that some leaped, others rolled over and over, others jumped onto the people, and still others climbed on the altar, so that there was great rejoicing and merrymaking. Then they took a great quantity of game and sacrificed the deer and large animals before the idol, tearing out their hearts with the same ceremony they used when sacrificing men. This done, they loaded the game onto their backs and returned with their idol in the same manner in which they had come and entered the city with all these things, rejoicing greatly and with much music of trumpets and kettledrums, until they reached the temple, where they replaced their idol with much reverence and solemnity. Then everyone went to cook the flesh from all that game and invited all the people to join them, and after the meal gave their usual performances and dances before the idol.[16]

Some practices were exotic, if not weird. Consider the fire-assisted hunt for monkeys in which a bonfire was lit, and into it hunters tossed a *cacalotetl* (raven stone), which bursts when heated sufficiently. The hunters would cover the fire and spread around it some maize seeds as bait. Mother monkeys with young on their backs would come for the seeds, the stone would burst, and the monkeys would flee, but not before hunters seized the young.[17]

Not least, there is no evidence that aboriginal Mexicans sought to extinguish fires set by nature. Carl Sauer, who spent many years of his storied career studying Mexico, once commented that he knew of no instances of hunting and foraging peoples (anywhere in the world) who actively suppressed landscape fires. Instead, they contained unwanted fires by substituting their own fires—preemptively burning places they wanted protected, replacing the wild fire with the tame, accommodating nature's fires as they did storms. They left more remote fires to burn on their own or incorporated frequently burned sites into the annual almanac of hunting and foraging. Those fires passed over the land like a spring breeze. In some places such fires, with minor supplements, were ample for people's needs.[18]

The paradox is that such burning could sustain, even enhance, the desired habitat. It renewed those grasses that, while green, fed the game, and, when yellowed or dormant, fed the flames. Exploring in northwestern Mexico in the 1880s and 1890s, Carl Hartman became "convinced that the continuous, immense forests here could never be destroyed by the Indians." Several times he noted that his Mexican assistants "through carelessness" loosed a fire into "high dry grass, and when it was windy, the fire would sweep over large tracts, burning all the luxuriant grass and shrubs as well as very young trees, but never destroying any trees with trunks thicker than a man's arm, only blackening the trunks and perhaps burning off the lowest [branches] of some trees—leaving the forest, on the whole, quite intact and green."[19]

In the Sierra Madre Occidental Hartman described how "these Indians, the pagans as well as the Christians, keep up the custom of burning off the grass all over the sierras during the very driest season of the year, April–June"; how "the fires are continually burning day and night all over the mountains up to the highest crests, leaving the stony ground, blackened and barren, but the forests stand tall"; how those routine fires left "still more convincing proofs of the indestructibility of the forests." Regular controlled burning prevented high-severity burning.[20]

FARMING

Hunting and gathering were typical throughout ancient Mexico, but the core fire practices, like core land use and central cultural obsessions, were related to farming. Mexico was one of a handful of Earth's great hearths for plant domestication. In farming, the duality of fire and water found a common cause. They were the linked strokes in the two-cycle engine that was Mexican agriculture.

The earliest migrants came with domestication as part of their heritage. Fire, after all, is a prototype for taming first nature into second because, more than a tool, it is a practice and a relationship. Its agricultural expressions are many and graded, more analogue than digital. Eventually some 48 species of Mesoamerican flowers, grasses, roots, fruits, and nuts were tamed; some, like the bottle gourd, agave, cotton, and rubber, were a source of material, while others were foods. The tomato, the peanut, the guava, the avocado, the chili pepper, the bean, maize, and manioc—all began replacing natural selection with selection by humans. The transition from gathering mostly wild foodstuffs to harvesting mostly cultivated ones took several thousand years. Interestingly, it seems the process began largely amid the more seasonally arid environments before spreading throughout Mesoamerica. One of fire's roles was to re-create sites that emulated the lands of a cultivar's origin.[21]

No threshold cleanly divides agricultural economies from aboriginal ones, or their associated fire practices. From a fire perspective, the distinction is that agricultural usage adds manipulation of fuels to control over ignition. Under aboriginal conditions, people must rely on nature to slash forests or drain wetlands and create conditions that allow fire to spread. In agricultural geographies, people prepare the fuels, and in this way alter and enlarge the patches and pulses of burning. It's easy to see how one style can grade into the other at a site, or over a span of time. The people of the Gran Chichimeca (or Northern March, to use Sauer's term) in the mostly arid north practiced horticulture and had gardens and fields. The people of the empires (or Southern Culture Hearth) in the mostly wet south complemented farming with hunting and gathering wild foodstuffs and medicines.[22]

The civilizations of Mesoamerica were based on squash, beans, chili peppers, and—critically—maize. Maize likely descended from teosinte (*Zea* spp.) found in the Río Balsas region of northern Guerrero. Seeds have been discovered in nearby archaeological sites from as early as 8750 BP. The process by which wild *Zea* evolved into modern maize was surely messy and varied. But over millennia, seeds and the skills to grow them spread into North and South America as well as the Caribbean. So, too, the process of transitioning from mostly wild plants to mostly tame ones, and from seasonal nomadism to sedentism, was gradual, and almost everywhere incomplete. But among the constants of Mesoamerican cultivation was the interplay of water and fire. Like all plants, maize needed water, and being a shrubby grass in a wet-dry climate, teosinte surely experienced fire.[23]

The core practice was *la roza*, or in its full expression, *roza, tumba y quema*—cutting and burning. The slashings prepared the fuels; the burning converted now-inert biomass into nutrients, drove off pests and weeds, and opened a milpa to a plant that on its own could not occupy the site. A firebreak around the plot helped hold the flames to the field. The two essentials required for the practice were fire and fuel (or, if returning to a site, fallow). Fray Juan de Torquemada, a Franciscan missionary in central Mexico and a leading chronicler of his time, described the basics, which he characterized as occurring "everywhere in the mountains and in the warm lands." The farmers would cut the trees and shrubs and burn the clearing (*roza*), then sow among the ashes. "The seeds grow most abundantly, without much labor." Still, "the land that is sown one year is not resown again in four or even six years, until the shrubs have sprouted again, and cover it, amid whose shade the land moistens again, and it is reformed for another sowing."[24]

Variations on this foundational practice were unbounded, as were local terms applied to them. They differed according to soil, exposure, past history (and hence combustibles), cultivar, duration and intensity of cultivation, and general social context. Roza-tumba-quema had its antecedents in a nature that grew profusely after burning. What farmers did was direct those ecological processes toward their own ends: with clearing and burning it was possible to transplant cultivars from their sites of origin to an alien field. A milpa involved not only a domesticated plant but the domesticating of a site.[25]

This basic formula could be exported to most landscapes of Mexico. It was used in pine-oak forests and chaparral shrublands, mountain slopes and floodplains, the Sierra Madre Occidental, the central volcanic plateau, the Gulf lowlands from the Huasteca to the Yucatán Peninsula—only the driest, coldest, and most remote settings denied the practice. It was a spear tip of first-contact colonization, and it was a method of repeatedly reclaiming sites known to be fertile after a suitable period of rest. It could be supplemented by weeding and successive cultivars, and, after the Conquest, by grazing. A roza site could last two years or ten before being abandoned and later revisited. Depending on the local flora, the period of fallowing could last years or decades. In such ways the land was reworked over and over across centuries and millennia, as useful plants were promoted and annoying ones removed, a patient weeding and sowing that remade Mesoamerican first nature into Mexico's second nature.[26]

The milpa coexisted with varieties of farming that made water, not fire, the informing presence. The *chinampa* of the Valley of Mexico is the best-known example, but there were similar projects in other places where water was sufficient to flood a site, at least seasonally. There were horticultural gardens of plants and trees; *huertos* of cacao, of fruit trees, of avocado, of fruiting cactus, of maguey. (Hunting and gathering meanwhile continued.) Most agricultural fields were polycultivated and required broad-spectrum treatments, like watering and burning, which could advance multiple purposes. Without water nothing could grow, and where it was possible to control water rather than rely on the vagaries of rain, farmers did. But Mexico was the creation of drying as well as wetting, which made fire and water, yin and yang, the tag team of Mexican landscaping. Even where water did the primary work of fertilizing and fumigating, somewhere in the chain of cultivation fire would appear.[27]

With maize as its mainstay—as defining for Mesoamerica as rice for the great civilizations of Asia and wheat for those of the Mediterranean—the agricultural economy of Mesoamerica, like its climate, stabilized. It spread and settled wherever it could across the crumpled terrains of Mexico, from tropical lowlands to semiarid mesas and highlands both tropical and temperate. It persisted through tidal swings of climate, of peoples, of empires, of ideas and institutions. Not until Mexicans turned to burning lithic landscapes (in the form of fossil fuels) rather than living ones did the fire economy of Mexico undergo a revolution as profound as the one that led to systematic cultivation.

In most of Mexico the milpa became the pith of the anthropogenic landscape, certainly its best-known emblem, as fundamental to Mexican geography as its mountains and monsoons. It did not always demand burning, but usually it did, and those fires were the most visible sign of roza-tumba-quema. Fire became the expression of all that observers liked and disliked about the system, or became the reason they disliked it. In the early 1970s the American ecologist Howard Odum famously noted that in the modern world a potato was made partly of oil. In pre-Columbian Mesoamerica, maize was made partly of ash—an observation that became literally true when lime was scarce and ash replaced it in tortillas.[28]

Those fires structured society as much as they did land. The hearth fire is a domesticated fire, and so became more than just a tool to be picked up and dropped at will; instead, it defined a relationship. It had to be kindled, tended, fed, trained, and its wastes cleaned up. It demanded that society organize around its needs, especially the ceaseless quest for fuel. Through it came the ecological alchemy that transformed dross landscape into food; with it proceeded the rituals and ceremonies that connected people to a spiritual and symbolic world; around it came the stories that formed the basis for explaining the origins and prescriptions of human life. It even entertained people as they gathered, mesmerized or telling stories, around a campfire, or traced the flicker of flames at night flowing sinuously over hills, or watched embers from field fires rise like slow fireworks to merge with the stars.

Yet for all its presence, so common and expected was fire that it went unseen. The Old God was not depicted holding up the world, only a brasero, the humble hearth that organized the day and that, disseminated through agriculture, organized the landscapes of Mexico.

FIRE AND WATER, PEOPLE AND PLACE

Fire prepared, water nourished. Without water, nothing could grow. Without fire, fewer resources could be extracted and land became less habitable for humans. Fire readied a thirsty land to receive the life-promoting water; fire helped hunt, fish, gather, plant, weed, and harvest; fire transformed raw tubers, seeds, and meat into food. Even lakeside vegetation was burned, "as early as the beginning of the Holocene," to enhance overall productivity. Not every activity involved fire, and fire was rarely alone in its actions, but like the Old God it seemed to always be around.[29]

In general, northern Mexico favored more hunting and foraging, and southern Mexico more cultivated crops; but the interplay between water and fire was complex and syncretic, and the borders between an aboriginal and an agricultural livelihood were porous. Some of the fire practices, and the peoples who used them, are featured among Mexico's best-known ethnographic studies.

SIERRA TARAHUMARA

The northern borderlands, the Gran Chichimeca, served as the source regions for the successive repeoplings of Mexico, much as the steppes did for Europe. They also apparently hosted the ancestral Desert Culture that served as the *ur*-font for the more elaborate civilizations that flourished amid the lusher landscapes of central and southern Mexico. For their inhabitants, sustenance followed from the cadences of water and fire.

The Rarámuri live amid the Sierra Tarahumara, the semiarid mountains, gorges, foothills, and mesas where southwestern Chihuahua meets present-day Sonora and Sinaloa. Apparently pressure from Spanish settlements, especially mines and their associated farms, ranches, and pueblos, pushed the Rarámuri northwest and southwest during the early seventeenth century. They had a flexible, synchronized economy whose parts could expand or contract as needed and in which the Rarámuri moved seasonally, a kind of aboriginal transhumance between mountain and valley, exploiting a natural larder of plants and animals as rain and drought made possible or necessary.[30]

Rains collect mostly along the flanks of the central plateau and the Sierra Madre Occidental, and among the sky-island mountains that dot the plateau. Most of the rain falls during the summer monsoon (70 percent); the rest, during winter nortes. The floodplains of streams made suitable soil for a classic digging-stick agriculture of maize, beans, and squash, supplemented by endless foraging, hunting, fishing, kitchen gardens, and scattered milpas in suitable niches beyond streams. Some spotty irrigation existed, but water on the landscape could not be easily manipulated.[31]

Fire could. The Rarámuri followed the classic formula, burning lines and fields. The fields were sites with good soil and water. They were cleared in February and March, the debris was burned in May and early June, and the site was planted after the first rains. The fields were generally cropped for three years, then abandoned and, after a suitable time, reused. In more forested areas trees would be girdled by having their bark stripped off in swaths around the trunk; they would dry for two or three years, then were felled and burned during the summer, when temperature was at a maximum and humidity at a minimum. When Carl Hartman visited in 1890, he recorded that the practice occurred "on a level place in the forest where the humus is rich and generally near some stream," across "a few or perhaps 20 acres or more"; the dead trees were left for "two or three years," and then "the whole mass of trunks and broken branches" was set on fire and "burned to ashes." Likely these were new sites, since it's far easier to burn small woody debris than large trunks. Most sites left mature pine and juniper and targeted oak for slashing and burning, since it would subsequently coppice into usable firewood.[32]

Carl Lumholtz gave a tidy account from his travels, also in the 1890s. "In the arroyos and barrancas . . . ," he writes, "the Indians use the ancient mode of agriculture, still in vogue among remote natives of Mexico and called *coamillar*. They cut down the trees, clear a piece of land from brushwood, and leave it in this condition until just before the wet season sets in." The agricultural consequences varied by locale. "In the sierra," the land could yield "good crops for three years in succession"; "in the broad mountain valleys and on mesas," the same field could produce annually for 20 to 30 seasons; "in the barrancas," where soils were better and wetter, a field could survive only two seasons before weeds "suffocated" the maize. The acquisition of goats in the colonial era added animal manure to maize, which boosted production in a generally marginal area for crops.[33]

There are a few reports of burning for other, specific causes. They burned to hunt rabbits and squirrels; and if the heritage of the Desert Culture survived, they probably burned to prompt fresh browse for white-tailed and mule deer and turkey, to promote berries and edible and medicinal native plants, to ease the harvest of piñon nuts, and perhaps to drive and harvest insects like grasshoppers. Snags attracted birds. Afterward, hearth fires transmuted the raw into the cooked. Otherwise, the Rarámuri burned, as explorers and missionaries recorded, "to encourage rain." In template-true fashion they burned in the lead-up to the rainy season. And not least, in an economy of use or lose, fire was a means of establishing ownership—the slashings and burnings testified to invested labor that conveyed legal rights to the products.[34]

That they burned widely and often is incontestable. The earliest Spanish accounts testify to the practice; Lumholtz lamented the problem of finding grass for his horses and pack animals from April to June due to the far-ranging fires—"the Indians at that time (May) always burn the grass, and the entire country is wrapped in smoke"; "Only by accident is some little spot of grass spared here and there"—and the practice was reported in the latter twentieth century. A burned land was clearly a better land. Having the land burned was so universally helpful that it was not necessary to kindle flames for specific purposes: the environment was fire-rich, and it was enough that what could burn did burn. What the Rarámuri burned, lightning fire couldn't, and each kindled source mopped up what the other missed. Equally to the point, apart from burning ahead of lightning fires, the Rarámuri made no attempt to keep fires from spreading wherever fuels, terrain, and wind might take them, across ridges and over the course of months.[35]

There is a complementary record in the trees, some of which can be scarred at the base by fire; once scarred, the cavities burn easily again, leaving a minimal chronicle of burns (not every fire leaves a mark). In the early twenty-first century researchers sampled three sites around Ejido Pino Gordo with trees that dated back to the early nineteenth century—and, equally relevant, had not been logged. When combined, the three sites yielded a mean fire interval of 1.9 years for all fires, which lengthened to 7.9 years when the domain contained 25 percent or more of the sampled trees. Roughly two-thirds of the scars occurred in the dormant season, likely in the lead-in to the annual rains. The trees confirmed what travelers had long witnessed: allowing for local variances in a significantly textured terrain, the Sierra Tarahumara burned as often as it was able. Researchers writing in 2011 regarded Pino Gordo as "the least interrupted surface-fire regime reported to date in North America."[36]

Even within the Mexican context, with its stubborn persistence of Indigenous peoples, the Rarámuri have remained notably isolated until recent times. Although missions introduced other cultivars, especially fruits, livestock, and tools like the plow, the core economy persisted largely as it had been in preconquest times. Not until the Ferrocarril Chihuahua al Pacífico railway finally crossed the region in 1961 were the mountains fully open to commercial exploitation, which quickly drove Indigenous fire practices into select refugia (by 1990, 97 percent of the timber harvested in Chihuahua came from the Sierra Tarahumara). As much as anywhere, the Sierra Tarahumara presents a likely lens on pre-Hispanic agriculture in the northern regions, and certainly one on how Indigenous fire practices could hybridize with contemporary pressures. (Traditional roza burning has, for example, been adapted from beans to poppies and marijuana.) Again and again, the Rarámuri have absorbed the novel into the ancient.[37]

Among those Rarámuri who accept Catholicism there is a practice of double baptism. The usual sequence in Christianity begins with a physical baptism by water before advancing, symbolically, to a rebirth by spiritual fire. The Rarámuri reverse that order. A fire baptism confirms membership in the community of Rarámuri. A handful of smoking cobs "lightly draws small crosses" around the child's head, burning away "threads growing from the head and up into the sky," symbolically severing a kind of spiritual umbilical cord and birthing the child to the world. A second, water, baptism brings the recipient into the church, a link to the outside world. The order is exactly the same as appears in maize cultivation and much else in the Sierra Tarahumara economy. The two baptisms are life-cycle rituals, abstractions into the sacred realm of what makes the profane world work. Fire prepares, water nourishes.[38]

MAYAN MILPA

Between the Sierra Tarahumara and the Yucatán Peninsula lies a steep gradient, an order of magnitude of difference, both environmental and cultural. Where the mountains have an annual rainfall of 80 millimeters, the lowlands receive 100 to 200 millimeters. During the mission era of the eighteenth century, the Rarámuri had an estimated population between 20,000 and 35,000; during the Classic Age the Maya probably had 60,000 people just at Tikal. A similar distance separates their material cultures. The Rarámuri had grass huts and moved seasonally; they transmitted knowledge orally; and apart from selective adoptions, they remained remarkably unchanged across centuries. The Classic Age Maya erected vast stone cities of monumental architecture, settled karstified lowlands and volcanic mountains, devised a complex calendar and system of glyphs, and participated in the rise and fall of Mesoamerican empires. Yet both landscapes experienced regular cycles of wetting and drying, and both cultures shared an economy of maize and a common core of fire practices.[39]

The peninsula is a limestone lowland, heavily karstified, lightly hilled, that curves northward in its eastern extent. Rain falls mostly from May to November, with peaks in June and October, and it thins from the great cordillera in the west to the coastal plains, with the northern littoral being significantly drier. Tropical evergreen rainforest prevails, with the dry coastal swath grading into tropical deciduous and thorn forest. Onto these rhythms come deeper swells of drought and deluge. Hurricanes pass in a matter of days. Droughts can last for decades. Historical records testify to major episodes between CE 200 and 300, 820 and 870, 1020 and 1100, and 1530 and 1580, with other dry spells spiking in 420, 930, and 1800.[40]

Managing water was half of Yucatán agriculture. What mattered was not the volume of rain but how it interacted with soil. A warm, wet climate slowly dissolves the Yucatán's limestone caprock, eroding out a subterranean drainage system of underground caverns, rivers, and sinkholes that accentuate the dry season. Some depressions are broad and shallow (*bajos*), some smaller and seasonal (*aguadas*), and some deep (*cenotes*). On suitable slopes, terraces help hold water. Underground subterranean cisterns abound. Seasonally available, the gathered water, sometimes directed, could support crops. There were terraces, channels useful for aquaculture, and raised fields alternating with canals above the water table—a variant of chinampa agriculture. Instead of bringing water to the garden, the garden was brought to the water. The productivity could be high; the site, reusable. But intensive agriculture is tricky in limestone-soiled tropics; and especially in the drier, poorer swaths it is believed that Indigenous populations existed by supplementing through trade what they could produce locally—trading salt for food, for example. Most crops were watered with rains.[41]

The other half of the agricultural equation was fire—in its core form, roza-tumba-quema ("la tecnología básica"). Like limestone soils everywhere, the ones here tend to be shallow and nutrient scarce; the Yucatán's good soils, like its surface cavities, are patchy. Cultivation requires a nutrient-rich deposit, which comes from burning, which follows from vegetation on the ground, parched by dry spells, thick and continuous enough to carry combustion. Such circumstances occur naturally on a spectacular scale when hurricanes crash into the peninsula and the subsequent debris dries and ignites. It happens routinely where humans slash and kindle the forest. (After Hurricane Gilbert, swiddeners swiftly moved into debris fields and created milpas.) Without burning the slashing is pointless—in fact, an encumbrance. The burning is what transmutes dead biomass into crops—what breaks down wood into nutrients, what drives off pests and fungi, what fertilizes and fumigates. Without a good fire, a milpa is worthless, a lump of inert biomass. In his 1843 *Incidents of Travel in Yucatan*, John L. Stephens described how "in the dry season, generally in the months of January and February, a place is selected in the woods, from which the trees are cut down and burned. In May or June the corn is planted," along with beans, squash, and chilis, the universal diet of Mesoamerica. A second planting might follow, in which, as Benjamin Norman recorded, simple burning was "the only preparation that the soil received prior to sowing it."[42]

The milpa was a constructed site, not ecological graffiti carved by axe and spray-painted by flame. Each patch was small, part of a sprawling mosaic of field tiles. After two crops the site was abandoned, and in the Yucatán these sites were left for 15 to 20 years of fallowing (in richer soils, closer to the sierra, the fallow stage might last 4 to 7 years). In addition to maize, beans, squash, and chilis, arranged so that each could aid the others, trees were often planted, an example of agroforestry.[43]

The milpa was endlessly adaptable: the Maya were not its only practitioners, only the best studied. It could be applied in rainforest, in pine-oak woodlands, in chaparral; it could be used on karst plains, on mountains and mesas, and in barrancas. Old sites could be revisited on a regular cycle. Large, useful trees might be left for shade, fruit, reseeding, and erosion control. The usual protocol had the site burned two to three days before seeding in the ash, and as soon as possible before the rains arrived. Plots were burned from the outside in or, on slopes, from the top down to dampen escape. Control was bolstered by natural or cut fuelbreaks. Neighbors assisted.[44]

Still, surprises happened: the rains arrived too soon or too late, winds blew sparks far and wide, smoke obscured the site. (In his survey of

Mayan civilization, veteran archaeologist J. E. S. Thompson described how a Mayan acquaintance died while burning his milpa, perhaps due to "a change of wind," or being "overcome by smoke," or a triggered heart attack.) From time to time, fires escaped across the countryside. In words that echo Old Testament prophets like Ezekiel and Joel, the Books of Chilam Balam speak of fire loosed upon the land. "The hills shall burn, the ravine between shall burn. The fire shall flare up at the great *sucte* tree. It shall burn at the sea, on the beach. The squash seeds shall burn, the squash shall burn, the *macal* shall burn."[45]

Throughout not only Mexico but Mesoamerica, across the southeast and the Ohio Valley of North America, even into lower New England, variants of the malleable milpa were the basis for settled life.

The milpa was not the only cultivated patch in the landscape. There were kitchen gardens for vegetables, seed nurseries, and spices; there were "forest gardens" (small orchards) for fruit trees (such as avocado, papaya, guava, and breadnut), legumes and tubers (like manioc), and cacao; there were sacred groves, left in something like their native condition and a source of special medicines and toxins; and there were created forests (*pet kot*), fashioned by favoring prized species and weeding out less useful ones, in sites sometimes enclosed by stone walls. Between the "natural" forest, the pet kot forest, the forest garden, the strips of woods left around slashings, and the circulating swidden of patches cut, burned, and left fallow, there swirled a great diversity of habitats and species. Guiding the entirety was the hand of the Maya. One modern observer has speculated that "if all these [plants] are added up, we may find that *all* of the flora of the Maya area was used!"[46]

Over time—across centuries and, in places, millennia—the Maya selected for those species they found valuable and against those they had little use for. This was a cultivated landscape, though not one Europeans with their notions of plows, fixed land ownership, and monocropping recognized (a distaste amplified by the dislike among European elites for fire-based farming in any form). With population collapses the guiding hand relaxed, or receded, leaving the landscape untended. By the time Enlightenment naturalists surveyed the scene in the nineteenth century they found not a wild forest but a feral one.

The Yucatán held more than woods. Under the right conditions repeated burning could convert a fallow plot to grassland, suitable for an abundant fauna including turkeys, tapirs, peccaries, and deer. Once burned, a site was more easily burned, and once grasses got established, the biota could flip into savanna—and tropical savannas, the world around, are notorious for near-annual fires. Where winds and terrain align, the grassy patches and their fires can be extensive, and in dry times flames can escape into the surrounding countryside. (What J. G. Myers observed in 1930s Guyana is endlessly echoed throughout tropical America: "I have never seen in South America a savanna however small or isolated or distant from settlement which did not show signs of more or less frequent burning.") Whether for livestock or hunting, such patches were the grassy equivalent of woody milpas. They fit into a general mosaic. Still, for the casual observer, seasonally smoky skies could obscure that landscape mosaic so intricate with food and fallow.[47]

Fire was a physical as well as a biological technology. The demand for fuelwood was incessant (the regrowing of woody fallow was a good source). Maize and meat needed to be cooked; whatever grew in water-based agriculture went into the pot. Torches likely assisted spearfishing at night; then the fish needed to be smoked. Boiling seawater over fires could speed up salt production; cooking limestone made mortar; forges smelted ore and hardened digging sticks and spears. By amplifying humanity's overall firepower, pyrotechnologies strengthened the grip and furthered the reach of human land use.[48]

Fire and smoke attract attention—even casual visitors may remark on their presence, especially the paradoxical juxtaposition of dry flame and tropical forest—and for travelers versed in European agronomy and forestry, fire was long condemned for tethering people to a traditional agriculture that was slovenly, inefficient, and ecologically damaging. Early archaeologists often speculated that pervasive swidden was a cause of the Maya's dramatic collapse. (Even firewood gathering for urban use was implicated in ruinous erosion on hillsides.) If the Mayan milpa was a foundational practice, modernizers of all stripes agreed it would have to go before the peninsula could advance. Yet these perspectives say more about the viewer than the viewed.[49]

In recent decades the Mayan milpa has undergone a reconsideration, moving from a custom to ban into a technology to learn from. Not a few observers have noted that Mayan agriculture thrived better and with less long-term harm than have European-derived replacements. Instead of denouncing Mayan agriculture as decimating a primary forest, they have celebrated the biodiversity that the intricate mosaics of milpa and fallow support, and the cultural sensibility that matched the nuances of sites to particular practices, in contrast to clearcutting, bulldozing, and generally razing the landscape in order to cultivate monocultures. Converting to plantation crops for export was a form of biotic looting, like conquistadores stripping temples to send bullion to Spain. Done right, the Mayan method was sustainable. It might not serve a global market for grown commodities, but it was a model for managing nature reserves. The ancient Maya used the land, as a recent review put it, "in a magnificent way." This pronouncement joined a general reassessment, at least among those sympathetic to Indigenous people, about the value of polyculture farming, and, by implication, of a polycultured society.[50]

The Classic Age of Mayan civilization is best remembered through its monumental cities—an impressive feat for any era. Yet those cities were supported by extensive agriculture, and both city and countryside were held together by water and fire. Together, they made the mortar that fused those stone monuments into an urban habitat, and they made agriculture possible in a place that would seem, by the principles of European agronomy, to have little potential. The templed towns and milpa-dappled countryside were, in a sense, mirror images, two versions of a built landscape: one of limestone, the other of plants and animals. Both were linked; both survive, if in ruins.

If archaeologists have tried to reconstruct the one, ethnobotanists have tried to reassemble the other. "The present floristic composition of many rain forests in the Maya area," as a study noted, "is a result of ancient silviculture practices." When one half of the system collapsed, so did the other, both overrun with a feral flora.[51]

VALLEY OF MEXICO

In central Mexico the two traditions, north and south, fire and water, meet on roughly equal terms. The high plateau punctuated by peaks attracted good rain in the wet season. A lake, Texcoco, filled the valley, which was a sprawling basin without an outlet. Cities clustered around the lake and on islands. Tenochtitlan, the Aztec capital, had three grand causeways to join the isle to the mainland. It collected not only tribute but the varied traditions of foraging, hunting, fishing, and farming that fed the population of Mexico from the Chichimecs to the Maya.

In his chronicle of the conquest, Bernal Díaz recorded the largesse available to Moctezuma: "For each meal his servants prepared him more than thirty dishes cooked in their native style, which they put over small earthenware braziers to prevent them from getting cold." Every day "they cooked fowls, turkeys, pheasants, local partridges, quail, tame and wild duck, venison, wild boar, marsh birds, pigeons, hares and rabbits, also many other kinds of birds and beasts native to their country, so numerous that I cannot quickly name them all." (Others did name them: shrimp, snails, salamanders, tadpoles, river oysters, turtles, turtle eggs, waterfowl from ducks to herons, moles, boars, snakes, iguana eggs, maguey worms, and grasshoppers.) Díaz dined on "maize-cakes, kneaded with eggs and other nourishing ingredients." He was brought "some of every kind of fruit that grew in the country." He sipped a frothy chocolate. Nor were these all the foodstuffs available. The lake had fish, and around its marshes were chinampas flush with tubers and vegetables, probably enough to feed Tenochtitlan.[52]

This is hardly the subsistence economy of primitive peoples. Few nations in Europe could match it for variety and volume. The gist, however, remained constant throughout the realm and beyond: maize, beans, squash, and chilis. Mexicans needed fire to prepare for planting, water to grow crops, and fire again to prepare the harvest for eating. But there were lots of other foodstuffs around, a biota cycled by wetting and drying and burning. The creatures that were hunted for meat lived on browse freshened by fires, or in habitats full of edges and patches, open spaces and covert, dappled by fires; even marshes could burn and have their niches renewed for waterfowl. Expansion into new lands proceeded by fire and axe. It is estimated that 15 percent of the central plateau was cultivated, concentrated in valleys and plains.[53]

The built environment, too, relied on mortar, burned from limestone. Timber might come from long-fallow sites, hauled away before the remaining slash was burned preparatory to planting. Roofing thatch from marshlands could be encouraged by proper burning. Within those constructed places people cooked their food, heated and illuminated with braziers, and gathered for stories, rites, and meals.

The two essentials were water and fire, both managed to make a world more habitable by humans. Water is a physical substance; its place on the landscape was modified with cisterns, small-scale irrigation, hillside terraces, ponds, raised-bed gardens, even a dike through Lake Texcoco in 1449 to help control flooding. Fire is a reaction. It can be called into being at will, it integrates its surroundings, and it propagates, which means both that it has to be fed constantly and that it can spread over large areas and across weeks and months. It could be altered in ways both crude and subtle. Most critically, it involved forests.

Woodlands typically supplied the necessary fuel, whether the fire free-ranged across mountains or nestled in braziers; and from Classic times there were restrictions placed on the exploitation of critical woods. Too-vigorous clearing on hillsides, whether for timber or fuelwood, could lead to erosion and unhinged watersheds; too much surface fire could prevent regeneration. In the thirteenth century a Chichimec prince, Nopaltzin, regulated fire-setting by permit. Tarascan kings appointed forest guards to oversee timber harvesting. Nezahualcoyotl, the king of Texcoco, established forest reserves with severe penalties for illegal cutting (later revised to accommodate firewood collection). When a park was created at Chapultepec—the oldest in the Americas—it centered on a woodland.[54]

Lives flourished and passed, tribes came and went, empires rose and fell, but sustaining them all—the substratum upon which human society depended—was an underlying rhythm of water and fire. Too much or too little water led to floods and droughts. Too much or too little fire led to ruin or famine. Even the grandest empires announced their arrival and departure with flames—the burning of temples and archives, the proclamation of a new order with ceremonial fires. The bundled reeds that marked the passages of the Long Count would be burned, but then they had arisen from previously burned reeds. The core cadence endured.

LINES OF FIRE HISTORY, FIELDS OF FIRE GEOGRAPHY

The grand transect from north to south is an abstraction of thousands of smaller transects from coast to altiplano, valley to mountain, nook to niche. All reflect various combinations of wet and dry, with each site working out a balance between water and fire. In the arid north, economies had more fire, but needed water to thrive. In the humid south, they had ample water, but typically needed some fire to function. The variations were myriad; chinampas needed no fire directly, and classic milpas only seasonal rains. The proportions shifted around a yearly almanac, and across years, decades, and centuries.[55]

Prior to European contact, an abundance of terraces, canals, weirs, and basins moved and stored water, stabilizing flow. But that system was complex, depended on abundant labor, and operated within a social system as syncretic as its agriculture. For central and southern Mexico, water commanded the most attention. After contact, the labor pool collapsed, and with it the ability to maintain an intricate hydraulic infrastructure. Fire likely became more prominent as a means to extensively farm, herd, hunt, and forage. One didn't replace the other—water and fire. They just shifted their proportions.

In passing through the countryside, Cortés observed that not a palm of land was left uncultivated. That intensive tending turned many landscapes into ecological hearths, vast braseros made of terrain and vegetation. That was fire's context. But as landscapes turned rough or unruly, so did fires. Likely, they became more visible. Undoubtedly, they assisted in midwifing a new world.[56]

NEW FIRE, OLD GOD

The populating of Mexico spilled along coasts, over passes, across valleys. The originating Desert Culture elaborated into scores of tribes, each with its language and consolidating homeland. At the time of the Spanish conquest some 14 major language families flourished. A counterpresence, this time cultural, emerged around 1400 BCE to partly consolidate that fragmentation.[57]

The civilization was known as Olmec. It featured urban centers, distinctive pottery and clay-fired figurines, monumental sculptures carved from basalt (massive heads are the best known), a controlled jade trade, and the idolization of the were-jaguar and Maize God, and it probably devised the Long Count calendar. The Olmec cultural model spread across the region as people had earlier, carrying embers from the Olmec hearth, again adapting and modifying to suit local circumstances. The Maya, spreading throughout the Yucatán and the western highlands. The Zapotecs at Monte Albán, centered in Oaxaca. The people who raised Teotihuacan, a city so impressive it became the new font of power and emulation. The Toltecs, at Tula, claiming the mantle of the Teotihuacanos after their collapse. "There is now little doubt," write Michael Coe and Rex Koontz, "that all later civilizations in Mesoamerica, whether Mexican or Maya, ultimately rest on an Olmec base." Or that, after the seventh century CE, the center of the informing civilization rested in Teotihuacan, its successor, to which all the efflorescence of the classic eras seems to point—the Rome of Mesoamerica.[58]

At the time of Spanish contact, the prevailing power was Aztec, or more correctly a triumvirate, a Triple Alliance of states in the Valley of Mexico, dominated by the Aztecs. The Aztecs had followed the traditional path to power, a Chichimec people trekking southward into central Mexico. They were a minor tribe, for a while serving as mercenaries. Steadily, they stabilized their position, consolidated power, founded a capital city on an island in Lake Texcoco, looked to Teotihuacan and the Toltecs for inspiration. As the seat of power moved, the dual temples to water and fire at Tenayuca were replicated in the Templo Mayor at Tenochtitlan.

Throughout the process of migration, invasion, and diffusion, local gods, like local practices, were added and adumbrated. Pantheons became syncretic, practices merged and sifted. So it proved for the Aztecs. If the basis of the Aztec economy, maize, depended on water and fire, so, it seemed, did the basis of their political power. Their symbol for war was a feathered coyote that spewed water and fire.

Like the Old God, fire was all around, related to much else, as unexceptional as sunlight and rain. It was present in quotidian life as heat and light, as a technology for cooking foodstuffs into meals, clay into ceramics, ore into metal, and as an inevitable accompaniment for rituals. It had its arc of practices through the day, not least the incessant need to gather firewood. It had a longer arc through the life of individuals—present at birth, marriage, and death (cremation was common among Teotihuacanos and Aztecs). It tracked the stately cadences of the calendar, the years bundled like reeds before burning. Fires started eras and ended worlds.

And it marked the cycles of conquest. Old capitals, once overrun, were burned—their temples scorched by fire, their codices and records razed by flames. Apart from practical destruction—fire can self-propagate in ways that an axe cannot—the symbolism could be widely understood. The same methods that prepared a new milpa for planting could prepare a new empire. In 1427 a great battle left the Aztecs supreme. The new Aztec king, Itzcoatl, and his principal counselor, Tlacaelel, celebrated with a ceremonial book burning. Rolls of deerskin, books of maguey fibers that folded like accordions—all went into the flames. With them of course also went the stories, the dynastic genealogies, the former glories, and the anticipated destinies, whatever was not recorded in stone. If a burning temple was the sign of triumph, burned archives were the means to confirm its legacy.[59]

For decades Tlacaelel continued, Coe and Koontz write, as "a kind of grand vizier to the Aztec throne through three reigns, dying in 1475 or 1480." He promoted major reforms and a new history appropriate to a people who no longer scurried around the edges and shadows of empire but had created one for themselves. "For this it was necessary to rewrite history, and so Tlacaelel did, by having all the books of conquered peoples burned since these would have failed to mention Aztec glories." As Camilla Townsend comments, "There was an old riddle in Nahuatl: 'What is the scarlet macaw leading, the raven following?' And the answer was *tlachinolli*, a conflagration."[60]

The Aztecs didn't invent the practice: it was what a newly triumphant ruler did. Nor did they invent the ritual of a new fire. But they remade them in their own image, made them a core celebration of their power and origin myth, and relocated the altar to their imperial heartland. The cycle of New Fires marked the epoch of the Aztecs.

The first New Fire had been ceremonially lit in 1351, announcing the arrival of the Aztecs. The fourth occurred in 1507, and for it Moctezuma II had a new throne sculpted, complete with solar calendar, to commemorate the advent of a new sun and world. It was the last. A dozen years later, the world of the Aztecs ended. Their temples and palaces were slashed, the chronicle of their achievements burned. A new sun rose from the east, a new empire pushed through the ashes, a new god presided over the established religion. Such was the cycle of fire and civilization. But beneath them persisted, unruffled, so embedded in daily life as to be banal and invisible, the Old God that still lit braziers, forges, and fields, the routines by which life in Mexico persisted.

THIRD SUN

FIRE AND SWORD

NEW FIRE

TEMPLE OF HUITZILOPOCHTLI

The first bad omen. Ten years before the Spaniards first came here, a bad omen appeared in the sky. It was like a flaming ear of corn, or a fiery signal, or the blaze of daybreak; it seemed to bleed fire, drop by drop, like a wound in the sky.

—*Florentine Codex*

THIS TIME *the invaders came not from the north but the east, and not by means of long treks over desert and mountain but across the sea. The conquest of Mexico was a further if extraordinary extension of a Reconquista that began in Europe and then sprawled across the Americas. The year that Christopher Columbus sailed for "the Indies," 1492, saw the fall of Granada and the final expulsion of the Moors from Iberia. Already Portugal had begun colonizing Atlantic isles, notably Madeira, the Azores, and the Canaries, and had doubled the Cape of Good Hope, announcing the passage to India and the Spice Isles. Portugal and Spain had sparred in Iberia, then down the African coast of Guinea, then in the Canaries, which Portugal finally ceded to Spain, and then over rival routes to the east. What might otherwise have simmered as dynastic squabbling in Iberia or Morocco went to the West Indies, and from there to Mexico, the coasts of Mesoamerica, and South America.*

The newcomers came for God, gold, and glory; carried sword, cross, and torch; set into motion global transfers of flora, fauna, diseases, and humans; and reorganized the peoples, waters, landscapes, and fire regimes of Mexico. They razed the old order—this went quickly. They rebuilt a new world on the ruins—this took decades, stretching into centuries. Their arrival in the

Antilles and then the mainland was a shock wave that kindled sparks that surged and smoldered, crept and swept across the landscapes of Mexico.

But while readily outfitted to conquer, Spain was ill prepared to rule. The Reconquest of Iberia had taken seven centuries, with institutions adjusting accordingly; even the Canaries took a century to subdue. The conquest of Mexico took three years, and institutions, already well beyond the limits of routine communication, lagged and proved inadequate. Events far outstripped the ability of authorities to contain them. The conquest was good at looting, poor at cultivating. As Bartolomé de Las Casas observed, "For all the Spaniards, young and old, in the Indies, are occupied in pillage, some openly, others secretly and stealthily." Not until 16 years after Columbus returned from his first voyage did Spain establish the Council of the Indies as a governing body for its island empire.[1]

Most of Spain's overseas empire was forged, or foreshadowed, in the painful conquest of the Canaries. The Antilles, especially Hispaniola, offered a microcosm of what was to happen environmentally to Mexico. Thanks to introduced disease and forced social breakdown, the human population (island Arawaks) crashed. What had appeared to Las Casas as a veritable "beehive of people" was, within 40 years, shorn of its Indigenous inhabitants. What had been cultivated fields went to seed; what had been savannas overgrew with scrub, or survived as pasture; what had been a functioning society fractured into shards. Pedro de Córdoba, the Dominican provincial, reported in 1518 how disease, ill treatment, and deaths of despair had stripped Hispaniola of its teeming peoples. "San Juan has been destroyed in the same manner, and in the same way Cuba, Jamaica, and Tierra Firme are being ruined. There are in this island at present ten or twelve thousand Indians, male and female. By common consent there will be none left in four to five years." Then smallpox struck. The biota that had flourished at the time of discovery, leading Columbus to declare that "Hispaniola is a wonder," was broken and refilled with new cultivars, feral fauna, imported slaves, and invasive weeds. In many respects, the island never recovered.[2]

Yet Mexico was also different. It was larger, a thinning continent, not an outright island; it had social and ecological resiliencies the Antilles lacked; it had a reigning empire, a deep political order, not congeries of villages; and it possessed disciplined military power, not farmers lacking weapons, armor, or a warrior ethos. At the time of the Spanish conquest, Mexico's ruling peoples had themselves come to power by conquest. They could expect to overwhelm the Spaniards by sheer numbers as well as organized might.

Neither group anticipated the other. Columbus never accepted that he had discovered a novus orbis *instead of the Golden Chersonese of the Indies, and the sudden vision of Tenochtitlan silenced even hardened conquistadores; Arawaks, Maya, Aztecs, and scores of others had no inkling Spaniards existed. Sooner rather than later, the rival groups overcame any initial surprise and worked to achieve their ends. Still, the conquest came with stunning speed. Looking back, searching for foreshadowings of what only became apparent in hindsight, Nahua informants told Bernardino de Sahagún about eight omens that had appeared a decade before the Spaniards and portended a dreadful future. Half of them involved fire. There was "a great column of flame which burned in the night, shooting out such brilliant sparks and flashes that it seemed to rain fire on the earth and to blaze at daybreak." The temple of Huitzilopochtli burst into flames and burned to the ground. "As an omen and miracle which boded evil," lightning struck the temple of Xiuhtecuhtli and kindled a fire that burned it down. Comets blazed even when the sun shone; "fire streamed through the sky," racing by threes "from the west to the east with great haste and violence, shooting off bright coals and sparks of fire."*[3]

Whatever happened in Mexico, it seemed, however much the sword would dominate the saga of conquest, it would contain some element of fire. Flame even immolated the temple of the fire god.

> The second bad omen. The temple of Huitzilopochtli burst into flames. It is thought that no one set it afire, that it burned down of its own accord. . . . And now it is burning, the wooden columns are burning! The flames, the tongues of fire shoot out, the bursts of fire shoot up into the sky! . . . They could not put it out, and the temple burned to the ground.
>
> —*Florentine Codex*

IBERIA MEETS THE MEXICAN MOSAIC

The encounter between Spain and Mexico began with survivors of shipwrecks off Yucatán. In 1518 Juan de Grijalva reconnoitered the coast and paused at what would become Veracruz. In 1519 Hernán Cortés left Cuba, landed at Veracruz, and began his march to Tenochtitlan. Pánfilo de Narváez followed months later, with orders to arrest Cortés, but Cortés defeated him, absorbed his troops, and returned to Tenochtitlan, only to be driven from the city and then return again to lay siege to and destroy the Aztec empire and its capital.

The feats of arms are famous. Less widely appreciated is the role not only of Tlaxcalan allies but of a *pestilencia,* probably smallpox, apparently introduced by a black Moor with Narváez and carried from Spain via Hispaniola, which ravaged the Aztecs while they fought the revanchist Cortés. The conquest was not simply a matter of arms and empire, but of microbes and ecology. Alfred Crosby famously termed the shakeout that followed "the Columbian Exchange" (and later, "ecological imperialism") and described the "portmanteau biota" that accompanied Europe's explorers, merchants, settlers, and soldiers. The Conquest and its aftershocks refashioned not only Mexico's society and institutions, but its habitats.[4]

Conquest continued for decades as expeditions marched west to the Pacific, north toward the Chichimecs, and south to the Mayan lands and beyond. Three years after the fall of Tenochtitlan, Cortés led a floundering foray to Honduras. In 1540 Francisco Vázquez de Coronado departed to trek as far as Kansas. By the end of the century Juan de Oñate y Salazar established colonies at the headwaters of the Rio Grande in New Mexico. The northern frontier, clustered around mines and missions, was still expanding at the time of Mexican independence. As conquest continued, so did the collateral reorganization of lands, peoples, and institutions. Each new encounter was both a provocation and an opportunity.

The Spanish conquest remade Mexico. It remade earth through mines and abandoned terraces and fields; refashioned water through the loss of old waterworks and renewed drainage schemes; rearranged living landscapes by reorienting the scale and character of agriculture and pastoralism; and reconfigured ignitions, both those that were set and those that were withheld. Only air was spared. While the stories of the Conquest inevitably emphasize the sword, with fire a kind of reserve force to destroy what had been seized, fire was not simply a collateral consequence. It was the means to conjure up a new world.

In sketching the "personality of Mexico," Carl Sauer determined that the "two most important things to know" were "the patterns of life that existed before the coming of the white men and the changes that were introduced during the first generation or two of the Spanish period." That was as true for its lands as for the lives of those on them.[5]

WHAT WAS EXCHANGED

The exchange involved disease, plants, animals, and people. From the Old World the New got smallpox, cholera, yellow fever, malaria, measles, whooping cough, bubonic plague, typhus, tuberculosis, typhoid fever, and other pestilencias, all of them unknown to the Indigenous people. In return, the Old World apparently got syphilis. The transfer of flora was better balanced. The Americas shipped maize, potatoes, manioc, tomatoes, tobacco, cacao, paprika, American cotton, chilies, pineapples, avocados, and scores of other fruits and vegetables. Europe reciprocated with wheat, barley, rice, sugar, indigo, bananas, wine grapes, citrus, pasture grasses, and other staples, along with a cavalcade of camp-follower weeds from Kentucky bluegrass to the dandelion. Much of the exchange was deliberate, much accidental; seeds could travel incognito and survive for long times.

The interchange of fauna was more uneven and more visible. The New World had domesticated dogs, turkeys, guinea pigs, llamas, alpacas, and the Muscovy duck. The Old World was a Noah's ark of domesticated livestock that ranged from chickens, cats, and dogs to horses, pigs, cattle, sheep, goats, oxen, and burros, not to mention the untamed stowaways like anopheles mosquitoes and Norway rats, or the abundant pathogens that afflicted each of the newcomers and came with their hosts, and which in some cases could cross to native species.

The upshot was an ecological scramble for the Americas, a scale of biotic reorganization not seen since the end of the Pleistocene.

THE PROCESS OF EXCHANGE

It's easy to emphasize the shock of the encounter. Almost overnight, the dominant empire in Mesoamerica lay in ruins. The Indigenous population, slammed not just by war and social upheaval but by waves of infectious diseases, plummeted for a century. New species challenged and often crowded out the old. Livestock seemed to explode across landscapes like gas molecules into a vacuum. With land and people so tightly bound, it's tempting to assign the same collapse to Mexico's physical geography that was felt by its social demography. A world ruled by fire and sword overturned a previous one ordered by fire and water.

But the real world was messier. Both Spain and Mexico gave and got. Like exotics everywhere, some newcomers grew exuberantly, and some fell on stony ground. Thomas Whitmore and B. L. Turner II observe that "the scale of environmental transformation of Amerindian agriculture has not always been fully appreciated, the scale of environmental degradation associated with Spanish introductions has been overstated at times, and the contrasting ideologies of nature between the two cultures has been oversimplified." Besides, nature held the strongest hand at the table, and it couldn't be bluffed, trumped, or intimidated. Landscapes could not be replaced as completely as a ruling class.[6]

Some of the reassembly came with a plan—a template of sorts hauled across the Atlantic from Iberia. With its Mediterranean climate, an agriculture adapted to well-defined wet and dry phases, and a broad central plateau (Madrid was the highest capital city in Europe), Spain came with a complement of preadapted plants, animals, and institutions, further field-tested in the Canaries and then the Antilles. It had experience with agricultural monasteries, with commercial plantations (especially for sugar) on island colonies, with the reapportioning of lands seized from heathens, and of course with landscape fire, which was ineradicable throughout the Mediterranean world, basic to much of Spanish agriculture, and critical to transhumance.

Mexico, too, had preadaptations. One ruling caste extracting tribute had replaced another, with marginal differences for farmers and craftsmen. The land tenure that emerged built on existing systems. Polyculture farming was accustomed to dealing with many cultivars and could test and selectively absorb new ones. Many customs and traditional uses persisted and were allowed as necessary. Fire as a catalyst for cultivation and hunting was well understood, and could adapt from maize to sugar, and from hunting deer to herding cattle. The big disruptions were religion, in which a single jealous deity with a chorus of local saints replaced a pantheon of hybridizing gods; a population collapse that decimated Indigenous Mexico; and an irruption of livestock.

In places, Europeans and Indigenes occupied separate scenes. Both Spaniards and *indios* grew their traditional crops. Wheat could flourish in valleys; maize thrived on hillsides. Wheat went to market, maize to families. Mediterranean-derived cultivars withered in tropical lowlands. Wheat could survive occasional frosts that blighted maize. Commercial crops were cultivated as monocultures, traditional foods as polycultures. Europeans developed sugar plantations and left to Indigenous people the cultivation of cacao. Mexicans wanted maize, Spaniards wanted wheat—each cultivar was fundamental to their sense of themselves. The newcomers also wanted wine, which led to vineyards; Natives held to maguey fields. The newcomers craved sugar, which served as sweetener, preservative, and medicine, as well as a kind of cultivated biotic bullion that had bankrolled colonization in the Atlantic isles.

Mostly, the two systems mixed. Spanish agronomy absorbed Mexican elements, and Indigenous cultivation accommodated Spanish cultivars. In place of citrus and fruit trees, Spanish farms in tierra caliente planted guava and avocado; indio farms incorporated bananas, melons, onions, and sugar. Farms in tierra templada mixed sheep, pigs, goats, and European chickens with turkeys and Muscovy ducks. The chickens could fold easily into an agriculture accustomed to turkeys. Cattle and sheep could not.

In the end, New Spain caused more upheaval for Mexicans than for Spaniards, but at all levels, save the elite, the outcomes tended toward the hybrid—a mestizo society, mestizo landscapes, and inevitably mestizo fires.

LIVESTOCK, GREATER AND LESSER

Livestock were a special case: they transformed Mexico from a land largely farmed into one mostly herded and ranched.

Mexicans had raised dogs and turkeys and organized hunts on a landscape scale, but they had no equivalent to flocks of sheep and herds of cattle that could number in the thousands, even tens of thousands. New plants could find niches in the existing agricultural mosaic; new animals could not. Their sudden irruption into Mexico spread like a contagion. To some they symbolized loss, a degradation of productive farmlands into habitats fit only for marauding flocks. For others they could symbolize a newfound abundance, as valuable as bison herds to Great Plains peoples. Even today, livestock, notes Sergio Guevara, is "a principal source of protein in Mexico," the country's "most extensive form of land use," and a practice "of great economic, social, and environmental significance."[7]

Still, livestock required a restructuring, whether deliberately or by accident. Small farms incorporated small numbers of cows, goats, horses, and oxen for milk, wool, meat, and draft. Large herds demanded extensive land, a good fraction of which resulted from once-cultivated lands that were now deserted amid the demographic collapse. As the population of Mexicans imploded, the population of *ganado mayor* (horses, cattle, burros) and *ganado menor* (pigs, goats, sheep) exploded. Abandoned fields and once-irrigated terraces became pastures; oak and mesquite woodlands once harvested for acorns and mesquite beans and foraged for edible and medicinal plants became mast for pigs; northern grass- and shrublands too arid for expansive cultivation shifted from the hunting of rabbits to the pasturing of cattle and sheep. Loosely herded flocks could overshoot the capacity of the land to absorb that shockwave before eventually stabilizing. The role Tlaxcalan allies served in the Conquest, livestock served for the remaking of the Mexican landscape.

The transition occurred on many levels—the most fundamental being that from Spain to Mexico. Each country had a high, dry central meseta dotted with mountains and well suited to grass, a pastoral economy, large estates, and seasonal migrations across plains and mountains. Spaniards carried centuries of experience codified into law, tradition, institutions, gear, and practice, dating back to Roman times. Along with cattle, horses, sheep, and pigs came the whole apparatus of their management. Instead of bringing forage to flocks, the flocks were brought to available forage in complex arrangements. In Old Spain legal grazing occurred on *prados* (meadows), *pastos* (pastures), *ejidos* (communal lands around towns), *dehesas* (a pastoral-silvicultural biome), *montes* (rough woods and shrublands), and *rastrojo* (stubble and fallow). Since the thirteenth century the Crown had granted the shepherds' guild, the Mesta, a monopoly and special priorities over other land usage for their far-ranging transhumance.[8]

It was a settlement pattern tested and tweaked over centuries, not only by environmental conditions but by the Reconquest, and then by settlement in the Atlantic isles and the Antilles, before final export to New Spain and the other new colonies acquired by conquest. All this made large-scale ranching remarkably preadapted to Mexico. It was flexible enough to accommodate the peculiarities of the New World; its demand on labor was light; novel grasses replaced familiar ones; burning to keep down the woody scrub of abandoned terraces replaced burning to clear out the shrubs on dehesas. The use of transhumance at many scales helped retard overgrazing at any one site, especially as flocks and herds swelled in the early decades. It was an extensive system, not unlike shifting cultivation, that could scale quickly as lands many times the size of Andalusia fell to the conquistadores.[9]

Difficulties arose because the old buffers that decree and custom, land and climate, had erected in the various pastoral landscapes of Spain did not exist in Mexico and would have to be re-created. Even classic controversies relocated: the conflicts over access to common pasture and woods around towns, over livestock allowed to feed on stubble and unfenced fields where crops and gardens grew, over transhumant flocks moving from one seasonal pasture to another. Livestock in Old Spain existed within an agricultural system that reconciled herding with farming within a fabric of customs and law that dated back to the Hispania of the Roman empire. In New Spain a clash immediately flared between sedentary Indigenes and pastoral newcomers. All the old conflicts had to be renegotiated across the Atlantic. An offspring of the Mesta appeared. Officials readjudicated the old quarrels between cultivated fields and open-range grazing.

All this occurred at a time when the preexisting agricultural landscape of New Spain was disintegrating as its tenders succumbed to introduced diseases and the ills of conquest. The swarms of sheep and cattle, many effectively feral, could seem to the conquered to be another of the plagues visited upon their unhappy land. Then the population explosion of ungulates stalled and declined, as the first wave of servant species exceeded the capacity of the land to accommodate them, as new limits were established, and as haciendas created a context to contain the swarms. The easy grasslands played out, like placer mines. Further expansion would require the ecological equivalent of hard-rock mining, and it took the form of transhumance.[10]

Meat joined maize as a staple of the Mexican diet, hides joined bullion as an export to Spain, and wool joined, then often supplanted, native cotton for clothing. The horse was no longer the preserve of the privileged; it was as common among laborers as the plow and digging stick. A milk cow, an ox, a pig, and some chickens could boost the productivity of small landholders, while extensive pastoralism argued for large holdings, whether communal or private, and helped underwrite the rise of haciendas. Mexico never experienced mass human immigration; instead, it had livestock, whose arrival remade its social and biotic ecology.

A traveling larder, moreover, was ideal for frontiers. In central and southern Mexico the attraction was a pool of labor from Indigenous populations. In northern Mexico the pull came from missions and mines. The relative scarcity of labor made livestock the primary means of rapid colonization. Along with Iberian-forged, Mexican-adapted practices, that meant a transfer and fusion of fire. Fire-forage systems did for pastoralism what fire-fallow systems did for cultivation.

A HYBRID CONQUEST

In the end, conquest led to adaptation, partition, syncretism, and hybridization. Even where conquest was seemingly total, as at Tenochtitlan, the new order built on the structure and with many of the materials of the old—novel agronomies appeared like cathedrals rising from the ruins of Aztec temples. Indigenous farmers learned to accommodate wheat, peaches, sugar, and especially cattle and sheep; Spanish colonizers learned the virtues of maize, cacao, tomatoes, and avocados. As Karl Butzer observed, "In effect, the agrosystem had to be disassembled, each element tested more or less independently, after which the viable components would be reassembled."[11]

It took time, gradually emerging over the course of three centuries of Spanish colonial rule. The new ruling caste had its own mandates, system of tribute, and ideals of governance, but the ordering of people and land resembled less a blow of steel on obsidian than an intergrowing of new and native plants on a shared lattice. Conflicts were not cleanly resolved but subject to negotiations in which nature, too, was one of the bargainers. The process continued through the demographic collapse of the sixteenth century and the population's slow recovery through the seventeenth. The lightened human pressure even unburdened some of the landscapes damaged by intensive preconquest cultivation.

The kaleidoscope of Mexican history turned, and tiles scraped, abraded, struck, chipped, and flipped, all of which threw sparks into fields, pastures, mines, towns, and lands depopulated by disease and discord.

KEEPERS OF THE FLAME

At the core of the crisis were people. It was people who made and received the shock of conquest, whose ideas informed the environmental makeover, and whose labor would make it happen. Ecologically, people were the keystone species of environments, and a species monopolist over fire, the prime movers who refashioned Mexico after the Conquest, not always well, often ironically, and sometimes catastrophically. The bonding of land and labor, fused in agriculture, has reliably informed Mexican history. "In truth," observed Manuel Gamio, a founder of Mexican anthropology, "the agrarian question has been the most powerful factor in the development, past and present, of the Mexican people." The same holds for Mexican fire history.[12]

Demographics, then, were the largest component of the Columbian exchange. Since there were never enough Europeans, and since African slavery, while dominant in the Antilles, never rooted to the same extent in Mexico, the demographic saga hinged on the Indigenous Mexicans. New Spain depended on its Native peoples to supply the labor to farm, trade, and mine, and furnish new souls for conversion. At the end of the sixteenth century Viceroy Luis de Velasco II confessed, "The estates, buildings, plantations, mines, and herds, the monasteries and religious orders,—I do not know whether it would be possible to maintain them or improve them without the service and aid of the Indians."[13]

Yet the Indigenous population crashed. Plague after plague swept over the land—New Spain had major epidemics in 1521 (smallpox), 1531–32 (measles), 1545–48 (typhus), 1550 (mumps), 1563–64 (measles), 1576–80 (typhoid fever), and 1595 (measles)—while social organization disintegrated in the face of war, relocation, and forced labor that was often indistinguishable from serfdom. Social resilience dissolved, and the population plummeted. Some studies suggest that the worst plagues involved indigenous hemorrhagic fevers and rodents, now liberated by the breakdown of social and environmental order and enhanced by a multidecadal megadrought. When the sixteenth century ended, Mexico's Indigenous people had decreased by some 90 percent. The collapse was a world-order event, felt even by the global carbon cycle and perhaps contributing to the Little Ice Age. Over the course of the sixteenth century the great dying left most of the heavily populated regions of Mexico vacant.[14]

The tending of the cultivated landscape decayed in proportion to the loss of cultivators, with an estimated 30 percent reduction of land under an agropastoral system. Cultivated lands fell from 250,000 square kilometers to 50,000 square kilometers, of which 10,000 were actively farmed and 40,000 in fallow. Pastoralism claimed 125,000 square kilometers. Some 75,000 square kilometers passed out of cultivation altogether, reverting to a quasi-natural state occupied by hunters and foragers. Much of the vacant countryside filled with livestock.[15]

One variable, land use, chased another, population, and vice versa. The process was complicated by the division between newcomer and native. Viceroy Velasco memorably phrased the issue as a Mexico divided into "two Republics," one of Spaniards, the other of indios, saying that "they are so repugnant to each other . . . that it seems that the conservation of the former always means the oppression and destruction of the latter." Their relationships were further complicated by the requirement (which helped legitimate the Conquest) that the Spaniards bring civilization—in this case, Christianity—to the heathens.[16]

The initial response to rebonding land and labor was the encomienda, a loose adaptation of feudal institutions evolved during the Reconquest. It granted Spanish conquistadores land and Native labor. Within twenty years, however, the Crown clawed back power lost to first-wave freebooters. The New Laws (1542) abolished slavery and limited the scope of encomenderos by declaring the Indigenous people subjects of the Crown rather than of the would-be barons. Later, the repartimiento, a requisition of Native labor, replaced the encomienda, until it too was phased out, though compulsory labor never wholly disappeared as old habits found ways to reincarnate themselves. In the meanwhile, by happenstance, a third Republic arose, a hybrid middle, the mestizo, which eventually grew to be the largest in number. The three classes of people roughly align with systems of land tenure.

By the time the shakeout congealed into a new matrix, as much remained the same, or only slightly modified, as was different. Spanish titleholders controlled perhaps a quarter of the land, almost all of which lay in central Mexico, with satellite clusters around mines. Perhaps 45 percent of New Spain remained legally under the ownership of Indigenous peoples. The conquerors replaced the Aztecs as a ruling class and imposed their laws and mores; yet a little under half the country remained under the *costumbres* of the indios. Some 30 percent of land remained outside the active sphere of either, neither conquered nor regarded as inhabited, and so belonged to the Crown as *tierras baldías*.[17]

Like land without water, land without labor was worthless. Over the course of three centuries, from the sack of Tenochtitlan to independence from Spain, colonial Mexico sought to reconcile land and people. The fact that indios and mestizos furnished the bulk of labor, and the necessity of allowing those peoples to continue in most of their customary ways, meant that traditional fire use persisted, even if, like its practitioners, it was scorned as primitive and perhaps pagan. That tension persisted throughout the colonial epoch, and beyond.

HACIENDAS, RANCHOS, PUEBLOS

Land grants by the state, or legal recognition for land already held, were a primary means of organizing the colonial estate. Elites oversaw large holdings, originally encomiendas, later haciendas. A few were massive—de facto fiefdoms like that enjoyed by Cortés, the Marqués del Valle de Oaxaca. The Indigenous people had pueblos, which were traditional lands in villages and surrounding commons. The small farmer and landowner had ranchos. The reality was finer-grained, of course. Grants of land were many and varied—for agriculture, for pastoralism, for towns, for missions, for mines, for sugar plantations, for all manner of economic exploitation, which necessarily also constituted a social order either by feudal arrangements, coercion, salary, or some variety of peonage. Their development helped domesticate otherwise free-ranging livestock. The resulting texture of land inscribed the fire regimes of Mexico.

The arrangement favored the big. The resulting hacienda was a microcosm of Mexico, an updated latifundium, as self-sustaining as possible, as much entail and estate as agricultural factory. A typical hacienda had grazing, farming (often with some irrigation), woods for fuel and timber, and a workforce living within or adjacent to it. Arable fields lay in the valleys and floodplains; pastures, on the slopes. An hacienda might emphasize meat, wool, sugar, wheat, or maguey; it might be owned by a religious order, an aspiring capitalist, a grandee, or the Crown; but all were large, all integrated people and land, and all were critical tiles in the political-social mosaic that comprised Mexico. Apart from mines and cities the hacienda was where most people worked, where most people lived, where and how most people organized land. Which is to say it defined a social and political order, or, as Molina Enríquez put it, the hacienda was not simply a business enterprise ("La hacienda no es negocio"). The hacendado was the hidalgo of Mexico. The hacienda, "a legitimate successor of the large holdings in aboriginal Mexico," informed land use in Mexico up to the twentieth-century revolution.[18]

The ranchos began mostly with immigrants from Spain—an estimated 240,000 in the 300 years prior to independence; at the peak of immigration in 1601–25 some 4,450 people arrived annually. The number of ranchos rose after the reforms of Benito Juárez broke up church holdings, communal holdings of Indigenous peoples, the organized pueblos of the mestizos, and much public land. The rancheros constituted a rude rural middle class. There were far more ranchos than haciendas. In 1923 the ratio of rancheros (47,939) to hacendados (8,245) was 6-to-1, and in many places closer to 10-to-1. But the bulk of land and of power belonged to the hacendados.[19]

Critically, there were also lands held collectively, either as incorporated villages, unincorporated settlements of Indigenous people, or legal properties invested in groups or the state. Many originated in pre-Columbian times and were reconstituted by Spain as pueblos. Their core were lands known as *ejidos*, a term originally from Spain that referred to communal grazing lands. Such lands surrounded villages or incorporated towns, were held in common, and provided an inalienable estate that could provision villagers. The pueblo was for Indigenous Mexico what the hacienda was for Spaniards.

MISSIONS

The missions, along with their associated presidios, were a primary institution of expansion to the north, part of the storied "spiritual conquest" of Mexico. As with agriculture, with which they were linked, missions tried to redirect existing populations to new purposes—in this case, new forms of worship, which necessarily led to new forms of social organization.[20]

Religion is not easily separated from social life generally, and missionaries regardless of order—Franciscan, Dominican, Jesuit—believed that Indigenous people would be less likely to convert if their old lives continued. Tribes who lived by hunting and foraging were less amenable to the rituals of Catholicism than were sedentary farmers; they would have to be settled and taught a new economy to support their new confession. So while their intentions were less overtly commercial, and while the clergy were frequently fervent spokesmen for their parishioners, by forcing their congregations into another social pattern they changed their lives and their relation to their homelands. The mission was an hacienda adapted to frontier conditions. It could be just as disruptive to Indigenous life, and Indigenous fire regimes, as mines and markets.

The church owned considerable land. Many estates were endowed to support religious orders—the Jesuits proved particularly adept at organizing Native lives and turning haciendas to profit. They had estates devoted to sugar, ranching, and milling. New missions came with an ark of cultivars and livestock. Even far-flung Alta California converted by means of cattle, sheep, wheat, and peas as well as by means of sacraments, votive candles, and Masses. Some 17 years after its founding, the Mission of San Juan Capistrano held 430 cattle, 305 sheep, 830 goats, 40 pigs, 32 horses, and 1 mule; 36 years later the mission had 31,000 animals including 14,000 cattle and 16,000 sheep. These figures did not count the boundless livestock that escaped into the countryside. The production of wheat, maize, barley, beans, and peas expanded comparably, along with grapes, fruit trees, and olives. In their cultivation, as in their architecture, the missions were patches of Spain, channeled through Mexico, transplanted to the frontier. Over that same time period the Indigenous population at San Juan Capistrano declined 74 percent. The pre-mission landscape unraveled.[21]

The missions typically viewed Indigenous fire practices as pernicious, either forming part of pagan ceremonies or enabling a seasonally nomadic life, and on occasion as an outright threat. Removing traditional fire practices was a way to enmesh Native people with a new economy, a new religion, and a nominally more advanced civilization. But the disruption, particularly if accompanied by a population implosion, could end up worsening the firescape. Rewilded landscapes encouraged wild fires.

MINES

Mining mattered to Mexico. Mexico's precious metals were what most bound its economy to Spain's empire; its silver went to the Philippines and China as well as to Madrid. The insatiable demand for more put pressure on Mexico's landscapes beyond what Mexico could have mustered on its own. Mining broke down and pushed northward the old frontier with the Gran Chichimeca, creating an arena for mixing peoples and creating mestizo Mexico. On the eve of the war for independence, Humboldt estimated that Mexico was responsible for half the world's stock of precious metals. Bullion financed the Mexican state, if unwisely. So much flowed that the export unhinged Spain's (and Europe's) economy no less than Mexico's.[22]

As Spaniards replaced Aztecs, so silver replaced jade. The reserves of bullion gathered by the Triple Alliance were soon looted. The search for more sparked expeditions to all points of the compass. Great mines were quickly located in Zacatecas, and then beyond. Mexico's silver age lasted from 1530 (or 1550) to 1630–40; it revived a century later before again cycling down. The mines became magnets for settlement, often in places that otherwise offered little sustenance for farmers. The mines brought in people, and people needed to be fed; farms and especially herds flourished to supply them.[23]

Mines by their nature are extractive: they had a similar effect on everything around them (including people). They were especially hard on woods, which supplied timber for towns, haciendas, and mine shafts, and fuelwood and charcoal for those same purposes as well as smelting. The pace of clearing far exceeded the tempo of regrowth, and where grazing was common, the fauna browsed down reproduction into scrub. The mines and their biomes became equally exhausted—yet another example of a migratory economy, a shifting cultivation, in which slag heaps like mineral fallow littered the countryside. New mines replaced depleted ones in a leapfrogging frontier. Mexico's mines forged a new economy of Mexican fire.

The mines were a fire force-multiplier—an ecological distortion that reshaped the pyrogeography of their larger settings. Unlike milpas and pastos, the mines themselves did not burn, or not intentionally, though deadly fires did occur. Alexander von Humboldt recounted how the Pachuca mining district, "one of the richest of all America," had been "wholly abandoned since the terrible fire which took place in the famous mine *del Encino*" when its wooden scaffoldings were consumed by fire and "the greatest number of the miners were suffocated before being able to reach the pit." A similar conflagration shut down the mines of Bolaños.[24]

Yet the mines depended on applied fire as much as agriculture did; most mine technologies were pyrotechnologies. Mining meant the cooking of stone. Torches lit shafts and adits. Heating and quenching helped crack veins; the extracted rock was roasted prior to milling. Reagents were fired before use. The ores were smelted, or heated to quicken the amalgamation of silver by mercury; the mercury was itself burned away; the metal was again heated to assay and refine it. Without fire even Mexican mining was reduced to artisanal placering and prying out jade and turquoise. Lime for building involved a less complex process, but one equally reliant on burning. One of the great works of Renaissance scholarship, Agricola's *De re metallica*, published in 1546, the same year the foundational mine in Zacatecas opened, illustrates profusely the techniques of the time and their reliance on pyrotechnologies. All those fires, however, demanded enormous quantities of firewood, or, better, wood processed into charcoal, which furnished a higher, more constant heat.[25]

Like a slow-moving fire (imagine a fire in peat), the mines burned away the fuels of the surrounding countryside, leaving the biotic equivalent of pits. A daily scrounging for firewood was boosted to industrial-strength dimensions. It has been estimated that six square kilometers of forest land was cleared for each kilogram of silver produced. Like missionaries, prospectors, and herders searching for new sources to exploit, *carboneros* ranged in ever-widening gyres to find biomass that could be converted into charcoal. By the 1590s they had radiated out almost 400 kilometers from the mines of San Luis Potosí. From 1558 to 1804 the mines affected roughly 315,642 square kilometers of woodlands, "a bit more than the surface of the contemporary states of Poland or Italy." Adding the inevitable informal economy of mining boosts the deforested landscapes to 391,650 square kilometers, or "20% of the surface area of contemporary Mexico."[26]

The mining frontier experienced two surges. The first commenced shortly after the Conquest, with the fabled finds at Zacatecas, Guanajuato, and San Luis Potosí. Discoveries continued northward along the great belts of the cordillera and its volcanoes, into Durango, Chihuahua, and Sonora. A second surge began two centuries later, after it was learned that mercury could separate silver from gangue, thus making accessible ore that was otherwise too refractory. This second wave accounted for an estimated 70 percent of land clearing during the colonial era. The frontier founded some 450 mining communities of various sizes. They all lived by their fires.[27]

The mines were an unquenchable maw for firewood and for that twice-burned progeny, charcoal. Charcoaling was an ancient art, typically conducted on site, shifting with the availability of wood—a kind of milpa that grew a special fuel rather than maize and beans (oak was preferred both for its heat content and its ability to coppice). But more intensive measures led to special kilns to ensure a better product—ovens for baking fuel. They were a common feature in the pyrogeography of mining. Even travelers not noted for sensitivity to natural history could not fail to see them. Frances Calderón de la Barca commented on "the fires of the furnaces of Angangueo, a mining village, at the foot of some wild hills. We rode past the huts, where the blazing fires were shining on the swarthy faces of the workmen."[28]

The carboneros' quest for fire stripped their surrounding environment of woods, and even of combustible yuccas and bunchgrass where wood was scarce. It was a kind of ecological strip mining, worse than a fire because it denied good fire where it was needed and enabled bad fires where they weren't wanted. Charcoaling hauled biomass away and burned it in kilns and mounds that did nothing to recycle and renew. Instead, the practice typically exposed the countryside—almost all of it hills and mountains—to erosion and reconstituted those landscapes in fundamental ways. Often grazing became the primary agricultural use, and sometimes the only one possible, which further pushed the landscape into novel forms, best suited to cattle and sheep. If intense, especially in recurring drought years, the heavy grazing could retard reforestation.

The mines, in brief, became black holes that distorted the geography around them, a visitation from outside the ecological order that set into motion a wave train of aftershocks, irreversible interactions, and unintended consequences. Their ripples helped dismantle the hunting and foraging economies of the Chichimecs, forcing them to migrate or relocate into missions. The carboneros had an effect on Indigenous people similar to that of commercial buffalo hunters in the North American plains. Market farming reconstituted indigenous agriculture, while grazing could influence landscapes far from the mine itself. Collateral economies like towns added to the stress.[29]

The Indigenous people were not the only ones who saw their landscapes and livelihoods trashed. Like an open-pit excavation that consumes the community around it, the massive stripping of the landscape undercut the mines themselves. Typically, mining projects failed either because they exhausted their ores or because their shafts flooded. In Mexico, they were threatened because they lacked timber and fuel. Eventually, the state had to intervene with edicts to protect the indispensable mines, which translated into sparing the woods from axe and fire.

TOWNS

Then there were the towns. Pre-Columbian America had a flourishing urban geography and culture, ranging from villages to imperial capitals. Some cities and towns continued, some were rebuilt with new materials, many were relocated or reconstituted. In brief, the built landscape showed the same trends that characterized the cultivated landscape. Perhaps surprisingly, towns, too, were firescapes.

Monumental cities were made of stone, and Spaniards also insisted on stone, which required lime, which meant the cooking of limestone. Brick, too, was baked in kilns. The older structures, as one nineteenth-century visitor observed, typically had walls "of stuccoed bricks of great thickness." That made fires "of rare occurrence, and, indeed, it would be nearly impossible to burn up one of these dwellings. If a fire does occur, it is almost always confined not only to the building in which it originates, but even to the room where it first makes its appearance." But most structures were not so immune to flame, and while the stone and stuccoed brick did not burn, building interiors stuffed with combustibles did; even major edifices had interior scaffoldings of timber. Roofs required wooden beams and were covered with shake shingles; furniture was made of wood. Indigenous houses, if not adobe, were fashioned from scrub, thatch, and wood, but whatever their exterior, their inside held ample fuels. The built environment was typically the surrounding natural environment reconstituted. It was subject to the same stresses that led to fire. There was plenty of hazard.[30]

The risk of ignition matched it, since every form of interior lighting, heating, and cooking involved open flame and adjacent fuels. Only a remarkable degree of social control kept structures from burning down with the regularity of the surrounding scene. Workshops held hearths, forges, candles, and torches; flame and fuel lived nearly cheek by jowl, separated only by their human tenders. When the social order broke down, so did control over fire. Fires were set; fires were no longer tended; fires broke out; fires rampaged. There was a chronic background count of fires, with occasional flare-ups and big breakouts. The built landscape, too, had its fire regime.

Not intentionally, yet inevitably, the town was a firescape. It had abundant fuel. It had designated lines and fields of fire to define and contain the flames. It had smoke that settled at night and rose during the day like tides. Without their pyrotechnologies people could not live, but when they packed themselves, their fuels, and their fires together more tightly, they boosted the capacity for fires to escape and cause serious damage. Only so much of the natural landscape could be tamed by cultivation—enough to contain wildfire. But in principle everything in a built landscape could be prescribed and designed to promote the working fires of hearth and forge and to prevent the wild fires that slipped their technological leashes.[31]

Especially as the Native population plunged, settlement became more diffuse and urban conflagrations less likely. Occasionally fires slopped over their confines, as they did in milpas and pastos. But the fundamentals pointed to a keystone species, people, doing routine burning in braziers and kilns, and engaging in that most mundane of fire practices—the ceaseless search for firewood, a browsing for combustibles. Their foraging cleared the forest floor, pruned oak coppice, and trimmed low branches. Where sheep metabolized grass, fire could not. So it was also in built and cultivated environments: where people cleared or burned off the surface fuels, wildfire could not.

OLD FIRE, NEW FIRE, MIXED FIRE

Mexico being Mexico, any single process, event, or shock—whether a Little Ice Age or a conquest—will break down into particulars, like white light refracted through a prism. With fire, Mexico's tangled terrain and rooted diversity fractured and slowed the process of conversion before the parts were reassembled into hybrids and mosaics. Spain had its fire heritage, Mexico had its. After they met, digging stick against plow, milpa against transhumance, an odd amalgamation resulted. Thanks to the sword, Spain ruled, but the need to rely on Indigenous Mexicans for labor meant traditional practices (other than those associated with religion and politics) prevailed. Mexico was mostly governed by customary fires.

Fire as fire was not novel for Spaniards. Almost every expression of Mexican fire had an analogue in Iberia. Spain had a long-extant agriculture based on water and fire—irrigation in dry lands, a fire-fallow system throughout. Spanish mining relied on fire to process ores and work metals. Spanish Catholicism had religious ceremonies that included fire and sometimes featured it. Towns burned. Victorious armies set fire to temples and fallen cities. What Spain added was livestock, which, carried to Mexico, transformed fire for hunting and habitat on a subcontinental scale. Even Renaissance humanists like Gabriel Alonso de Herrera, whose *Agricultura general* was published in 1518, three years before Cortés sacked Tenochtitlan, included burning as a source of fertilizer, weed and pest control, and stubble removal, all with appropriate citations from the Ancients. The practices of what was termed "la quema del monte" persisted almost as long in Spain as in Mexico, complete with local terminologies and obligatory condemnation as an "archaic" system of cultivation.[32]

In brief, for Spanish newcomers fire was not strange. It just occurred in peculiar settings. What the Conquest did was to restructure, sometimes on the order of landscapes, the context of those fires. Conquistadores and their chroniclers recorded exotic religious fire ceremonies and fires that destroyed towns, but they did not record quotidian fire in Mexico's homes and habitats any more than Miguel de Cervantes did with Don Quixote in Spain.

Old fire habits persisted where European agriculture was not possible, where intensive cultivation, mostly based on water, was too difficult because of the site or because labor requirements could not be met, and where the old ways were not shattered because the land was too remote, Spaniards too few, or depopulation less severe. New fires accompanied transplants like sugar and livestock. But mostly odd ensembles of fused fire emerged, in what Birgitta Leander has called a *mestizaje ecológico*, which not only resembled but resulted from a parallel mestizo society. Inevitably, mestizo fire practices and regimes emerged.[33]

FLAME OLD

Fire records are sparse throughout the colonial era, from the sixteenth-century conquest to the nineteenth-century war of independence. Typically, fire was chronicled only when it destroyed buildings, accompanied an uprising, or threatened critical assets. Otherwise, there was no more reason to record the burning of milpas, montes, and pastos than to document the lighting of candles and cooking fires. Still, some observations do appear, and some reasonable guesses are possible.

Many old practices persisted. Traditional harvesting endured in most places, along with its various forms of burning. Even amid the wreckage of maguey fields, many peoples like the Otomís were able to survive by hunting and foraging. In Zacatepec, a version of fire hunting, a surround, occurred on the scale of two or three leagues. The grass was lit all around, which prevented escape, while drums, horns, shouts, and whistles frightened the animals toward installed nets. The circle tightened. Eventually, the hunters loosed their weapons, which killed and wounded many animals while others were trapped in the nets or caught and killed by hand. So great were the numbers of creatures taken that Viceroy Mendoza could not believe the reports and arranged for a personal demonstration. Reportedly, some 11,000 Otomís formed a great circle over five leagues long and repeated the hunt, with the viceroy himself and Fray Toribio Motolinía, one of the original twelve Franciscan apostles, as witnesses. So prodigious was the hunt that Mendoza ordered a great part of the catch released, which still left 600 deer and mountain goats dead, more than 100 coyotes, and abundant skunks, hares, and rabbits. In commemoration the plain was named Cazadero (the Place of the Hunt).[34]

The greatest persistence was the most mundane. The milpa remained the axis of postconquest agriculture and was present nearly everywhere either as a primary form of cultivation or a supplemental source on less productive lands. In places like the Yucatán Peninsula, European plows were useless and most European crops withered. Conquest broke existing societies, but old methods of farming had no real alternatives. In his summary of Indigenous life, Fray Diego de Landa noted that "Yucatan is a land of less soil than any I know, being all live flat stones with very little earth"—clearly unsuitable for a plow. But none was needed. "When they work the land, they do no more than gather the brush and burn it before sowing." After the rains came, the growth was "marvelous." That sentiment was seconded by Antonio de Ciudad Real, who accompanied de Landa on his second trip to Yucatán in 1573, and then returned in 1584–85. "It seems impossible to be able to grow maize" in such a land, he concluded, because the seed was literally cast among stones, but the Maya needed "only to set fire" and then sow in the ashes, which "served as dung." They "cultivated the earth with fire." The better the burn, the better the maize.[35]

Even where alternatives were possible, they were often implausible due to the collapse of labor. Especially in the depopulated terrains of postconquest Mexico, fire could spread over abandoned terraces and into areas previously informed primarily by water. Writing from central Mexico nearly a century after the Conquest, Juan de Torquemada noted that "all the highlanders and those from the hot lands" made milpas "on the mountain slopes and barrancas." They cut, then burned and sowed among the ashes, and reaped a "most abundant" harvest "without much work." Rainfall was sufficient. The site could be sown for "one to four years" before being left fallow. The practice was, he affirmed, "very common" throughout the mountain region.[36]

The milpa was a center pole that held up the diverse tent of indigenous agronomy, a practice that continues today. Often the milpa serves as an outfield to more intensive infields and gardens, granting a measure of resilience. Oscar Schmieder describes a syncretic arrangement typical of part of Oaxaca: "The Mixe agriculturist has some permanently cultivated fields, of small extent, occurring in patches in the rugged country. These are the *terrenos de aradura* and are highly prized. Too small, in the main, to support a family, the cultivator also makes temporary use of low-grade land. The vegetation is cut down, burned when dry, and the land is planted to corn and beans amid the charred stumps; each field is the property of a single cultivator." The Zapotecs developed "a more complex system of land holding, for a group engaged in making a common clearing, which was then divided into lots." Mixe families "moved frequently." In sum, the Mixe and Zapotec territories exhibited "a vastly complicated pattern which results from the mosaic of numerous small land holdings which are either cultivated or have been lying fallow for varying lengths of time." The Chiapas Highlands developed complex relationships between mountain and valley farms.[37]

A few years later Oliver La Farge observed south of Los Tuxtlas that the landscape divided into two forms: "The lower parts are covered with thick jungle; the higher, whether from clearing or through a change in soil, are open grass and oak country." The "jungle" could support four crops a year—corn mostly, supplemented with melons, beans, papaya, pineapple, and sweet potatoes, along with gourd trees. It was "cut and the dead wood burned on the field, after such wood as may be useful is carried off." "All trees and branches are cut in small pieces, and left to dry in the baking sun. When thoroughly dry, they set fire to them, and in a short while the ground is clear." (He thought the method "very wasteful.") He says nothing about the oak savannas, which were surely anthropogenic, but others reported regular burning of similar landscapes farther south, most likely for hunting and habitat.[38]

Not every fire was neatly contained. On the border of Chiapas, John Lloyd Stephens rode through a village only to find that "at a short distance beyond we were stopped by a fire in the woods." His party tried to detour but were balked by flames, and retiring along their initial route they found the fire quickening. "The flames came creeping and crackling toward us, shooting up and whirled by currents of wind, and occasionally, when fed with dry and combustible materials, flashing and darting along like a train of gunpowder." Their trail contoured along the ridge; the fire rushed from below. "The clouds of smoke and ashes, the rushing of currents of wind and flames, the crackling of burning branches, and trees wrapped in flames, and the rapid progress of the destroying element, made such a wild and fearful scene that we could not tear ourselves away." The flames threatened the village. Stephens's party waited two hours, then moved cautiously ahead. "The brush and underwood were burned away; in some places were lying trees reduced to masses of live coal, and others were standing with their trunks and branches all on fire." A hut lay in white ash. "Our faces and hands were scorched, and our whole bodies heated when we emerged from the fiery forest."[39]

FLAME NEW

The Conquest added new fires. They clustered around the transformed landscapes made by mining, new-crop farming, fast-assembled towns, and, above all, livestock. The mines were a mess, making an omelette of fire regimes. Hasty settlements with footloose populations lacked the social control and building standards that fire prevention required. And New Spain had its fire quirks as much as the monkey-hunting Mexicans did.

Sugar was a multi-fired crop. Sugarcane plantations in Veracruz, Cuernavaca, Cuautla, and Izúcar de Matamoros became seasonal fields of fire—fire set prior to harvest, fire used to clean post-cutting debris, fire in the waste from milling. As with field fires, these, too, could escape. Frances Calderón de la Barca watched from the "great hacienda of Cocoyoc" as the "sugar-house, or rather the cane rubbish, took fire, and the great bell swung heavily to and fro, summoning the workmen to assist in getting it under." Calderón's party "stood on the balcony, which commands a beautiful view of Popocatepetl," and watched. Most observers were less enthralled by the spectacle.[40]

The most expansive novelty was the various indirect consequences of the menagerie of livestock Spain shipped to the New World. The resulting pastoral economy operated in three fire regimes. In humid lands fire was essential to freshen grass and fight back the ever-pressing woods. Old grass was inedible grass, old prairies were vulnerable to being overrun by revanchist woods. Fire had helped create most of the savannas; routine fire sustained them. Pastoralism first took over old hunting grounds—herding replaced hunting. Then the desire for more forage created more grass.[41]

In arid lands, the effect was more complicated. Burning could control brush, but only if there was enough grass to feed a hot, timely fire. That required a lush rainy season or weak grazing pressure. An explosion of ungulates could strip the land of its grasses, which meant no available fuel, which meant no fire. How intense was grazing? It depended on local history and conditions. The vast, untended herds probably behaved like herds of other ungulates, wandering in search of forage, leaving old grasses alone and then being enticed back after the dry, unpalatable grasses were burned. Their numbers were limited by forage and predation, not least predation by the peoples of the Gran Chichimeca, who could regard feral cattle as analogues to bison or deer. The Sánchez Navarros, who had a latifundium in Coahuila that boasted thousands of sheep, accepted a certain level of attrition from Apache raiders as they did losses to drought or freak snowstorms. The immediate fire threat came from heavy winter rains that could temporarily fluff up combustibles. The longer-range threat came from overgrazing, which shifted the biome from grass to woody shrubs.[42]

The simplest solution was to move the herds to seasonally available pastures. Transhumance was an ancient practice, known in Iberia in two forms, one that crossed the central plateau and one that tramped up and down mountains. With so much of Iberian pastoralism transferred to Mexico, it would be hard to believe that the same burning patterns that accompanied transhumance in Spain did not also transfer. Certainly, they did in nineteenth-century North America, where Basques accustomed to burning in the Pyrenees fired mountain pastures throughout the cordillera as they left in the fall. With more fuel and fewer constraints, burning in Mexico must have been routine.[43]

In Spain (as elsewhere throughout the Mediterranean Basin) the practice came with hundreds of years of protocols and constraints. The routes of travel (*cañadas*) were defined; the interaction between flocks and adjacent fields and gardens was closely monitored by custom and statute. Escaped fires incurred penalties. Similarly, the dehesa—a pastoral-silvicultural landscape typical of Extremadura and Andalusia—was the product of centuries of juggled accommodation between woods, meadows, and herds extending back to Roman times. Its oaks provided shade for sheep, mast for pigs, and timber and cork for people; its understory fed sheep, cattle, and goats. Burning helped keep the scrub down and freshen the grass, but this occurred within an intensively managed scene where each activity provided a check on others.[44]

Little of this was initially true in New Spain, and like so much else that was hauled across the Atlantic, pastoralism had to be relocated, adapted, and enforced. Flocks trampled gardens, and fires surely romped where not confined by close cultivation and watchful residents. Many of the early irruptions of cattle and sheep occurred with little assistance—in fact, they occurred because there was so little effort to restrain them. The mariners of the great voyages had habitually deposited goats and seeds on newly discovered islands and left them to propagate, often to the ruin of the landscape (St. Helena became the most notorious example, a place finally fit only for deposed emperors). Livestock ran amok in the Antilles. In Mexico they had more room and fewer leashes. As Alfred Siemens observes of the lowlands around Veracruz, "Cattle were not so much herded as hunted"; the *desjarretadera*, a curved knife on a pole, replaced the lariat. Most likely the same disintegrating landscape that invited livestock also invited fires, and all could become wild together.[45]

Eventually a Mexican Mesta was established to provide some oversight and regulation. The shift in settlement patterns toward haciendas and latifundia, large enough to accommodate herds, not just family milk cows and pigs, imposed some order and husbandry. Herds too immense to control shrank to manageable sizes, with rotating pastures and systematic burning replacing overland treks and their trailing fires. Still, the primary check was the most elemental: the fact that stock and flame competed for the same vegetation.[46]

FLAMES FUSED

Although old and new could find separate niches, the story is mostly one of their interaction: competing for the same site, absorbing elements of each other, adapting techniques to new conditions, synchronizing two fires into one. The interplay occurred not only in forest and field but in politics, law, science, even national identity. Fire to harvest mesquite beans or hunt rabbits could interfere with the cycle of pastoralism; fires that escaped milpas could harm fuelwood gathering by carboneros; what one group regarded as a traditional and necessary burn could be viewed by another as malicious and destructive.

Some observers understood (even if they didn't like) the way old and new collided. From Monterrey, Alta California, in 1774, Captain Fernando Rivera y Moncada wrote in his diary how "there came a fire from the west that was burning the forage of the countryside, and as it neared the Presidio, the soldiers, servants and even I went out to fight it, not because of the danger to the houses but to preserve the grass for our animals." They succeeded. Then the weary captain turned his ire to the cause of the commotion: "The heathens are wont to cause these fires because they have the bad habit, once having harvested their seeds, and not having any other animals to look after except their stomachs, they set fire to the brush so that new weeds may grow to produce more seeds, also to catch the rabbits that get confused and overcome by the smoke." (A French traveler in 1841 noted that "frequently such fires smolder for several months, and spread from one end of the province to the other.")[47]

In brief, the Spaniards had one regimen of burning and the local tribes another, and as long as the two economies—one primarily of hunting and foraging, the other of plow farming and pastoralism—coexisted, their fire practices would clash. Nine years later California governor José Joaquín Arrillaga responded to "various complaints that have reached me about the serious damage that results from the fires that are set each year in the pastures by Christian and Gentile Indians" with a proclamation "to prohibit for the future . . . all kinds of burning, not only in the vicinity of the towns but even at the most remote distances, which might cause some detriment, whether it be by Christian Indians or by Gentiles who have some relationship or communication with our establishments and missions." To enforce the edict, commandants of presidios were ordered "to take whatever measures they may consider requisite and necessary to uproot this very harmful practice of setting fire to pasture lands," while the missions were charged to "do their part in instructing the Christian Indians not to commit such transgression."[48]

It's not hard to imagine similar scenarios playing out throughout the far-flung mission system, with added leverage provided by demographic collapse as diseases swept away Indigenous people and their close-tended flames. Fire didn't vanish—it couldn't, and even newcomers didn't want all uses banned. They just wanted fire practices suited to the New Spain they imagined in the Americas. They might fear wildfire as a challenge to social order, but fires as such they could hardly abolish, nor did they want to. So it's hard to know how much success the fire prevention campaigns had—probably some close to missions where troops and neophytes could be rallied against fires succeeded. It's hard to believe, too, that much could happen in remote sites, or that the basics could change before the local tribes were converted not just to Christianity but to another economy, or before the friction of two peoples abrading against one another eased.

What happened in Monterrey, Alta California, might stand for all those clashes, not recorded, in which former fire practices, perfectly suited to place and society, were no longer acceptable to the newcomers who wanted the Indigenes to eat domesticated wheat rather than grass seeds and to get meat by raising cattle rather than hunting rabbits. The sedentary fields, with prescriptions for burning fallow, were the complement of the sedentary societies the missions wanted; the Indigenous people were more inclined to migrate through the countryside, burning in specific times and places as opportunity allowed. The introduced herds and flocks of the Spanish needed winter fodder; the Native people wanted the grass burned in time for the winter rains. Being built to burn, California had its own ideas as well.

The same might be said for colonial Mexico overall. The best protection against bad fires was a settled landscape, with close cultivation, fuelwood gathering, and good fires. Instead, Mexico got a depopulated countryside, former farms and hunting grounds gone to seed, and newcomers who had different norms about what constituted good and bad fire and only marginal capability and time to enforce proscriptions. Like its people, colonial Mexico's fires could neither be banned nor wholly contained, and they were ever likely, when circumstances permitted, to break free in pyric riots. Its fires, in brief, looked a lot like Mexico itself.

From the time Cortés landed his first fleet, Veracruz was Spain's primary port of entry to Mexico. That applies to livestock as well as Spaniards. Here two seemingly incompatible ways to occupy land met, clashed, and eventually accommodated. It helped that Spain imported a system of cattle ranching that was in some ways preadapted, and it helped that Mexico had a fire-primed landscape. The Veracruz story was, if not a paradigm, a prototype, and perhaps an ecological parable of how two fire worlds met and sorted themselves out. As conquest segued into settlement, fire and hoof replaced fire and sword as the primary ways and means of the new order.[49]

Most Mediterranean transhumance involves seasonal relocations between valley and mountain, a pattern common to Spain, most prominently in the Pyrenees, and readily transferred to the Americas. Spain added another pattern as well, which evolved out of the seasonal campaigns of the Reconquest that carried their provisions with them in the form of herds and flocks; this, too, traveled to the New World in the form of long cattle and sheep drives. And there were local variants in Spain. The most significant for Mexico emerged in Las Marismas, the seasonal marshes along the Guadalquivir River south of Seville. The classic Extremaduran model of transhumance herded mostly sheep; the Andalusian model that embraced Las Marismas herded cattle, which moved between hills of oak-pine forest that greened in the winter (and were burned in late summer) and lowland marshes that were accessible in the spring as flood waters receded. In the marshes the cattle were left to themselves, which required mounted herders to move them back to the hills.

In Mexico, the marshy lowlands at Veracruz were bordered on the coast by tall sand dunes and toward the interior by a piedmont. The earliest cattle arrived from Hispaniola and were soon herded between upland and lowland—to the piedmont as it greened up from rains, to the lowlands as they dried. The rhythm was identical to that in Andalusia, except the seasons were reversed, and the oak-pine forests of the piedmont were burned prior to the summer rains. Interestingly, a similar cadence may have existed in preconquest times in which farmers, in Andrew Sluyter's words, "might well have cultivated wetland fields during the winter and piedmont terraces during the summer."[50]

With depopulation those hillside terraces and lowland ditches fell into ruin. Conquest replaced maize with meat, Huastecan farmers with Andalusian herders, and burning for milpas with burning for pastos. The pattern spread along the Gulf coast, with much of the land already receptive thanks to centuries of clearing and burning. Spanish colonists added livestock as leverage to broaden and maintain the impact. Later, land reform after the Mexican Revolution again partly reversed the conversion, allowing "ejido sugarcane to compete with hacienda cattle."[51]

What didn't change was the continuity of fire. In 1571 the priest at Veracruz recorded fires set to sustain "meadows," annually kindled around Christmastime. In the 1850s Carl Sartorius reported how vaqueros, after the herds had been driven to the lowlands, would fire the uplands "partly to destroy the clouds of tormenting ticks and tarantulas, partly to call forth a new crop from beneath the ashes." Over the centuries, populations ebbed and flowed, *estancias* grew, shrank, and reorganized, and the particular character of fire evolved with them. Its purpose changed, the timing and manifestations of its presence changed, but fire persisted. In the absence of burning neither farm nor pasture was possible, especially in the lowlands, where dense shrubs and woods would soon reclaim clearings.[52]

The Conquest had provoked an agrarian shift from milpas to pastos, which caused burning to alter its timing and scale. The Mexican Revolution would later stimulate a return to the land by the landless, mostly mestizos. As the large haciendas were broken up, a mixture of people moved in, and with them a stewpot of farming, herding, and fallowing. But among the big and the small landholders, even where the soil was fertile for crops, of which an impressive variety thrived, pressures pushed toward *ganaderización*, the dominance of livestock. This remained true after places like the Sierra de Santa Marta, south of Veracruz, were gazetted as biosphere reserves. Burning for pasture overrode other reasons for and patterns of burning.[53]

As Juan Revilla, a resident of the Sierra de Santa Marta, described, "The paddocks are burned in the month of May of each year, three months later, when the grass reaches one meter in height, the cattle are introduced. In August the grass blooms. Cattle eat and disperse the seed in their feces in the paddock. Cattle are removed between September and October, to allow the grass to reproduce by seed. In November, the cattle are reintroduced to the paddock and the cycle is repeated." It is a rhythm that would have been recognized 250 years earlier, and a trend of land use that recapitulates what followed in the hoofprints of the Conquest.[54]

The shock and scramble of conquest was followed by an exercise in syncretic ecology. One response was to replace small-scale cultivation with large-scale plantations, multicropping with monoculture, and subsistence hunting with livestock herding, and to match a shrinking pool of labor with less intensive varieties of agriculture. Abandoned land created opportunities for novel agro-ecosystems; it also created vulnerabilities when combined with weather systems beyond human control. Years were too hot or too cold, too wet or too dry. Epidemics continued to abrade human populations, while epizootics swept over cattle, sheep, and pigs. A society that had adapted to a broadly temperate climate and a political order challenged by Chichimec raiders met environmental extremes and invaders from across the Atlantic.

Another arrangement replaced small-scale milpa farming with large-scale maguey plantations, or what Bradley Skopyk has termed a *metepantli* system, in which agave was cultivated on hillside terraces. The maguey—a native agave—was distilled into pulque, an alcoholic drink. A state-sponsored system oversaw the resulting commerce. Then the new landscaping on friable earth met the climatic shock known as the Late Maunder Minimum, which prompted an outrush of soil erosion and new rounds of land desertion in the Valley of Teotihuacán and the Zahuapan River basin in Tlaxcala.[55]

In many places the mass wasting buried some fields and wetlands even as it created new sites for fields elsewhere. With less intensive cultivation, a result of depopulation, the land regrew with grasses and shrubs—from one perspective, simply weeds, the biotic debris of abandonment; from another, fallow, a fresh and more expansive version of the fuels required for fire-catalyzed farming. In a naturally fire-prone land, such vegetation was certain to burn. Viewed from the first perspective, fire burned like carrion feeders in a rewilded landscape. Fire free-ranged because fire was destined to burn, and, ever an opportunist, it burned when dry spells, wind, and spark met. From the second perspective, those fires appeared as cultivated flames, part of reinitiating agriculture. They were a means to reassert control through the hallowed principles of close tending.

The most contentious response was to convert arable to pasture. On this topic the general narrative is known: places without sheep, cattle, or pigs now had them in abundance. But there were no ecologists to record the enterprise, certainly no one ready to code the story in a language easily translated into today's categories. So the details remain murky, along with the meaning and moral that can transform narrative into fable and apologue, which seems to be one of the end points, because an irruption of sheep can stand as a literary proxy for the Conquest itself.

An example is the classic study of the Valle del Mezquital by Elinor Melville. When Spain first encountered the valley, it was an agricultural mosaic, irrigated and tended by the Otomís. By the time Spaniards completed their possession, it had become a semiarid landscape full of cacti, mesquite, invasive grasses, and agave, with a spotty capacity for agriculture, ruled by newcomers, and mostly occupied by sheep. The details of this transformation matter—the particulars of place, the particulars of process. The Valle del Mezquital was not uniform and itself ranged along a climatic frontier between the temperate and the semiarid; its conversion came and went at varied rates and times. But the end points are known, and a rough chronology is possible thanks to records that document the transfer of land tenure from Otomí to Spaniard.

The first phase (1530–65) saw expansion out of the Valley of Mexico into the Valle del Mezquital. By the 1570s the land was saturated with sheep; "intensive small-scale pastoralism" supplanted "intensive irrigation agriculture." An era of consolidation set in (1565–80), then a final takeover (1580–1600). By Spanish tradition grass was held in common, its use regulated by long-established custom; but the eruptive sprawl of the Conquest overran landscapes before any Mexican alternative to such customs could be established. Pastoralism resembled a gold rush, with placers of rich forage on relic fields and a mother lode of rough pasture in the mountains. The early outrush had gushed everywhere, like the outflow of a thunderstorm.[56]

As grazing pressures intensified and conflicts between flocks and farmers flared, pastoralists looked to Spanish models for relief. Seasonal transhumance moved flocks between pastures, unburdening the best sites from the unbroken stress of tooth and hoof; haciendas (more like latifundia) absorbed grazing within a newly consolidated pattern of agricultural use; and a Mexican Mesta organized pastoralism, with some control by the state, to prevent hostilities between Otomí and Spaniard and between regulated and uncontrolled grazing. Such institutions served an environmental role similar to the political one the establishment of land grants and legal tenure served, as the Crown reined in the freebooting conquistadores who were responsible for so much of the damage.[57]

By the seventeenth century, however, so suddenly (in places) had the sheep swarm irrupted, so completely (again, in places) had it overgrazed the former regime, so far had lime mining deforested hillsides, so thoroughly was the hydrology of the region disrupted, that it was hard to imagine that the Valle del Mezquital had ever been anything but the hardscrabble countryside that it had become. Sheep had not simply seized a niche, the way grapes might replace tomatoes; they had reshaped the habitat. Not until industrial-scale irrigation with transport across watersheds appeared did the valley reclaim a heritage of intensive farming. The outcome was partly a matter of scale. Just as maguey went from being one component of a biota to a monoculture, so livestock for milk, meat, wool, hide, tallow, and the plow—all embedded within a larger farm ecosystem, the faunal equivalent of a kitchen garden—had metastasized into swarms that descended on the land like locusts. Those Indigenous people who felt the onslaught most keenly regarded the teeming flocks as an addition to the "ten plagues" visited upon them since the Conquest.[58]

And fire? The official records are mostly silent. It is implausible, however, to believe that fire did not flourish early on abandoned lands, that in drought years (like the great *sequías* of 1543, 1576, and 1594) it did not range widely, that during the expansion phase it was not used to promote fresh forage and stimulate the conversion from farm and monte to pasture, that it did not find a niche in transhumance as old pastures were burned when flocks left for new ones, that, until grazing overreached and left little for flames to consume, it was not a seasonal presence like the monsoonal rains. Along with microbes and fungi, herbivorous insects and small mammals, fire contributed to the decomposition of living landscapes. It just blazed openly where the others combusted in unseen metabolism.

Conquest by ganado mayor and menor had no single cause or singular effect. It resulted from both pushes and pulls—pulled by stubble and fallow, by grassy reclamation of abandoned fields, by oak savannas, and by grasslands novel and ancient; pushed by the inexorably increasing flocks, by markets for meat and fiber, and by the peculiar logic of sheep production in which wool could be cropped annually, as milk could be taken from a dairy cow, without having to kill the animal, unlike with cattle, whose hides, tallow, and meat demanded culling. For a brief period, unbounded forage was available and normal constraints were missing. As with Spain's military conquest of Mexico, it took decades for the state to corral the free-ranging into the herded.

Eventually, the scene calmed. The new regime was as much a matter of politics, laws, and scholarship as of ecological equilibriums. It persisted even through the War of Independence and the serial wars and unrests that followed. Not until a reconstituted form of combustion, and not just new species and societies, arrived in the late nineteenth century did the fundamentals of fire alter.

THE FIRE THAT FOLLOWED THE SWORD

What were the ecological, and hence fire-ecological, effects of the Conquest? It depends on specifics of time and place, and, because people were the keystone species, it depends on how the Conquest affected people's numbers, their social organization, the character of their agriculture, their connections to the outside world, and the ways in which they adapted to the newcomers, human and otherwise—as well as how both Indigenes and Spaniards understood what was happening.

Early on, a year before the New Laws, King Carlos had declared that the natural realm of New Spain—its forests, its waters, its pastures—was to be held in common, thereby granting Indigenous people free and open access. The indios were subjects of the Crown and deserved protection not only of life but of livelihood. That edict was reaffirmed by Philip II. Yet officials distrusted the means by which much of that livelihood was possible. Bishop Alfonso de la Mota y Escobar denounced agrarian burning and charcoaling: "The native peoples' use of the milpa to cultivate land causes great harm [to the forests], as does their harvest of firewood and conversion of coppice to charcoal." But the Natives could also protest against Spanish forest abuses that harmed their livelihood. When villagers from Tlalmanalco complained about overcutting, the Spanish viceroy vowed to enforce regulations, not only for the villagers' sake but also to protect construction timber needed by Mexico City.[59]

In reality, the indios had access only so far as their claims did not harm the Crown's need for wood for mines, ships, structures, and fuel. The mines, upon which the finances of the state depended, soon highlighted the conflict. In 1550 the disruptions around Taxco created a wood crisis that led Viceroy Mendoza to ban burning. The goal was not the protection of mine and town so much as the regeneration of the forest upon which the future of both mine and town depended. A similar logic applied to forage. In the 1570s officials denounced inappropriate fires as "an injury to the conservation of pastures for livestock" and "a vice for no profit or purpose." Guards were enjoined to watch for outbreaks. Penalties included a year's exile and stiff fines (for Spaniards) and "100 lashes and exile for three years" (for a mestizo, mulatto, Morisco, or Indian). The provocation is not specified, so it is not clear whether the burns were arson, protest, carelessness, or traditional burning that propagated where the hacendados and alcaldes didn't want them to go.[60]

That ideals and practice clash, that the wishes of the state and those of its subjects disagree—such conflicts are to be expected. What matters is how in colonial Mexico the state justified its decisions and how fire figures in that discourse. Ecologically, milpa and *carbón* were far less damaging than the aggressive mining, logging, and herding introduced by the Spaniards. Such destruction was not only a knock-on consequence of conquest but a proxy for it, and had to be justified by a similar logic. What Spain used to argue the legitimacy of its New World empire was that it brought a higher civilization to the indios. That assertion mandated attention to the welfare of the Natives, but it also allowed the authorities criteria to justify their decisions when conflicts arose, as they inevitably did.

What made fire special was that, for Europe's intellectual elites, open flame symbolized primitivism. Humanity's use of fire separated it from other creatures, but the character of that use sorted out higher from lower societies. Their ubiquitous use of fire in circumstances unwelcome to the state indelibly branded Mexico's indios as backward, and was a visible cause of that backwardness.

This perspective was present in the Renaissance, but it became overt in the Enlightenment. Like Christianizing the heathen or, later, bestowing upon the benighted the Reason of science, finding alternatives to flame was a duty of the state, especially when it ruled a civilization at least equal to itself and founded on a very different understanding of the world and humanity's place in it. The state was willing to overlook landscape fires set by Native people as long as they didn't threaten critical infrastructure; the church was keen to condemn fires used in pagan rituals while celebrating those baptized into Christian liturgy.

Here was the principal occasion where vernacular fire concerned elites. Fire was not just an agricultural technique, an ecological process, or an ethnographic curiosity, all of which could be invisible, the white noise of the everyday world. Fire could serve as a conspicuous index that placed Mexico along a scale of the world's civilizations. That index, reincarnated again and again in new avatars, has continued since the Conquest, through the epochs of colonial Mexico, independent Mexico, postrevolutionary Mexico, and industrial Mexico. It continues today.

FIRE IN THE MIND

For something arguably so axiomatic to Mexican geography and agriculture, fire remains remarkably absent in the written record, even in the gaze of a Europe that had its own tradition of fire myths, fire-informed natural philosophy, and classical references (of the 12 Olympian gods, 2 were devoted to fire—Vesta to the hearth, Vulcan to the forge). The easiest explanation is that observers document the unusual, not the pedestrian, and that scholars develop only those themes they are trained in or are compelled by circumstances to ponder. Outside of religious ceremonies, Mexican fire at the time of the Conquest was not sufficiently compelling as either a problem or a curiosity.

Much of the New World was as perplexing as it was unexpected: its very existence questioned inherited learning. It had unusual flora and fauna, some useful, some merely exotic. Its terrain held natural wealth in ores and an abundance of volcanoes. It beckoned the curious; natural and human history needed intellectual order akin to what the New Laws and land tenure review were bringing to politics. Above all, its teeming peoples were a challenge and a puzzle. Where did they fit in the great Chain of Being? Did they have souls? What was the proper relationship between discoverers and discovered?

These were issues critical to the Spanish New World. They were discussed within a government that was still creeping toward the status of an early modern state, within the ecclesiastical revolt known as the Reformation, and amid an academic tradition still infused with late medieval scholasticism and grappling with a Renaissance. When observers turned their eyes toward landscape fire, they saw it through the spectacles of the scholarship of the day. It might be regarded as an element, in a symbolic sense, but oxygen would not be discovered for another 280 years after Tenochtitlan fell. Fire only mattered when it blew out of its quotidian rhythms and threatened towns, pastures, forests, and mines, or when it illuminated pagan religious ceremonies.

THE RENAISSANCE ENCOUNTERS MEXICAN FIRE

Mexico's silver age was a golden age for an educated elite who experienced the New World. Reports and books, some of immense length, emerged not only from the dominant missionary orders—the Franciscans, Dominicans, and Jesuits—but also from inspectors sent by the Crown, Renaissance scholars, and even the conquistadores or their secretaries. *Historias* of the Indies and the indios were full-spectrum CAT scans of the encounter: military, missionary, and mental, encompassing plants, animals, ores, mountains, peoples, beliefs, rituals, languages, politics, anything and everything that would explain what the Spaniards had found and how better to exploit and govern that new dominion. The first *colegio* in the New World was established in Mexico City in 1536. The first printing press followed in 1539.

Scholars were educated in the classics of Greek and Roman antiquity, in rhetoric and logic, and in theology. Their first instinct was to tinker with existing categories and schemas and appeal to the ancients for models. Scholars looked especially to Pliny the Elder, the great encyclopedist of Rome, for inspiration. But they had Livy and Herodotus on political history and geography, Theophrastus on economic plants and fire, Virgil's *Georgics* on agrarian life, and Aristotle on the four elements and much of everything else. With regard to fire they had commentaries that mixed ancient sources with contemporary examples like Gabriel Alonso de Herrera's *Agricultura general* (1518), which considered farming, Vannoccio Biringuccio's *Pirotechnia* (1540), which surveyed the panorama of fire-catalyzed technologies, and Agricola's *De re metallica* (1546), which dealt with mining and metals.

The genuine mystery, and most awkward discovery, was the Native Americans, who were critical for labor and whose conversion provided a justification for the conquest. When the educated elite commenced their volumes on *la historia general y natural de las Indias*, the Native people typically dominated the discussion. Virtually all mentions of fire occur within larger commentaries about Indigenous practices and ceremonies.

The primary sources were written by men like Bartolomé de Las Casas and Bernardino de Sahagún, who were formally educated and personally engaged in the events they describe. The Dominican Las Casas exposed the cruelties of the conquerors; the Franciscan Sahagún, the crushed society of the conquered. Las Casas had arrived early in Hispaniola and saw the havoc firsthand; he argued the case for the Native people that led to the New Laws that gradually ended the encomienda and its echo, the repartimiento; he served as bishop of Chiapas, and he wrote tomes on the history, both human and natural, of Spain's Caribbean and Mexican possessions. Fire appears only incidentally in these writings, in relation to sacked villages, tortured prisoners, and cooked meals, and as a means of cultivation. His *Short Account of the Destruction of the Indies* (1542) was a polemic. His more exhaustive compendium was not published for centuries.

Bernardino de Sahagún had a Renaissance scholar's facility with languages but a missionary's unwavering zeal. He learned Nahuatl, established the Colegio Imperial de Santa Cruz de Tlatelolco, and over the course of five decades painstakingly compiled the 12 volumes of his *Historia general de las cosas de la Nueva España,* among the best-preserved records of the Nahua world at the time of the Conquest. More than anyone else he translated Indigenous knowledge and empathized with Native understanding. Yet his purpose remained the furthering of the missionary project. As he put it, he resembled a physician who studies a disease so that he might better treat it. He viewed suspiciously the rapid conversion of the Indigenous people to Catholicism, especially the apparent fusion of the Virgin Mary with Tonantzin, an Aztec goddess whose Nahuatl name means Our Mother, and the puzzling myths about the gods Tecuciztecatl and Nanahuatzin, who sacrificed themselves for humanity's benefit. Toward the end of his life, Sahagún fretted that the True Faith would wither under the blows of depopulating plagues and a deracinating syncretism.

The tensions and fusions that would define Mexico were well under way. But there was no appreciation for how they might express themselves in fire. As a ceremony of a competing cosmology, the New Fire merited close attention; the fires in braziers, fields, and new pastures did not. Such things were known in Spain; only their novel setting in Mexico differed. In any event, Sahagún's summa, originally published in Nahuatl with later sections translated into Spanish, failed to propagate outside the authorities, a victim equally to the closed colonies favored by Spain and the Counterreformation unleashed by the Council of Trent. The work disappeared, a lost manuscript almost from the time of its completion. A copy finally surfaced in 1793 in Florence (hence its customary title, *Florentine Codex*).

The real New World rival to Pliny and Herodotus—as he himself boasted—was Gonzalo Fernández de Oviedo y Valdés. Like his contemporaries, notably Las Casas, Oviedo had personally witnessed the encounter. He spent half his life in Europe and half in the New World. He knew Hispaniola and the Spanish court firsthand; the 35 years in which he wrote (1514–49) pivoted on the Conquest and its aftermath; and when he became the designated royal chronicler, he had access to all the official documents available. Like Thucydides, he was both participant and recorder; like Cicero, his writings helped inform the debates of his time. He regarded his own labors as monumental, as fashioning a new kind of history for a New World, as doing for the Americas what the ancients had done for antiquity. He boasted that his was the greatest such work since the time of Adam.[61]

For all that, there is little regarding fire. It appears in the history the way a candle might in a room—illuminating, but not itself worth mentioning; it's just there. So it becomes visible during the tales of conquest (*la guerra á fuego y á sangre*), as villages are burned; in descriptions of food preparation, from barbecues to sugar to cacao; in explanations of how canoes were fashioned from logs; in the methods of making fire (with echoes of Pliny and Vitruvius); in descriptions of charcoal and smelting; in dozens of allusions, such as when Fernández de Oviedo describes a color as resembling that of women's hair when it gets singed. And there are endless references to the many volcanoes of the Indies, the unquenchable fires that had lit the course of the Conquest from Tenerife in the Canaries to Hispaniola to the Valley of Mexico. But *fire* was not an organizing category of discourse. Often it appears as no more than shorthand for anything that produces heat and light.[62]

The text was written, rewritten, and revised unceasingly, yet while portions entered the political debates of the times, the entire 50-book opus was not printed until 300 years later under the auspices of the Royal Academy of History. Fernández de Oviedo had by then become himself one of the Ancients, kept alive in part as a source of historical data and as an honorific figure the same way Pliny the Elder continued to be revisited even after his encyclopedia had long passed its shelf life. For fire history he is relevant more for what he overlooks than for what he describes.

Instead, the telling observations belong to the Jesuit José de Acosta, who wrote his own *Historia natural y moral de las Indias*, trying to place the things of the New World within the matrix of the Ancients and Scripture. While he had "no doubt that a work could be written equal to those of Pliny, Theophrastus, and Aristotle," Acosta did not "find that vein" in himself, "nor would it agree with my aim even if I did." His grand goal was to advance the True Faith. His literary ambitions were more modest, and he was willing to encase them within the matrix of Renaissance learning, with references to Seneca and Virgil, Plutarch and Pliny, Josephus and St. Jerome, Aristotle and Holy Writ—a menagerie of references and a pantheon that must have seemed as baffling to Nahua audiences as the gods of the Aztecs were to Franciscan missionaries.[63]

With Pliny as a coarse model and Aristotle as an authority, fire deserved consideration as one of the four elements of the world. Apart from the occasional curiosity—fire-making by friction, cooking with hot stones in gourds, a fire drive embedded within a ceremony—Acosta concludes that "I see no special qualities of fire in the Indies that differ from those of other places." That observation might stand for all the *historias* that the encounter sparked.[64]

With one exception: Diego Durán, who, though born in Seville, was brought to Mexico as a young boy, arriving around 1540. He grew up bilingual and effectively bicultural in Texcoco. While most of his contemporaries studied the Indigenous peoples as they would a text, Durán knew them as neighbors, and empathized with them even as he came to loathe their residual paganism. A biographer noted that even when he spoke Spanish he was thinking in Nahuatl. Several of his contemporaries considered his works the "most exact" account of their subject and an indispensable reference.[65]

In 1556 he entered the Dominican order and became an ardent missionary, mostly in central Mexico. He commenced his monumental *Historia de las Indias de Nueva España* in 1571 and completed it in 1581. Its declared purpose was to advance the missionary enterprise by documenting the character of the Mexicans through their history, intellectual heritage, and religious beliefs and practices. He was dismayed at the persistence of pagan rites among the baptized, and he disdained his colleagues who fooled themselves and refused to see the shallowness of conversions because they did not understand the old ways and how the Native people had adapted them to Christianity. The spiritual conquest of Mexico had replaced one overt ruling caste and one religion with another, but they were grafts on deep rootstocks. Mexican religious practices tended to show a mixing of old and new similar to that of Mexican agriculture. Durán sought to sort through and expose the muddle. So thorough was his text that he was actually berated for possibly helping perpetuate the heathen rites and beliefs he exposed. Still, like so many tomes of that era, the *Historia* was not published for centuries and instead circulated in the works of others. Durán died seven years after he finished it.

The sprawling text consists of three broad sections. One relates the history of Mexico up to and through the Conquest. The second documents the religion of the Mexicans: their gods, rites, and calendar. The third was intended to continue the second, since the Devil was indeed in the details. Collectively they supported Durán's belief that "we who have been chosen to instruct the Indians will never reveal the True God to them until the heathen ceremonies and false cults of their counterfeit deities are extinguished, erased." Only when "all roots and stumps" have been extirpated can the Holy Gospel flourish. True conversion can only come by wiping out the "ancient idolatry and false religion," which is only possible if missionaries "understand the old religion profoundly."[66]

Because fire was so much a part of conquest, pre- and post-Spaniard, and because so many gods and rituals had dimensions related to fire, Durán's *Historia* is perhaps the greatest single source for fire's presence prior to European contact. Of course, his roster is rigorously selective: his passion was religious; he had no interest in fire as fire. He lists fires in feasts and processions, in ritual offerings and human sacrifices, and in the burned temples that signaled a battle's end. He records fires on altars, but not in fields. He speaks of the Divine Hearth attended by Aztec priests, but not the braziers that cooked the maize raised in burned milpas. What made his task tricky was what also made it essential: the fact that fire-accompanied or fire-themed rituals in ancient Mexico had echoes in pagan Europe, many of which Christianity had baptized and incorporated into its liturgy, and these could be easily confused with the vestigial flames of Nahua ceremonies, which is what Durán thought was happening and went unseen by his ignorant and complacent compadres.

Diego Durán's contribution to Mexican fire history is minor. He was assiduous in ferreting out fragments and expressions of Indigenous fire rites that had survived conquest and conversion. Yet in another sense he is a pioneer, foreshadowing all those later proselytizers who came to convert Mexicans to what they regarded as the best of civilization, whether this be Christianity, Enlightenment, capitalism, modernity, or economic development, whether by example or force, and for whom those inextinguishable flames were an enabler, emblem, and taunt. Substitute fire costumbres for fire ceremonies, and his successor missionizers follow in Durán's well-trodden path.

So it is that the paradoxes of Spanish scholarship might best be illustrated through the career of Diego de Landa, a Franciscan friar who first arrived in Yucatán in 1549 and promptly immersed himself in all matters Mayan so that, like Sahagún, he might better exorcise all pagan vestiges from the converted. His was, in a sense, an Inquisition-driven scholarship. His collective writings remain the fullest source of information about Mayan life and beliefs, but that is also because at Maní in 1562 he conducted an auto-da-fé that burned 5,000 idols and 27 codices, constituting the bulk of written lore about pre-Columbian Mayan civilization. For his apparent excesses, he was forced to stand trial in Spain before being exonerated and returning to Yucatán as bishop.

In his *Relación de las cosas de Yucatán* (1566) he mentions fire mostly within the context of myths and rites, and casually within milpa agriculture. Yet his larger contribution to understanding the evolution of Mexican fire is his justification for the Conquest, even when it involved compulsory conversion: "It is with reason, then, that Spain can glorify God in that he elected her among the nations for the remedy for so many peoples, and for which they owe her much more than they do to their founders or progenitors." The Conquest brought "justice, Christianity, and peace" to Yucatán. But if a "blessing" to the Maya, conquest was also a burden to the Spanish. They had obligations to oversee the uplift of the natives, even if that meant unremitting political control and paternalism by force. His apologia reads like a first draft on the White Man's Burden.[67]

It sounds to modern ears like legalistic hypocrisy, another version of the self-rationalizations that accompanied the Conquest, like the requirement that before battles conquistadores read a long proclamation (in a language unintelligible to its audience) inviting the peoples to yield to the Pope and the Most Catholic Monarchs of Spain for their own good and the good of their souls or else suffer war. They could accept the cross or face the sword. But the sense that rulers had a mission to civilize, that they should guide the ruled toward higher enlightenment, that force might be warranted to achieve that goal, and that fire was a handy marker of success—these assumptions persist. Only the nature of what constituted civilization has changed.

✵

Fire in Mexico was not fundamental to Spanish administration except where it accompanied disorder or destroyed buildings. Its role in landscapes and ordinary life called for no special attention. The great tomes themselves, even Oviedo's and Las Casas's, went unpublished in full for centuries. Spanish America, like Earth generally, had to be rediscovered by a second wave of European exploration, this time outfitted with the methodologies of the Enlightenment. The Renaissance triangulated novelty from geodetic markers laid down in the past. The Enlightenment redefined the character and means of knowing and resurveying *las cosas de Nueva España* based on a vision set in the future. Secularization passed over learning like Earth's terminator, dividing the enlightened from the shadowed. Discovery began anew.

The most defining contact between fire and formal learning during the Conquest and its immediate aftermath was, paradoxically, loss. The Aztec and Mayan libraries were burned. The tomes of Spanish scholars were buried.

THE ENLIGHTENMENT EXPLORES MEXICAN FIRE

As the conquistadores had used Indigenous guides and interpreters, so scholars and missionaries appealed to local savants. They learned Nahuatl, identified flora and fauna in Native terms, tested European cultivars in Mexican settings and Mexican cultivars in Europe, matched European texts against Mesoamerican societies, and in general agreed that they had to understand the Indigenous people's view of the world in order to govern and convert them to Christianity.

Their task was the classic one of the Renaissance: to translate from one text to another. They found texts, made texts, translated and glossed texts. They did in Mexico City what they did in Toledo. In the Valladolid debates that led to the New Laws, Las Casas and Sepúlveda appealed to Aristotle and St. Augustine. In assembling his *Historia natural de Nueva España,* which became the basis for the Mexican Royal Botanical Garden, Francisco Hernández argued for Nahuatl names for the plants but organized his text according to Pliny's *Natural History,* whose 37 books he was translating.[68]

Scholars operated with the tools at hand. Modern science, in particular, was a distant prospect. Nicolaus Copernicus published his solar-centric cosmology (*De revolutionibus*) the year after the ill-fated Coronado and De Soto expeditions returned, and 22 years after Cortés sacked Tenochtitlan. The great scholars of Mexico passed well before the early salvos of natural philosophy: Fernández de Oviedo died in 1557, Bartolomé de Las Casas in 1556, Bernardino de Sahagún in 1590, and José de Acosta in 1600. Galileo Galilei published the *Starry Messenger* in 1610, and Francis Bacon the *Great Instauration* in 1620. Isaac Newton, born the same year Galileo died (1642), published *Principia Mathematica* in 1687 and *Opticks* in 1704. Linnaeus released the first edition of his modern systematics, *Systema Naturae*, in 1735. Large portions of the *Historias generales* of Spanish America went unpublished during the lifetime of their authors, and they were rarely known outside Spain, which tended to hoard information as it did silver. Spain fought off encroachments by the northerners, whether Reformation or post-scholastic philosophy, as they did the plague of French and English pirates that infested the Caribbean. The massive botanical compilation of Francisco Hernández was treated as a state secret, until a fire in El Escorial destroyed the manuscript and its illustrations.

Yet fundamental changes in learning were under way. Novel discoveries challenged the authority of the Ancients. The Book of Nature competed with Holy Writ as a source of information and data; mathematics, with Aristotelian logic; experimentation, with literary exegesis; and naturalists, with missionaries. Britain and France supplanted Spain and Portugal as global rivals. A new age of exploration spanned the globe—as much rediscovery with new eyes as first-contact encounter. The Book of Nature was translated into the idiom of an emerging science; Linnaean systematics replaced Nahuatl. The eighteenth century's enlightened despots, looking to modernize governance, turned to those new sciences and the pursuit of "useful knowledge" to help revive the fortunes of the Spanish state and its overseas colonies.

Spain joined the other great powers in what swelled into a second great age of discovery. In 1761 it permitted Jean Chappe d'Auteroche to transect the country on his way to California as part of an international campaign to measure the transit of Venus. In 1769, in the next global effort to measure the transit, New Spain dispatched its own expedition to

Baja California under Joaquín Velázquez de León. Between 1785 and 1789 a series of regal orders established a Royal Scientific Expedition to New Spain that sponsored a Royal Botanical Garden and dispersed collecting parties to fill it. At the same time Spain planned and then launched in 1789 a five-year mission led by Alessandro Malaspina to challenge the voyages of British and French explorers, mostly in the Pacific. In 1791 Malaspina sent a party under Antonio Pineda y Ramírez overland through New Spain, from Acapulco to Mexico City and its environs and then on to Guanajuato; orders from Spain then instructed members of the Royal Scientific Expedition to join Malaspina in the Pacific Northwest. A year later other members under José Longinos Martínez traveled to Baja California, then continued north to missions in Alta California. Subsequently, Longinos Martínez explored south to Guatemala and Yucatán. All in all, it was an impressive endeavor to collect and interpret the natural wealth of Spanish America through the prism of the new sciences.[69]

But by the time the new reconnaissances were under way, landscape fire was vanishing from scientific scrutiny: the Enlightenment found its illumination elsewhere. The surveys amassed thousands of specimens—Malaspina tallied 10,622—but they were housed in herbaria, in gardens dedicated to useful plants (especially medicinal or commercial), in lists organized by artificial criteria. It was an exercise in identifying and classifying species, not landscapes and the processes that shaped them. The savants weren't charged with recording fire or understanding how it operated in the world, so they didn't. Worse, European agronomy, always suspicious of fire, determined that the use of fire was the divide between an agriculture based on Reason and one founded on superstition and tradition. Even Linnaeus in his journey to Skåne was forced by the Ministry of Agriculture to delete a passage favorable toward burning for pastoralism from his published report. Paradoxically, what people actually did on the land—what they had to do to cook, heat, light, farm, herd, mine—stood outside the realm of scientific inquiry.[70]

In Mexico, a country conspicuous even today for its pervasive burning as imaged by satellites, fire was, for intellectuals, invisible. A few incidental references appear. In his 1769 journey through California, Fray Juan Crespí recorded 17 days of active fires, often finding pastures "burned by the heathen for the purpose of hunting hares and rabbits." Longinos

Martínez observed that "in all of New California from Fronteras northward the gentiles have the custom of burning brush, for two purposes: one, for hunting rabbits and hares (because they burn the brush for hunting); second, so that with the first light rain or dew the shoots will come up which they call *pelillo* (little hair) and upon which they feed like cattle when the weather does not permit them to seek other food." Outside Acapulco the Pineda party observed that "the labor of this country is limited to three operations: clearing out the reeds, weeds and bushes; burning the land; and then planting it two continuous years, leaving it uncultivated during the third year." On the Cerro de la Brea pines were tapped for pitch and tar, both distilled over fires. But for travelers to document what fires were around them would have been like recording what their pack mules ate each day.[71]

In any event, most of the extraordinary labors of the expeditions were lost—ignored by the Crown, forgotten and unpublished, or sequestered as state secrets. Like Hernández's labors, the results were not recovered and published until the 1880s, then as historic artifacts, not contributions to the science of the times. The Royal Botanical Garden in Mexico City fell into decay. Alessandro Malaspina ran afoul of palace intrigues and was imprisoned. Spain was convulsed by the Napoleonic Wars, and Mexico by a war of independence.

The fundamental problem, however, was systemic. The stellar naturalist of the era, Alexander von Humboldt, had made his reputation with a five-year trek through Spanish America, including a year in Mexico. Yet he—the vaunted Second Columbus, the "scientific discoverer" of the New World—failed to regard fire as more than an incidental ornament or an environmental quirk. Even Humboldt, who sought to discover the laws that bound life to the physical world, did not see fire as one of the most elemental of ecological processes. This was especially true in Mexico, because fire was not among the "statistics" that he tallied or that the state cared enough about to routinely record. Still, fire was often unavoidable.

In the Canaries Humboldt noted how, "following a custom introduced by the Spaniards, though it dates back to remotest times, the shepherds lit the fires of Saint John. The scattered masses of fire and columns of smoke driven by the wind stood out from the deep green of the forests lining the peak." In the *llanos* of Venezuela he witnessed a scene in which "several parts of the vast forests that surround the mountains were on fire. The reddish flames, half hidden by clouds of smoke, stunned us. The inhabitants set fire to the forests to improve their pasturage and to destroy the shrubs that choke the scant grass. Enormous forest fires are also caused by the carelessness of the Indians who forget to put out their camp fires." He thought the "mountainous savannahs of the Cocollar and Turimiquiri owe their existence to the destructive custom of Indians burning the woods to make pasture land," creating a thick tangle that prevented reforestation. Outside Caracas during the dry season there was more: "to improve the land the savannah and grass on the rocks were set on fire." Down the Orinoco he "saw a part of the cane fields devastated by a fire spreading from a nearby forest. Nomadic Indians set the forest alight everywhere they set up camp for the night; during the dry season vast provinces would be in flames if it was not for the extreme hardness of the wood, which does not completely burn." On Isla Conserva he recorded how their campfire "attracted crocodiles and dolphins whose noise stopped us sleeping until we decided to put the fire out," while elsewhere they considered fires around their camp "indispensable to scare off jaguars." Such was "the wonderful fertility of nature" that Indian agriculture meant little more than "setting fire to branches" on a patch of land.[72]

It's hard to see why similar practices would not also characterize similar environments inhabited by comparable economies, although in Mexico proper, where political economy rather than nature's ecology was his purpose, Humboldt recorded only a variant of the milpa in which "the Indians of the district of Sola and Zimatlán establish their nopaleras [prickly pear cactus, fodder for the cochineal insects that produce a red dye] on the slope of mountains or in ravines two or three leagues distant from their villages. They plant the nopals after cutting and burning the trees which covered the ground." Mexican agriculture used fire much as Mexican mining did, as a pyrotechnology to smelt sugar, vanilla, and other plants into economic goods. If Humboldt was among the minority of travelers that thought flames worth recording, he nonetheless appreciated them as exotica, like electric eels or boldly colored macaws.[73]

Others filled in what Humboldt missed. Carl Sartorius in 1824 described for the coastal lowlands of tierra caliente how "the tall grass is burnt down." Ashbel Shepard observed the same in 1859, noting how, once "these plains are laid bare by fire," it was apparent that the "entire region was formed into terraces" complete with masonry walls. The rancheros hewed down "all the trees and bushes. The wood is allowed to dry for some months, and then set fire to. When the rain sets in, the grain is sown without the soil being turned up." Frederick Ober described the process adapted to henequen in the Yucatán. While the ground was dense with scrub, the surface was thinly vegetated, "owing to the annual fires that run over the country." Workers cleared a plot of land and let the slashings "dry for a season,—and the next, if the previous one has been dry," then set fire to them, after which laborers dug holes for the henequen, whose leaves could be harvested annually for 12 to 20 years. Carl Lumholtz recounted how the local people kindled a fire "in the dry reeds and grass" and a strong gale increased it "in magnitude and fierceness, and in the evening it threatened to destroy the houses of the Indians, when suddenly a change of wind [likely the passage of a cold front] to a northerly direction brought relief, and the fire died away after a few hours." Lumholtz had no doubt that the fire had been made purposefully, "for they [the Indigenes] were just at this time busily engaged in such work all along the river, in the usual Indian belief that smoke produces rain and clouds." E. B. Tylor, the folklorist, described how, while looking for Chalma, in a scene that might have been lifted from de Costa's account of the Tlaxcalans, "a light behind us made us turn round, and we saw a grand sight. The coarse grass on a large hill further down the valley had been set fire to, and a broad band of flame stretched right across the base of the hill, and was slowly moving upwards towards its top, throwing a lurid glare over the surrounding country, and upon the clouds of smoke that were rising from the flames. Every now and then we turned to watch the line of fire as it rose higher and higher, till at last it closed in together at the summit with one final blaze, and left us in the darkness." The northern savannas burned mildly or savagely depending on the grass available, as Julius Froebel discovered. In one instance, the "scantiness" of the grass allowed his party's horses to cross over the flaming front without pause. In another, where the grass was "long and dry," a spark nearly overran their camp and wagons and threatened "the lives of the whole caravan." (Escaped campfires were a constant hazard; the Boundary Commission after the Mexican-American War had several narrow escapes.) Others observed charcoal burning, and how campfires could explode across the cuttings and up slopes.[74]

Most such practices had European analogues, but Mexico had its pyric oddities: fires like those burned by travelers into the bases of trees as a makeshift hearth, then left to fend for themselves, or the methods of the milpa applied to prickly-pear cacti to furnish sustenance—indeed, "salvation"—for cattle in times of drought. Ethel Alec-Tweedie wrote how, "to encourage the cattle to eat sufficient prickly pear to live on in times of drought," the ranchers "cut it off the stem, make a fire, and burn off the worst of the prickles, which they do by holding the leaves over the flames for a moment; the wildest animals will follow anyone about in the tamest manner to obtain this specially prepared delicacy, which goes far to save their lives." John Audubon recorded an episode in his journal in which he explained that "those intending to run off the 'cabalgada' of a travelling or trading party, get a strong and vigorous horse, cover him with the skin of an ox, newly off, the fleshy side out, tie all the bells they have round his neck, and fastening an enormous bunch of brush to his tail, set fire to it, and start him with yells and shouts, through the camp of those to be stampeded."[75]

But there were occasionally observers who sensed the paradoxes of European law and practical life. The naturalist Hans Gadow in the early twentieth century told first of the casual vandalism abundant in the mountains, aggravated by "herdsmen" who were "in the habit of firing the dried tussocks of grass to produce new crops." This was, Gadow confessed, "a difficult question to deal with." Law forbade the burning, science denounced it, and the wealthy owner lived well away from the action, allowing a "sub-regidor to 'punish' the herdsmen." Yet the practice, while prohibited, was "hallowed by immemorial custom," besides which it "benefits the cattle" of the burner's own community. The conflicts at work were several—caste, class, abstract law versus customary practice, all conflicts that have haunted Mexico since the Conquest and haunt it still.[76]

By the late eighteenth century the shift was under way from early European missionaries like Toribio de Benavente, who were intent on exposing the cruelties of the New Fire rite, to naturalists or professional travelers versed in the expectations of their literate audiences, who recorded fire on the landscape as they would sightings of rare orchids or a jaguar.

But it is remarkable that even a scientific icon like Alexander von Humboldt, who was celebrated for his advocacy of the equality of humanity and who insisted that "European geographers are wrong not to admit to seeing things as Indians do, for they are the geographers of their own country," could not see fire as those native geographers did. The central European suspicion and prejudice toward fire among the educated, sanctified with the baptismal waters of science, was too deeply ingrained. An Indigenous perspective did not much matter for Enlightenment reformers of Humboldt's time.[77]

But it would matter when those views, refracted through French and German traditions of forestry, tried to impose their condemnation of fire onto a land that could not exist without it. Humboldt lectured that "by felling trees that cover the tops and sides of mountains men everywhere [he reckoned 'European planters' the main villains] have ensured two calamities at the same time for the future: lack of fuel, and scarcity of water." Fire and water—the dialectic that informed Mexico's biota and shaped how humans sustained agriculture. Reformers saw Indigenous fire removing trees; they did not see how those fires made and sustained the forest and, if done rightly, improved the watershed.[78]

It was never plausible to extinguish fire on the landscape; even the attempt could be profoundly disruptive. True reformation in the fundamentals would require another kind of fire. At the time of Humboldt's expedition its early tremors were already being felt. The discovery of oxygen redefined the scientific definition of fire; the development of the steam engine created a combustion chamber that quickly demanded a more abundant fuel than wood; the renewal of European imperialism offered a vector to carry both around the globe. Humboldt himself hinted at the change when he noted in passing that Mexico's mines lacked the coal needed to run the machines that could drain their flooded shafts.[79]

BINDING THE YEARS OF COLONIAL MEXICO

Few moments of history truly rupture the space-time continuum. The Conquest did. In the flow of time it was a cataract. No comparable break followed, not even the uprising in which Mexico overthrew Spanish rule, or, arguably, the civil war that was the Mexican Revolution.

The politics of the postcolonial era present a rolling thunder of upheavals, without a controlling class, idea, or institution. They boast a seemingly endless succession of *tumultos* and *pronunciamientos*, coups and constitutions, a French intervention and a disastrous war with the United States. Church and state quarreled; Indigenous people struggled with elites; tradition wrestled with modernity. The election to the presidency of Benito Juárez, a Zapotec from Oaxaca, marked a milestone of social progress and led to another attempt at updating Mexico's tangle of customs and land use. The dictatorship of Porfirio Díaz, a successor to the enlightened despots of imperial Spain, meant a reversion to autocratic rule, though outfitted with better hardware. In many senses, the revolt begun in 1810 culminated in the revolution of 1910. Yet much of the action over that century was more churn than progress. Creoles replaced *peninsulares*; then the "men of the north," like latter-day Chichimecs, this time with allies from the south, challenged the creaky establishment of the Porfiriato.

Between independence and the Mexican Revolution, fire had its own narrative, sometimes parallel and sometimes oblique to politics. Its surface story flickered with bonfires and torches, burned haciendas and villages, battlegrounds aglow with firearms. But its deep currents followed overall demographics, the politics of land tenure, and the ebb and flow of traditional agriculture, because those processes affected how many people there were and how they lived on the land. While liberal politics tried to transfer what had been commonly held into forms better suited to a market economy, and progressive thinkers wanted to replace what they regarded as folk superstition with science, it proved easier to transfer land tenure from common governance to capitalist ownership than to replace traditional fire practices with modern alternatives. Substituting Reason for Revelation did not alter attitudes toward fire. It only gave further justification to attack flame in hearth and field, not just in religious rites. *México profundo* persisted, still aglow with its customary fires. The deeper Mexican revolution had to await a conversion to fossil fuels.

In 1942 Carl Sauer published his famous observation about the personality of Mexico. The Mexico he had long studied was the outcome of people interacting with living landscapes—lots of peoples, in many physical geographies, with an abundance of histories. Yet they all shared common constraints imposed by a matrix of air, water, soil, and fire, and the ecological principles that prescribed what was possible and what was ruinous. The outcome was a transformation of raw nature into a second nature whose character would have been recognizable to Gabriel Alonso de Herrera or to Cicero.

Yet the decades of Sauer's research in Mexico coincided with the slow emergence of a new fire that fed on fossil biomass, the lithic remains of landscapes from the geologic past. Coal mostly, then oil and gas—these powered the engines and economies of modern societies. At the time of his essay, Mexico was pivoting toward an era of rapid industrialization that would reconstitute the country's fire history—a fire-powered revolution as profound as the Conquest. These new fires were not bound by the ecological give-and-take by which fire had shaped Earth's terrestrial landscapes for 400 million years. Lithic landscapes, once living, now fossilized, could burn day and night, winter and summer, through drought or deluge. But while such fires expanded the source of humanity's firepower, they could not similarly expand the sink needed to accommodate their effluent. That was for the future to learn.

Meanwhile, fire and iron replaced fire and sword as organizing principles. This new fire replaced Cicero's second nature with a third and sparked for both Mexico and Earth a fourth sun.

MAP 1 Physical map of Mexico.

MAP 2 Political map of Mexico.

FIGURE 1 Huehueteotl, the Old God.

FIGURE 2 New Fire Ceremony as depicted in the *Codex Borbonicus*.

FIGURE 3 The Aztec calendar stone, which shows time organized around the Five Suns and includes several fire deities.

FIGURE 4 Fire and water: the yin and yang of Mexican agriculture, each with its own god (Tlaloc for water, at bottom left; Xiuhtecuhtli for fire, at right). Also note the fire-scarred tree on the right, an unusual feature in fire art.

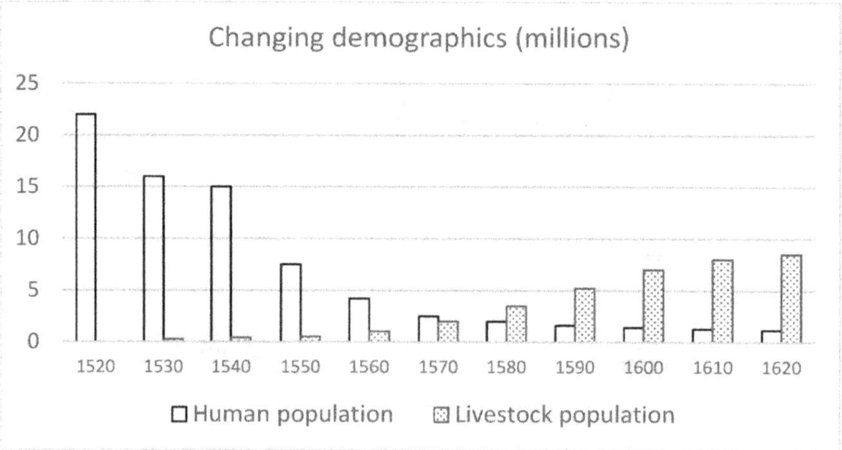

FIGURE 5 Competing populations of people and livestock in the sixteenth and seventeenth centuries: one empties, the other fills.

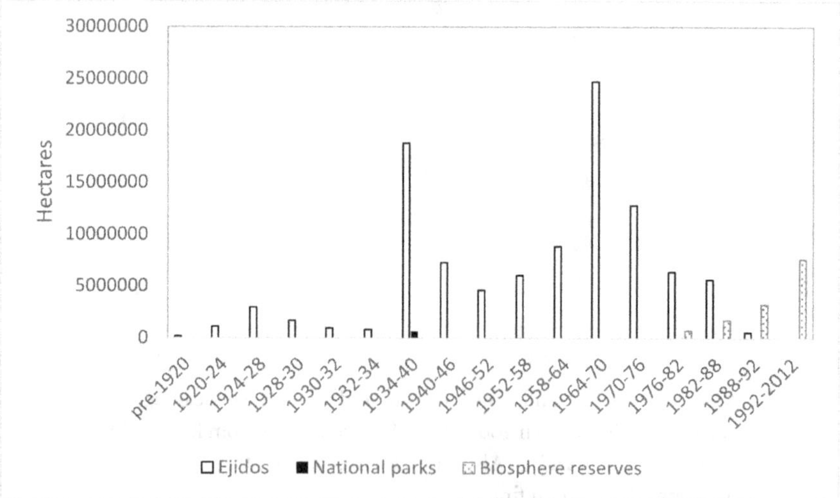

FIGURE 6 Living landscapes. Reforming land use: ejidos, national parks, and biosphere reserves.

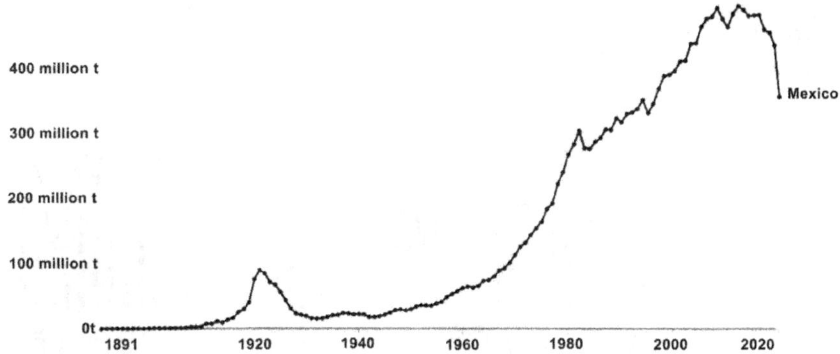

FIGURE 7 Lithic landscapes. Carbon dioxide emissions for Mexico, 1891–2020. The curve tracks the rises and falls of Mexico's industrialization. Note the impact of Revolution, Depression, Lost Decade, and even Great Recession and COVID-19.

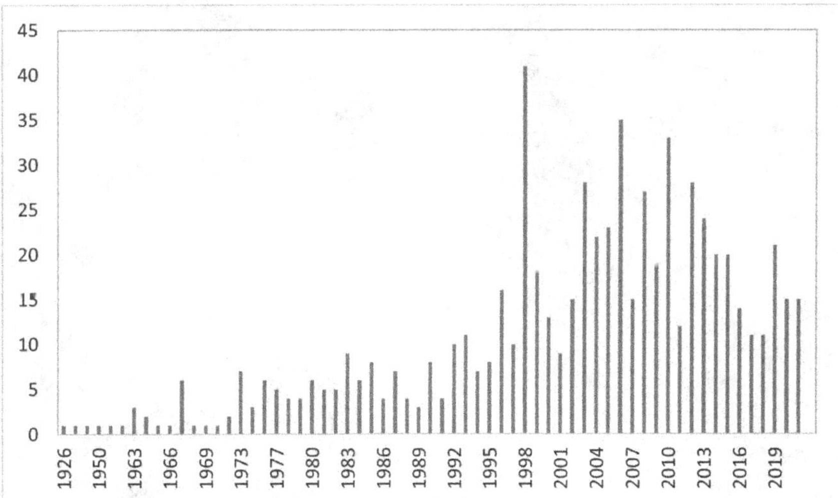

FIGURE 8 Numbers of fire-themed publications dealing with Mexican fire. Note, especially, the impact of the 1998 season.

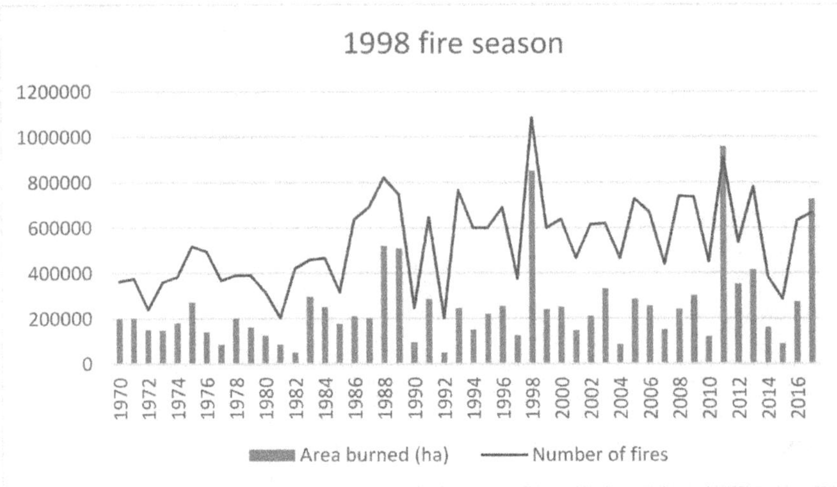

FIGURE 9 Fire profile: number of ignitions and burned area, 1970–2017. Note the spikes in the 1988–89, 1998, and 2011 seasons.

FIGURE 10 Three fires: the thrice-burned log at Desierto de los Leones.

FOURTH SUN

FIRE AND IRON

NEW FIRE

DOS BOCAS

ON JULY 4, 1908, *an oil well at San Diego de la Mar, on the tropical north coast of Veracruz in the Huasteca, abruptly cratered and then erupted into a historic gusher. Oil blew upward like a pyroclastic plume "until it blended into the clouds" and then fell in drops to the ground, where pooling petroleum met the hot particles from blasted boilers and caught fire. A mass ignition kindled a mass explosion. Two immense craters, later named Dos Bocas, marked ground zero.*[1]

A boiling, cloud-topping flambeau poured fire heavenward; its brooding pall spread down the coast for 200 kilometers. The flames, inextinguishable, burned for 57 days. Eventually the flames exhausted their reservoir of fuel, but not before an estimated 420 million gallons of crude had blown out of the geologic past and seared the present. It remains the largest oil spill in history. Its scars endure a century later, standing as an industrial echo of the Chicxulub crater that both ended an era and kindled a new one.[2]

The Dos Bocas blowout serves as a New Fire—macabre to the state of parody—for an emerging Mexico, as profound as the political revolution that roughly coincided with it. "Fire permeated hydrocarbon extraction from start to finish," notes Myrna Santiago, "to the point where the oil companies introduced a new season in the Huasteca, one that lasted all year: fire season." But industrial combustion did not stay on its derrick altar. Under the guidance of científicos *and capitalists it propagated to the four quarters of Mexico.*[3]

This new fire, like its predecessor, demanded that the old fires be extinguished before it replaced them. The process was slow, often tedious, and, this being Mexico, fragmented and incomplete. There were too many local points of resistance, too many niches that industrial fire could not penetrate, too many fire costumbres that it could not easily supplant. But once begun, the process seemed implacable. As it interacted with its settings, the new fire—here creeping, here flashing—refashioned the land and birthed a Fourth Sun.

There was no clear break between epochs, no undeniable symbolism that commemorated the passing of one fire-sparked Sun into another; unlike the Conquest, there was no sharp rift in which Spaniards replaced Aztecs, votive candles supplanted fire sacrifices, and wheat and cattle challenged maize and rabbits. The propagation of industrial fire spread more like missions on the frontier than Cortés's marches.

Globally, the arrival of fossil-fuel combustion provided a scale by which to distinguish developed from undeveloped nations, and elites fashioned intellectual arguments to support it. Now that fossil biomass provided fertilizers, herbicides, pesticides, and the motive power to pull plows, not to mention the transubstantiation of coal and oil into electricity—now that a new, machined fire was available—open field fires and *leña*-burning *hogares* looked more archaic than ever. Practices based on flame were denounced as primitive, and fire-fallow farming as atavistic; practices based on industrial combustion were hallmarks of advanced civilizations. How this new fire interacted with the others was little understood. It had no precedent in nature. At first, it simply meant power, money, and modernity. Historically, it wasn't present, and then it was.

It came to Mexico slowly, in fits and starts, and like so many introduced novelties it splintered as it encountered stubborn points of resistance, yet it advanced with an implacable power that eventually restructured the relationship between people and land as thoroughly as the Conquest had. It soon acquired a Mexican accent. Unlike Britain, Germany, or the

United States, Mexico held few reserves of coal, instead burning its traditional fuel, wood. Then it leaped into oil, of which Mexico had vast reserves. Rail linked oil to internal markets; commercial production for export commenced in 1901. Tampico served as Mexico's nexus to fossil-fuel civilizations as Veracruz had earlier to Europe. By 1922 Mexico was the second-largest producer of crude oil globally.

Eventually, the fires of revolution proved less powerful than a revolutionary fire. The Grito de Dolores of Mexico's revolution in firepower was the whistle of a locomotive.

THIRD-FIRE, THIRD NATURE

In 1917, as Mexico promulgated a new constitution, Fernando González Roa and José Covarrubias, members of the Madero cabinet, wrote that "the agrarian problem" was "the fundamental question which we must settle. About it gravitate all of the great problems which have agitated us in the past, which disturb us now and with which we will have to concern ourselves in the future." All Mexican governments, "revolutionary or otherwise," were "under the strictest compulsion to find a solution."[4]

They were writing in a long tradition. "Agrarian problem" was shorthand for the reconciliation of indios (and later mestizos) with Spaniards. The Native people would not disappear, and in fact were vital, but they lived in mostly traditional ways. How to grant them some measure of autonomy while still demanding their labor had, time and again, hinged on the question of land. If it remained in Indigenous hands, it would be used in Indigenous ways. If it was transferred to newcomers, it could be put to such economic uses as mines, pastures, and missions—or, by the late nineteenth century, factories, plantations for export crops, and modern cities, all eventually lubricated by oil.

Each change of regime had its characteristic transportation system. Pre-Hispanic Mexico had its "great arterial highway," as Carl Sauer labeled it, up the west coast, with transport by human porters. Colonial Mexico had its critical roads to bind the capital to seaports at Veracruz and Acapulco, and then to mines and missions; oxen, mules, burros, and wheeled *carros* were added. Still, into the seventeenth century goods moved mostly by packtrain; only the route from Mexico City to Zacatecas, Durango, and

Chihuahua had routine wheeled traffic. (Mexico had no navigable rivers.) Industrializing Mexico had rail. Locomotives opened new mines, planted new fields and orchards, carried new goods, and supported new urban centers. Those engines needed steam, and steam came from combustion. In Mexico's economy of nature, fire acted as a broad-spectrum ecological catalyst. In Mexico's industrial ecosystem it did the same, but sublimated into steam and refracted through engines.[5]

At first the discovery of fossil biomass resembled the discovery of another America, a new world exhumed from the geologic past. It enlarged the map of known landscapes, it expanded existing power relations. What initially complemented combustion-by-fire came to compete with it as fossil biomass's energy replaced flame, its petrochemicals supplanted ecological cycles, its mass of plastics and asphalts competed with primary biomass, and its machines, metabolizing the new combustion, created a mechanical fauna. It took a century for Mexico to feel the full impact, but eventually the second nature in which human wit and artifice had remade raw nature into working landscapes was itself remade into a third nature powered by a third fire.[6]

Whatever remade land remade fire regimes. But fire in living landscapes remained visibly fire. This time fire itself was remade as well.[7]

MEXICO'S NEW LINES OF FIRE

Mexico's early phase of industrialization rode on two rails that ran in parallel. One was land; the other, steam. Land supported fire in living landscapes—an ancient dynamic that came with internal baffles and barriers, both ecological and social. Steam, however, granted leverage, even on an Archimedean scale, to move societies. Where it burned wood, it quickly stripped the countryside, and the biotic feedbacks were soon felt. To continue, to leap beyond those ancient ecological boundaries, it had to burn lithic landscapes, fuels fossilized over geologic time and exhumed to burn in machines. It took the better part of a century for land reform and steam to merge and collude, like an ecological version of non-Euclidean geometry in which parallel lines can meet, but that convergence shaped twentieth-century Mexico. Fire was remade into steam, and steam reorganized land—together they introduced modernity and rearranged the traditional.

Land, the first rail—always in Mexico the question was about land and who would decide its purposes. The century of unrest between Mexico's declaration of independence and the Mexican Revolution was fueled, like a coal fire smoldering beneath the surface that from time to time kindles the surface, by conflict over land. Traditional owners—the Indigenous peoples, the mestizos, the villages with commons, and the small rancheros—preferred the countryside remain agrarian and under their control, even if conditions appalled visitors and officials.

But modernizers considered common resources a quagmire that recycled endless backwardness; they wanted collective land privatized, small farms enlarged, and the resources of those lands opened to whatever capital could be attracted. From the 1857 Ley Lerdo through the 1883 Ley de Colonización, land moved from public (state), communal (village), and ecclesiastical (church) control to private landholders who could more or less do with it as they wished. In return for surveying public land, private companies would receive a third of it, and an option to buy the remainder at a pittance. Like missionaries, who soon appreciated that conversion was not simply a matter of belief and baptism, but one of changing the social basis for beliefs, the científicos who proselytized modernity understood that their project was one of wholesale reconstruction of Mexican life and livelihood.

Meanwhile, the various communal lands were parceled into private, individual lots. Some were reassembled as *condueñazgos*, in which a version of joint tenure emulated the communal holdings of old. Much was scooped up by railways and mining and petroleum companies, or leased by them for nominal sums. The amount of land transferred was immense—52 million hectares went to the survey companies and hacendados, roughly a third of the national estate, as much land proportionately as was ceded to the United States in the Treaty of Guadalupe Hidalgo. Haciendas, creaky with outworn mores and economies, evolved into latifundia, while foreign companies eager for mines and oil acquired large blocks of the now-unencumbered lands either through purchase or lease.

The influx of capital came from outside the country, however, mostly from North America, and the wealth returned, on the same rails and ships that brought it in, back to its foreign sources. The mercantilism of Spanish colonialism returned in a more modern incarnation. The process began in the 1850s and was embedded in the liberal reforms of the Juárez administration, but it smoldered until the dictatorship of Porfirio Díaz. Commerce required transportation, but in mostly mountainous Mexico, lacking navigable rivers and canals, hauling goods and people was arduous and slow; the 1878 census tallied 9.4 million people, supported by 11.5 million donkeys and 1.5 million horses. During the Porfiriato, catalyzed and symbolized by a metastasizing railway network, the state transferred large swaths of public land to private ownership in the hopes that it would help galvanize a modern economic system. The liberal experiment succeeded in expanding production, and it increased wealth, though most of those gains went to elites or foreigners. Ultimately, the project ended in a revolution whose ambitions and slogans grew out of a 400-year-old contest over land.

The "agrarian issue" was a land tenure issue. It underpinned much of the agrarian revolt that stoked the Mexican Revolution, it forced President Carranza to issue a redistribution decree in 1915, and it was enshrined in Article 27 of the 1917 constitution, which affirmed national ownership of land and its resources and sought to reinstate a better equilibrium between rural people and arable land. In a variety of ways the constitution proposed a better balance between private and public, and between a traditional and a modern economy. The outrush ended in the administration of Lázaro Cárdenas, who renationalized the land's mineral wealth by appropriating foreign oil operations, and who gave political muscle to the ejido program that broke up large latifundia and parceled up public lands for transfer to small and traditional users. In 1946 the Partido Revolucionario Institucional (PRI) institutionalized the movement's political expression.

The ebb and flow of communal lands had followed demographics of conquest and the rise and fall of economic philosophies. Local or regional economies favored traditional burning. Global economies favored privatization and export, and relied on industrial combustion. The Mexican Revolution sought to reconcile not only land with the landless, but nation with world, and custom with modernity. For 50 years after the revolution calmed, and for 30 years after the Cárdenas regime, the adjustments proceeded, with surges and pauses, until the 1980s. The pyrogeography of Mexico aligned accordingly.

The second rail was steam, powered by new fuels—this was a novel source of seemingly unbounded power; not coincidentally were its inventors called

New Prometheans. The bald statistics almost speak for themselves. In 1877 railroads were feeble, extending only 400 miles, of which 71 were powered by mules. In 1880 they extended across 680 miles; in 1890, 5,974 miles; in 1900, 8,460. In 1910, at the start of the revolution, there were 12,000 miles of railroads. In 30 years the railway system had increased thirtyfold.[8]

The trains pulled a new economy behind them. Yet railways fed first on themselves. Construction galvanized the production of steel, fuel, and timber, and the organization of labor, including the creation of technical and managerial classes. Their ability to haul bulk freight made low-grade ores minable, plantation agriculture economical, and ranching for export possible. Cities connected by rail flourished. Like the money that financed the program, however, the primary effects went mostly outside the country: what the rails took in, they also took out. Mexico supplied raw goods and labor.

Mostly the project modernized Mexico's old social dynamic: it transformed Mexico more thoroughly than anything since the Conquest, and in truth, more than the Conquest. (*Harper's Monthly* even characterized the sudden onslaught of rail and capital as "the railway invasion of Mexico.") Together with land privatization, like two shears of a scissors, railways encouraged the large to absorb the little, and the big to become bigger. Hacendados usurped village commons, ranchos, and formerly public lands; foreign companies took over mines, oil fields, and factories. Elites tended to prosper; the agrarian population, parted from its land, sank into pauperism. Just as the loot of the Conquest went to Spaniards, so the wealth of industrialization went to elites.[9]

It took time to convert Mexico's fire economy from living to lithic fuels. The early engines burned wood—Mexico's legacy combustible. What had happened with the growth of towns and mines—the exhaustion of surrounding woodlands—now expanded to include anything within the reach of rail. The Díaz regime took notice and sponsored an inquiry. Alarm over a "fuel famine" arose, as increased combustion for the modernizing economy bit into existing stocks. A quick—seemingly self-evident—solution was to promote coal. But while Mexico held some coal deposits, mostly in Coahuila, they were sparse and fed regional markets, though, importantly, that included Monterrey. Mostly, Mexico imported coal, along with coal miners and, for that matter, the railways that burned and hauled it. Unlike the major industrial nations, Mexico experienced no lasting coal age. Coal's influence passed quickly, with production peaking in 1925.[10]

What Mexico had in abundance was oil. Not only did Mexico hold vast reserves along the Gulf coast, but it was easily found and more or less readily developed, and, as with so much of the Porfirian economy, oil linked with rail. In an era that favored monopoly, two companies came to dominate. In 1900 Edward Doheny journeyed from Southern California to northern Veracruz, found oil seeps everywhere amid the tropical forest, and founded the Mexican Petroleum Company. He persuaded the Mexican Central Railroad (also a U.S. company) to replace coal with fuel oil. In 1907 he established the Huasteca Petroleum Company, taking the name from the local term for the region around Tampico. Meanwhile, Weetman Pearson, an accomplished British engineer, began to acquire small holdings in the region before seizing large chunks of prospective land through "surveying." In 1908 he organized the Compañía de Petróleo "El Águila" S.A. The Díaz administration assisted the duopoly with laws in 1901 encouraging a wide-ranging reconnaissance, and in 1909 assuring them of subsurface mineral rights. In return Díaz personally received preferred shares from Doheny and saw his son installed on the board of El Águila.[11]

The oil rush overthrew in short order not only the ownership of the affected lands but their fundamental ecology. Tropical forest was transfigured into an industrial landscape, often a comparative wasteland; milpas and oil seeps morphed into haciendas and gushers; patch burns and savanna quemas turned into steam engines and blowout holocausts. But it did not broadly affect rural Mexico. Much as burros continued alongside locomotives and wood still heated hearths, so a Promethean economy fed by oil scarcely penetrated the countryside except insofar as its operations overturned local landscapes and drafted local labor. The oil went to railways and export. Not until 30 years after the Huasteca Petroleum Company was chartered did oil begin to orient its firepowers to improve the quotidian life of rural Mexicans. Until then Mexico's black gold, like the bullion looted after the Conquest, went elsewhere.

MEXICO'S NEW FIELDS OF FIRE

Railroads were not only a literal engine for growth; they furnished a model for a general energy transformation from wood to fossil fuels, and hence served as an index of modernity. What mines had done to surrounding forests, railroads and factories now did on a more prodigious scale. Locomotives burned wood in their furnaces and cast sparks into the disturbed landscapes to their flanks. Where logging and locomotives were part of a common project, stoking engines with wood bolstered demand and provided a partial outlet for the slash left by logging. Yet in the early years of steam, mines and forestry competed for the same fuel, and like mines, industrial forestry became not only self-supporting but self-consuming.[12]

Worse, traditional usage persisted so that instead of a "primitive" economy being replaced with a "modern" one, both often coexisted and compounded. Around Mexico City, as early as 1864, Leopoldo Río de la Loza noted the consequences. Firewood and charcoal powered furnaces as well as hearths, savaging the hillside forests, while continued landscape burning in the spring for pasture slowed regeneration. Echoing Humboldt, Río de la Loza argued that deforestation would lead to a crisis of fuel and water, and ultimately of habitability and human health. With railroads Mexico City could reach far beyond the Valley of Mexico and its peripheral forests into Michoacán and Puebla for fuelwood—and other urban centers like Guadalajara followed its example. The solution was to convert to coal.[13]

Unease over deforestation was hardly new. Aztec rulers and Spanish viceroys had pronounced edicts to protect woods from overuse. What was novel was an added, potentially bottomless consumption by industry. The modernizing project, however, demanded an uptick in energy sources, and mining fuel came naturally to a country whose global wealth had emanated from silver. What made the argument for coal so provocative was that the transformation it promised would both increase human wealth and improve forests. It would unburden forests as a source of combustibles for energy and equally replace primitive fire practices in landscapes. More coal meant more power, and more power, properly distributed, meant a wealthier society and a healthier environment. As the examples of Great Britain and the United States appeared to show, in the words of mining engineer Pedro Monroy, "el hierro y el carbón"—iron and coal—were the basis for industrialization.[14]

The Ministry of Development, Colonization, Industry, and Commerce (Ministerio de Fomento, Colonización, Industria y Comercio), established in 1853, acquired political power during the Porfiriato and became the point agency for promoting railways, finding coal, and overseeing forests. The ministry's director, Gen. Carlos Pacheco, argued vigorously that the nation's forests were at risk, facing "transcendental damages"; a new forest code, which the ministry oversaw, was adopted in 1881. A year later, "naturally alarmed" at what appeared to be an "impending" fuel famine, the government appointed a further commission to investigate. Their conclusion was what had become the consensus wisdom of the country's elites: that Mexico needed to replace wood with coal, and, further, that the railroads could make an inviting profit by simply carrying coal to other industries. Like Aztec priests carrying the new fire throughout the empire, not only would the rails run on coal, but they could spread the fuel wealth elsewhere. Industry could make more industry. Industry would remake society and land.[15]

Mexico's coal era lasted 30 to 40 years. Coal was critical for early rail and factories, but it was transitional. Mexico had ample supplies for start-up industrialization and could import more from the United States, but it did not have enough to sustain growth, and many forest-based industries preferred wood fuel since it fed their business. Coal production peaked in 1925; by 1922, however, Mexico was the second-largest world producer of oil. It imported coal, it exported petroleum. The substitution of *combustibles fósiles* for wood provided the template for a subsequent transition nationally from coal to oil. Both the Mexican National Railway and the San Rafael paper mill (the largest in Latin America) began their conversion in 1909. The steel mill at Monterrey soon followed, and then oil served as the primary power source for Mexico's major metropolises (by 1937 some 96 percent for Monterrey).[16]

These were modern fuels to run modern machines. But while they added power without decimating Mexico's woods, they did not abolish traditional burning. Firewood for heating and cooking persisted, and well into the 1940s it remained the "product of choice" for the 1.5 million inhabitants of Mexico City. In the 1930s it is estimated that the Federal District consumed 650,000 kilograms of charcoal daily and "thousands of kilograms more of firewood," perhaps as much as 25,000 hectares of woodland a year, hauled by an estimated 260 railroad cars per month. Even preparing kindling trashed forests. In his 1948 study *Rural Mexico* Nathan Whetten fulminated about "a further custom which tends to destroy thousands of trees"—namely, "slashing into the trunks with machetes for the purpose of securing splinters of pitch-pine fagots to be sold in the city markets as torches for kindling fires in the charcoal burners."[17]

Eyler Simpson, an American chronicler sympathetic to Mexico's experiment with ejidos, could barely contain his exasperation: "To supply Mexico City alone with charcoal for cooking and heating purposes involves the destruction—largely without plan, rhyme, or reason—of almost 5000 hectares of standing timber a year. And this goes on year after year despite the fact that it is well known that the Mexican charcoal burning *brasero* and *anafre* are among the most expensive and inefficient cooking appliances that the ingenuity of man has ever devised and despite the fact that other sources of fuel are available." What happened in the city happened on a vaster scale in the countryside: what wasn't burned in hearths and furnaces was fired for pasture, farming, and miscellaneous purposes on an "almost incalculable" scale (Simpson reckoned forest destruction at 400,000 hectares a year by fire alone). The conversion of railroads and factories to oil, a deliberate campaign by the state, showed how to reduce this drain on Mexico's natural estate.[18]

Success came with a lacquer of irony, with Mexico City the poster child for both. Once the city had converted to oil for energy, as Germán Vergara observes, it "witnessed a period of massive industrialization and uncontrollable urban growth . . . that devastated some of the very forests conservationists sought to protect by promoting fossil fuels." The story then "repeated itself time and again in other industrializing cities across Mexico." Similarly, the program to substitute fossil fuel in domestic use had unanticipated ramifications. The preferred fuel, oak, regenerated by sprouting, and thus the source woodsheds were typically coppices with firewood harvested as one might gather berries, or it was collected off the ground as dead logs, ground litter, and windfall, much as one might harvest acorns. One way or another the wood was destined to burn, whether in stoves or in free-ranging flames. Rather than furthering the deterioration of Mexico's forests, habitual firewood gathering and charcoaling provided a kind of combustion hygiene that sanitized the woods against wildfire. This experience, too, was replicated across Mexico's cities and villages. The small smokes of cooking fires were replaced by the noxious emissions of fossil-fueled power plants.[19]

THIRD-FIRE'S PRACTICES AND PRACTITIONERS

Conquest required conquistadores, missionizing required priests, ruling an overseas empire required scholars and clerks, and modernity demanded a corps of suitably trained technocrats. As Spain's enlightened despots had recruited naturalists, including Humboldt, to explore its American empire, so Porfirio Díaz assembled a cadre of men capable of replacing donkeys with railroads, firewood with oil, small farms with commercial plantations, and ecclesiastical mumbo jumbo with hard science.

"Order and progress" was the regime's motto. Order, as Myrna Santiago succinctly notes, was "entrusted to a special rural police force known as the *rurales*," whose "efficiency kept Díaz in power for the next thirty years." Their counterpart for progress was a "group of highly educated men who called themselves *los científicos*," who studied and promoted "the most fashionable American and European ideas." Among themselves, they organized professional and scientific societies. Within the Díaz regime, they focused especially on the Ministerio de Fomento as a bureaucratic mechanism for applying the power of the state to advance economic development. Together they oversaw an experiment to overturn traditional lore, practice, and ownership and replace them with the latest upgrades, and to do so with a "conscience." They applied the project to both city and countryside. They tried to counter the wreckage wrought by steam and capitalism with programs of state-sponsored conservation—the sustainability campaign of its day. They were determined to dispel the miasma of superstition, tradition, and ignorance that occluded systematic improvement. They sought to rationalize the syncretic maze that was Mexico and lift it to the highest standards of what they regarded as civilization.[20]

Building a railroad; financing a modern economy; creating the infrastructure to find, pump, and move oil; recoding knowledge of forests in ways that would allow for heavy harvests yet encourage regeneration for the future—these were not tasks that campesinos, clerics, and Indigenous elders were deemed capable of doing. So even after the Mexican Revolution swept Díaz and his cadre out of office, the científicos endured or were replaced with upgraded versions, which meant that the tension between the two castes persisted. Their successors even turned to anthropologists as a secular version of frontier missionaries to help convert the premodern peoples of Mexico.

Critically, the successors would also define forests and agriculture, and, through those two great landscapes of Mexico, the fires that seemed endemic to them. More than ever, flaming fire retained the stigma of primitivism.

Conversions on this scale take time, appear in spurts and pauses, and occur in patches. The conversion happened quickly where top-down political or economic power ruled; it happened slowly if spottily where local conditions prevailed, and inevitably displayed a degree of mestizaje. Ultimately, the authorities had to compromise with long-standing practices related to fuelwood and fire, as they had had to do since the Conquest. Despite proclamations and laws, officials tended to ignore firewood gathering and traditional burning—often not even recording these practices. They were too deeply embedded in the landscape and too little could be done to eradicate them.

Since the Conquest, Mexico had exhibited two civilizations, two castes, two land patterns, two realms of combustion. What Viceroy Mendoza identified as "two Republics" in his report to Charles V, Frank Tannenbaum observed for the Brookings Institution in 1929: "In fact, Mexico comprises two civilizations, living side by side but neither one fully conscious of the other." When the Mexican Revolution boiled over in 1910, 97 percent of the national estate was owned by 835 families, out of a population of 13.5 million. This partition was as true for Mexico's combustion castes as for so much else. The extremes continued, with an awkward mestizo middle. The railroad and the donkey, the hacienda and the milpa, the village and the metropolis, burning wood and combusting oil—their overlap was sparse and spotty.[21]

As with its political revolution, Mexico's industrial revolution was a drawn-out affair, never fully resolved, often violent and patchily felt, and profoundly significant. The old rhythms bent, and here and there broke, however, as industrialization applied another source of combustion, one that upset the interaction between people and land, sharpened the distinctions between peoples and economies, and polarized even fire. Yet instead of wiping away all the old ways, rail could "facilitate the survival of traditional, even archaic, institutions" such as the hacienda and reinforce the "regressive redistribution of wealth and income in Mexico." In Chihuahua, for example, railroads used wood for fuel and threw sparks into the countryside, where they kindled more fire on the disturbed lands; and when the market for wood stalled, locomotives burned yet more wood to avoid paying for coal. The new firepower intensified the polarized forces in Mexican society.[22]

Eventually, the Mexican Revolution and its aftershocks reclaimed the railroads, then the oil, and finally—first in constitutional principle, and then through decades of reallocation—the land. By reorganizing how its peoples occupied Mexico's bewildering mosaic of landscapes, the revolution rewrote fire regimes. Like a drop of oil suspended between oppositely charged plates, Mexico found itself precariously hovering between two realms of fire.

REVOLUTIONARY FIRE

The bond between the two revolutions was unstable—one created new tinder, the other decoupled the old connection between fire and fuel. Predictably, their impact splintered as they met the hard-textured realities of Mexico's lands and peoples. They arrived at different times in various ways with often unforeseen complications, like two good drugs that interact with unanticipated effects. A century after the Porfiriato began, most of Mexico had felt the impress of their messy mutual force field.

EJIDOS AND COMUNIDADES

The liberal reforms that followed independence and culminated in the 1883 Ley de Colonización had sought to privatize communal, church, and public holdings and so put them under market control. It was a kind of national enclosure movement. The Mexican Revolution intended to undo that process: it argued for a kind of reverse enclosure, breaking up what had been privatized (especially large latifundia) and returning it in the form of ejidos to communities and campesinos. Instead of the Ministry for Development directing privatization, a counterbureau, the National Agrarian Commission, oversaw a process of reversion. In 1915 the Ley de Desamortización gave legal grounding to the project. In 1917 Mexico's new constitution embodied the principle in Article 27.[23]

Getting rid of the Díaz dictatorship and giving land to those who worked it were the slogans of the Mexican Revolution. Writing in 1916, Manuel Gamio argued that "the possession and the exploitation of the land has been the fundamental cause of the bad economic condition experienced by the Mexican people from the time of the Conquest to our days." Land was the crisis that animated the revolution, and land's reversion was its most popular social goal.[24]

Common property took two forms, updates of ancient land-tenure arrangements. The *comunidad*, in David Barton Bray's phrasing, recognized "territorial rights of indigenous peoples first authorized during Spanish colonial rule." The ejido extended "land grants to landless farmers." (Etymologically, *ejido* derives from the Spanish term for village outfields employed primarily for collective grazing.) *Ejidatarios* could hold land privately or collectively. A century after the revolution set the process in motion, ejidos and comunidades constituted over half of Mexico's national estate. Over two-thirds of those lands—more than a third of Mexico's landed territory, and 60 percent of its forests—were classified as "community common property."[25]

The comunidad helped root people to traditional homelands. The ejido project would disperse much of Mexico's agrarian poor beyond their places of origin and inherited culture. People from central Mexico found themselves in the deserts and sky islands of the north, or in the valleys and coastal lowlands of the south. Ejidatarios from Michoacán and Jalisco settled in the mountains of Durango, while ejidatarios from central Mexico could end up in Chiapas and Tabasco. Ejido settlements could upset old arrangements, especially among Indigenous communities. Fire practices long adapted to a particular place could be relocated to other places that had known a very different regimen of burning. And since the easiest way to accommodate the new settings was often to raise livestock, fire had a new competitor for fuels. In the disruptions that followed, fires were both a cause and a consequence.

The reversions began in the 1920s, ceding enough to keep rural unrest from boiling over, but became serious under Lázaro Cárdenas, who redistributed 18 million hectares to 800,000 recipients, constituting roughly half of the country's agricultural lands and a fifth of its forests. In truth, a variety of communal lands existed (such as *monte comunal* and *tierras comunales*), but the ejido and comunidad had legal sanction and political symbolism. Impressive as they were—and their cachet rings with a brazenness equal to that of nationalizing oil—an even greater outburst of redistribution occurred during the mid-1960s to mid-1970s. By 1976, "79 million hectares had been granted to over 2.3 million heads of household and land was promised to about 850,000 more." By then there was little left to distribute (at least good land), and migration to the United States instead helped to dampen unrest. "Peak periods," as Susan Sanderson notes, "have corresponded to periods of economic recession, massive unemployment, and political instability."[26]

All this dissociated the newcomers' ability to burn from their inherited traditions and protocols for burning. In keeping with the paradoxes of Mexican fire, too often this meant newcomers could maintain traditional burning on the land in some places where it wasn't useful and remove it from others where it was. Fire history is not about whether or not there is fire, but what pattern fire assumes. The Mexican Revolution twisted the kaleidoscope.

Still, land is only land. Its quality mattered, and campesinos needed all the accompanying implements that made it tillable. The revolution had implicitly replaced the Porfirian mantra "order and progress" with "justice and progress." While land tenure promoted justice, bad land failed progress, as did land without the water, fertilizer, livestock, credit, and other requirements needed to sustain even minimal farming. Writing in the 1920s after the worst violence of the revolution had subsided, Frank Tannenbaum observed that even the free rural villages "have insufficient lands for their agricultural needs, and insufficient tools and animals to cultivate the lands they do have." They were "not only poor in worldly goods but primitive in technique." They "frequently lack sufficient lands for the use of plows, and where they have the land they often lack the animals. Where they do have enough land these are generally mountainous, forested, or broken and rocky." The upshot is that "they burn the forest and frequently, after one crop, leave the land fallow for six years." That observation was not a compliment.[27]

A change in tenure did not necessarily mean a change in fire practices. What mattered was not just land but how it was worked, which is to say, what kind of fire brought it into production. Modern agriculture found substitutes for fire—tractors in place of grain-eating oxen and mules, nitrogen fertilizers and herbicides for fire's recycling of nutrients and the purging of weeds. But fire was a full-spectrum catalyst, and replacing it demanded a suite of surrogates: the result, if done well, was complex and expensive and, like the later Green Revolution in agriculture, relied on the petrochemicals, draft power, and social infrastructure sustained by a fossil-fuel civilization. All of this meant that, outside high-productivity sites and commercial farming and ranching, agricultural burning endured.

The fire effects of ejidos thus varied by location, past history, and relocated cultures. A supporting infrastructure in the form of roads and logging, along with the change of population and fire culture—not just increasing the local population but injecting people from other places and practices—could break patterns of burning that, as recorded in fire-scarred trees, dated back several centuries, and almost certainly millennia in the Sierra Madre Occidental. The greater the number of ejidos, the smaller the number of fire scars. Surely agricultural burning continued, but perhaps on plots, not rambling as free burns over the hills, whose grasses fed cattle rather than flames. The emblems of a wrecked landscape followed. The peculiar climate and soils of the Chihuahuan sierras could, as Aldo Leopold observed, absorb fire, but not grazing.[28]

The underlying dialectic had changed from that of the Conquest, in which two civilizations competed and partly merged, to that of industrialization, in which two realms of combustion, the burning of lithic landscapes and the burning of living ones, competed and partly merged. Without roads there could be no effective market for ejidos' timber and increased livestock, but roads were paved with asphalt and required vehicles that ran on diesel and gasoline. The two grand realms of combustion interacted with unpredictable, mostly unknown outcomes. As measured by its fires, Mexico was part developed, part undeveloped, and part mestizaje.

OIL

An oil-informed landscape first appeared in Mexico along the northern Gulf coast (La Huasteca Veracruzana)—in the words of Myrna Santiago, "the first site of oil extraction in Mexico, the first site of oil production in the tropical areas of the world, and one of the most important sites of global oil production in the early twentieth century." One could also add, the site of the first major oil fires.[29]

It was one thing to have petroleum refined into fuels and burned in machines. It was another to have it flare freely like a monstrous Bunsen burner. The Dos Bocas blowout of 1908 announced an era of oil-powered wildfires. Locals remembered the cause of the conflagration as the falling of the ejected oil, descending like black oobleck, on "glowing ashes." That tableau, in which the new fuel was kindled by the residue of old fires, could stand for Mexico on the eve of revolution overall. A more formal inquiry admitted that "various versions" existed for the cause of the disaster but favored an account that pointed to gases that reached the boiler and blew into flame. In this sense the oil industry originated in self-immolation. Either way, the gusher turned into an anthropogenic volcano that blasted flame 1,200 feet in the air and burned uncontrolled for 57 days.[30]

All the usual tropes to characterize a conflagration applied, except that here the pall was soot, the flaming front did not spread across the landscape but burned down through millions of years of geologic time as gas and crude blasted upward, and the fire was unquenchable—was not, in fact, even approachable. The company, El Águila, was helpless. So was the Mexican army, which constructed a fire line "200 meters long by 15 meters wide and 8 meters tall," only to have it melt before the boiling oil. They were more successful at hacking a broad fuelbreak through the nearby forest. Meanwhile a flow of oil and simmering salt water streamed into the Carbajal River and then into Tamiahua Lagoon, and 50 kilometers later into the Gulf of Mexico. The havoc ended only when the well ran out of fuel. The lost oil was estimated at 10 million barrels. The lost landscapes, and presumably lost lives, were inestimable. The landscape bore the scars of the searing for decades. Its enduring monument was two gaping holes at the origin that lent their name, Dos Bocas, to both the drilling site and the holocaust it provoked.[31]

Blowouts were a fact of life. The primitive geology of the day meant that they were hard to forecast, and the primitive technology of drilling meant that they were difficult to cap. Inevitably, there were other blowout burns. In 1914 at Potrero del Llano, where a poorly run well leaked oil and gases, lightning from the summer monsoons kindled a lake of flames that boiled off a broad plume of roiling soot. Some 3,000 local campesinos were rallied and outfitted with shovels, then sent to contain the blaze as they would lay siege to a castle. They built fire walls, some five meters high, around the basin, then moved these walls inward to the point where a direct assault could be attempted by 20 pumps powered by gas-fired boilers, later supplemented by another 44 boilers used to shoot steam toward the cone of the well—the boilers were run by "gas taken directly from the ground." Outside the walls, workers patrolled with flat shovels to beat out any oily firebrands that might land and ignite the gases seeping through the abundant cracks. The water and steam failed, as they were destined to in an unconfined environment. But the fire was eventually restricted to the active crater.[32]

It was wretched work, done without protective equipment or much technical know-how. The men atop the walls were shielded by sheets of corrugated iron faced with asbestos. Eleven died, and six others suffered severe burns, when a wind shift blew gas over the flames and caused an explosion. Workers deserted, workers staged strikes. One lasted eight days, during which the company stationed "as many white men as possible," all armed, at critical sites. Campesinos were accustomed to fire, but what they faced at Potrero del Llano No. 4 was to quemas as a hurricane was to a summer thunderstorm. They demanded, and got, more money and better food.[33]

An attempt to suppress the flames by flooding the zone with mixtures of sodium bicarbonate, glucose, and water, along with some aluminum salt with acid, failed when gases rekindled once the chemicals had burned away. Next the company attempted a side shaft to intercept the flow of oil to the flames, a project assisted "with the aid of breathing-apparatus, as the atmosphere was irrespirable." Oil, however, destroyed the rubber critical to the apparatus. While the suits later proved useful for work in noxious environments like storage tanks, they were of suspect value for fire. Ultimately, the fire receded enough that a concrete cap could be placed over the borehole. In May 1932, on another monster flambeau, an airplane was enlisted to drop "flame-smothering fluid." The experiment was not repeated.[34]

Blowouts like Dos Bocas and Potrero del Llano No. 4 were the megafires of petroleum's fire regimes; some 12 oil wells caught fire. But they were monadnocks in a landscape of routine burning. As Myrna Santiago put it, "*flammable* was the operative adjective for the Huasteca in the oil era." Fire hazards were "omnipresent." Unsurprisingly, so were fires. Site-preparation fires readied the land for drilling and transporting operations, and all the infrastructure to support them. Cleaning fires mopped up after the endless spills that accompanied drilling. Some of the oil was collected in pits, and larger leakages in ponds, and then torched; in some cases, close to wells, the oil was allowed to harden until active drilling had ceased before burning in what resembled an asphaltic counterpart to peat. Wildfires were kindled by lightning as this primordial ignition source found fresh fuels in the cheap, wooden tanks (reinforced with metal or concrete) used to store chemicals, all flammable—the oil equivalent of slash piles. Close to towns, flames could spread to houses. (When an oil pipeline burst outside Mexico City on February 28, 1932, adjacent adobe structures were demolished to create levees, and water pipes were laid for more than a mile to assist with firefighting.) Endless fires spread into the surrounding countryside, often drenched with oil splatters that acted as an accelerant and enhancer. Flames tracked barges and tankers and followed spills into rivers like the Pánuco; in December 1910 and January 1911, the oil-coated Buenavista and Tuxpan Rivers burned after El Águila ignited the mess in an effort to remove it.[35]

A seasonally fire-prone landscape, long burned by nature and people in patches, metamorphosed into a fire-informed oil-scape, the signature regime of Mexico's industrial age. It was not just the flames that defined the regime, but the accompanying palls and gases, acidic toxins and penumbral wreckage. The oil industry polluted air, water, and earth—the battlegrounds and sacrifice zones for the country's combustion revolution. During the Dos Bocas blowout, reactions with sulfuric acid spewed out with the oil had caused gold-plated objects in surrounding towns to turn red and English silverware black. Still, as Mexico's rulers had tolerated the environmental and social havoc of silver mining for the sake of the added wealth and power it brought, so they now tolerated the collateral damages of oil.[36]

CITIES

Cities were made of the same materials as their surrounding countrysides (even stone palaces and cathedrals were stuffed with combustibles), so they were subject to the same rhythms of drought and wind, and often burned at the same time and in the same ways. The more humble neighborhoods were the most susceptible, given structures of thatch and wood, compact settlement, and an abundance of open flames for everyday use. Modernizing Mexico meant eliminating that threat from fire in its built landscapes. In principle, these could be constructed in ways that eliminated burning; in practice, houses, villages, and metropolises were not easily abstracted from their surroundings. Even the "scientific city" of progressives could burn just as well as Aztec temples had.

Traveling through the savannas of the "torrid region," Carl Sartorius came upon a hut on a hill. "There is neither brush nor grass about the house, and within a circuit of ten or twelve paces all is cleared." The reason, he soon learned, was that fire could spread from the oft-burned savanna, and two years previous, while the owner was away, the house "was burnt down, with everything in it." The clearing removed the threat. It was a common solution, around not only isolated ranchos but also villages.[37]

In major cities the issue was more complicated. Whatever their exterior material, the buildings held endless sources of flame within. Moreover, the flames from the countryside were themselves less threatening than their perceived effects on the wood and water that was essential for urban life. Nowhere was the concern greater than in the Valley of Mexico, which offered not only an urban cluster, the most advanced in Mexico, but a site that posed acute geographic challenges, beyond the capacity of traditional technology, and was surrounded by a majestic countryside rich with forests and natural wealth that cried out for rational management. Here city and country joined; a sound program of conservation could heighten the best of both and link them to mutual benefit. Through nature parks it could bring the countryside into the city, and then through social engineering it could project the new enlightened order of the city back onto the countryside. The Valley of Mexico distilled the national issues into a complex cameo.

The valley was a basin: flooding was an endless hazard that had inspired causeways, canals, and eventually a massive public-works project to drain it. But effective management required control over the sources as well as the sinks. The forested watersheds that contributed runoff were as vital as systematic removal of the excess flow. Control over water required control over forested watersheds, which demanded control over fire. And in ways that no one anticipated at the time, overcontrolling water—with all its unintended consequences, from dust to subsiding soils—could stand for the consequences of overcontrolling fires.

Unrest worsened conditions in cities, leading to fires and a breakdown in social capacity for fire control. The endless uprisings and tumultos that ran from the war of independence through the Mexican Revolution sparked fires, both aggressive and incidental. War with the United States, war caused by the French interventions, wars between political rivals—all saw houses and huts, haciendas and towns, burned. But fire was also a weapon of the weak, well suited to guerrilla war, harassment, and political protest. Though hardly milpas of the built environment, there was a macabre slash-and-burn quality to fired and refired scenes. Long before the oil boom, the Huasteca, for example, had seen villages, haciendas, and rancherías torched by one side or the other of the ceaseless conflicts, as Liberals burned Pánuco, Conservatives Amatlán and Tamiahua, guerrillas Temapache, and everyone swaths of the natural landscape. It was as though social friction was enough to cast sparks. Such outbreaks of fire were an expected part of insurrections.[38]

Yet just as the oil fires in the Huasteca introduced a difference in kind that demanded other methods of control, so modernizing cities created new hazards that required other than traditional means of prevention and suppression. As Mexico City swelled in the latter half of the nineteenth century, from 190,000 inhabitants in 1860 to 417,000 in 1910, as it transitioned to industries better adjusted to global markets and absorbed more kinds of fuels and appliances to burn them in, it realized that its habitual fire outbreaks were "occurring more frequently," as *El Monitor Republicano* put it in 1882, "to the point that the precautions being taken are useless." A fast-evolving city—worse, a city transitioning to fossil fuels—found that feral flame was outstripping the city's capacity to contain it, not just as the occasional conflagration but as a background count of quotidian wildfires. A modern city, promoted as a model of Porfirian Mexico, needed modern fire protection.[39]

As the number and variety of hazards and risks increased, compressed by the forces of modernity, the historic fire ecology of the city underwent a metamorphosis. There were new fuels like kerosene, coal, nitrate film, and other industrial chemicals; new ignitions like sparks from electrical wiring and locomotives; and novel arrangements that put candles and ovens next to explosive gases, railroads near public parks, and electric bulbs next to ceremonial ornaments. These were layered on such traditional sources as *pulquerías*, match factories, and even lightning (which did not bother to distinguish between a church steeple and a tall ocote). Some efforts to emulate European-themed modernity magnified the problem—for example, mansard roofs and wooden sidewalks. So it was, also, with the response to the threat, an effervescent, oft-confusing exercise in parallel processing by tradespeople, insurance companies, inventors, doctors, and technocrats, with no single guiding authority. Still, urban fire protection was generally animated by a sense of modernity summarized by what Mauricio Tenorio-Trillo has called the "scientific city."[40]

The 1882 sanitation code for Mexico City announced the future by regulating "all dangerous, unhealthy, cluttered, and crowded establishments in the capital" and "defining fire hazards as detrimental to public health, hygiene, and safety." It stipulated "explicit directions for restructuring the city based on the calculation of fire risks, linking health with social change." It proved hard to change behavior, and easier to move people, build anew, and install equipment. By the twentieth century, city plans incorporated fire hazards into zoning, building codes addressed fire risk, customary melees of ordinary citizens rallying to fight a fire evolved into professional fire corps, and technological innovation matured to help reduce hazards and bolster firefighting machinery. Firemen did for the chaos of urban fire what rurales did for unrest in the countryside. An infrastructure for fire coexisted with those for water and power. A rational approach to fire had seemingly replaced traditional lore and credulity.[41]

Inevitably, reforms came with some unintended consequences. Nothing represented a complete replacement, only a trade-off, with some trading up. Insurance and fire brigades encouraged a feeling of security that paradoxically could lessen interest in prevention (no one liked to change their behavior or attitude). Suppression—in fire as in politics—could only hold off a crisis for so long. Since fire derived its power from its surroundings, true fire protection resided in the built environment. That precept extended to the surrounding countryside.

The valences between town and country were many, but for the científicos of the city the most critical consideration was water. The Valley of Mexico had too much of the wrong kind and too little of the right. A modern city needed both draining and a reliable supply of freshwater for domestic and industrial use. Both floods and municipal water came from adjacent watersheds. Mexico City undertook heroic measures to remove what it regarded as excess water. But it needed to look to where the percolation and runoff came from, not just where it went.

A suite of vital springs, some in haciendas to the north, but most in parks such as the Desierto de los Leones, Santa Fe, and Chapultepec, supplied the bulk of the city's potable water—another reason to preserve those sites from development. The stability of the water, however, depended on the stability of the woods, and that depended on the practices of the campesinos who resided or worked within them. The logic that had brought parks into the city was turned toward projecting the character of the city onto the surrounding sierras.[42]

In the valley's peripheral mountains, fire was as much a feature of the forests as were the summer rains. Either the woods were burned for farming and pastoralism or they were removed as firewood and charcoal (or they would burn by lightning or accident). These traditional uses were so elemental and habitual that to control the practice required controlling the practitioner, and vice versa. It was partly a matter of the character of firewood, and partly a question of scale. As late as the 1950s Starker Leopold wrote that "no one who has traveled in Mexico need be reminded of the importance of firewood in the ordinary Mexican home. The pack trains of burros bringing *leña* or *carbón* from the mountains, the women gathering sticks to carry home in bundles, and the haze of blue smoke over each village in the morning are constant reminders that firewood, like water, is one of the basic necessities of Mexican life." Sadly, in the more populous places and the southern uplands, "the constant daily demand for wood has resulted in local depletion of the forests and in some places total annihilation of even the shrubs." Oak, the most desirable fuel, went first; then pine; then "other inferior woods" until they too were gone. "The only tree species to be seen on many of the hills around Mexico City is the introduced Peruvian pepper tree . . . , which because of its aromatic odor is considered unusable as firewood. All other trees have been cut."[43]

For the científicos fuelwood was Mexico's fire story in miniature. They had little confidence in the campesinos, whose traditional uses not only threatened the nation's material future but visibly advertised Mexico's backwardness. To foresters, who regarded themselves as engineers, fire of any kind was anathema, and to modernizers, as to European elites generally, open flame came with a stench of the irrational. If Mexico City was to become a model of modernity, its reach had to include its proximate environs.

The pivotal personality was Miguel Ángel de Quevedo. Thoroughly marinated in French forestry and hydrology, he viewed swidden, pastoral burning, charcoaling, and firewood gathering as intrinsically incompatible with modern life and economy. The transition away from wood for domestic use helped solve some of the forest-as-fuel issues, though the customary practice never ceased. Otherwise, Quevedo sought to regulate usage with laws, with gazetted public forests, and, as necessary, with guards. In 1901 he successfully leveraged a position on a public works commission to advance a program for urban parks. Within a decade Quevedo's project boosted the number of parks from 2 to 34, and the amount of urban land in parks from 2 to 16 percent. By 1909 he and a small cohort of colleagues had created the foundations for state-sponsored forestry not only in the valley but in Mexico overall.

That same year a small group, with Quevedo as president, created the Junta Central de Bosques del Distrito Federal to urge for further forest improvements. Three years later it became an official organ of the state as the Department of Public Works. So convincing were their arguments, and so compelling were early successes—such as a tree nursery at Viveros de Coyoacán whose stocks lined new boulevards, planted over dusty lake beds created by massive drainage projects, and reclothed barren hillsides—that President Díaz supported the cause. In 1908, with French assistance and a cadre of teachers, a Mexican forestry school opened at Santa Fe. The next year the Junta Central de Bosques was rededicated as Mexico's first forestry bureau (*servicio forestal*).[44]

The goals were both to halt ongoing depredations and to rehabilitate previously deforested landscapes. The classic agenda called, first, for protecting existing forests by restricting usage and establishing formal reserves, and, second, for reforesting denuded or damaged lands. These were the goals of state-sponsored conservation in Europe and throughout the European imperium, including settler societies like the United States and Canada. They told the world that Mexico was modernizing;

they made Mexico City as much a paragon of land management as it was of urban development. Mexico City offered, it seemed, proof of concept for a national program.

Political forestry meant more than timber. Forest influences extended to air, water, and earth; forests could dampen floods and droughts, retard soil erosion, reclaim abandoned lands, drain malarial swamps, and enhance the purity and power of springs. Ultimately, forestry represented a rationalization of land based on putatively scientific principles, or at least notions agreeable to modern economies. Unfortunately, forestry and public waterworks were separated (unlike in France), and the Mexican Revolution soon uprooted both forestry's grander schemes and its nursery. Fire and water remained joined, however, in their shared landscapes, the ambition they inspired to control them, and the paradoxes that excessive control created.

Until that political conflagration, the practical problem was the fourth element, fire. The campesinos refused to give it up. They continued to burn in habitual ways and to harvest firewood and charcoal. Plumes of smoke mocked the presumption of modernity—literally obscuring the curated view. Foresters' training demonized fire, not only for what it was but for what it made possible for those who had no respect for the greater good of the woods, those whose deplorable ignorance and poverty allowed no environmental conscience. These people had to be regulated, they had to be educated. Their fires burned up Mexico's future.

The Mexican Revolution overturned that relationship. Up to a point the common people had rights to land, or at least its use. Parks and public forests had to compete with ejidos, the values of *ciudadanos* with those of campesinos, traditional use with protected watersheds and industrial use for timber and paper. The effort to abolish fire was more likely to spark an insurrection. Decrees and laws might regulate access to woods, but fire restrictions could make those places biologically inaccessible. In reality, fire would persist alongside fire prevention campaigns. Each of the two castes had its own fire.

What emerged was a hybrid program, a *paternalismo ecológico*, in the phrase of Christopher Boyer, by which the científicos would instruct and oversee appropriate use. A 1926 forest law (influenced by Quevedo) sought to reconcile the ambitions of foresters with the 1917 constitution by means of a compromise in the form of ejido forest cooperatives.

Predictably, some of the first of these bordered the southern periphery of the Valley of Mexico in the Serranía del Ajusco, and soon afterward came others in San Nicolás Totolapan, Magdalena Contreras, San Bartolo Ameyalco, San Salvador Cuauhtenco, San Mateo Tlaltenango, and San Lorenzo Acopilco, all united in a federation of forest cooperatives under the direction of a forest engineer.[45]

After a fashion, Mexico City did become a model, but not always in the way intended. In the valley the "two civilizations" of Mexico met and managed to work through an accommodation. "Fire forced different groups, through varying measures of conflict and cooperation," notes Anna Rose Alexander, "to grapple with their hazardous environment and assert their interests in discussions about how best to confront it." The same held for the rural countryside. If the resulting compromises pleased no one, that is the nature of compromise. The alternative was revolution, and in the end revolution is what Mexico got.[46]

The outcome was littered with paradoxes. Trying to control Mexico's geographic heritage with heroic engineering works could temporarily improve but would ultimately destabilize its matrix of earth, water, air, and fire. Social engineering underwent a similar experience, as attempts to manage traditional societies under the direction of properly educated elites (whether priests, foresters, or anthropologists) repeatedly stalled.

Still, discourse, pronunciamientos, and exemplary demonstration sites mattered only up to a point. The city had its own momentum and ramshackle logic, which overrode the most enlightened plans. Reducing the middle suite of problems left the Valley of Mexico with extremes. The city's sprawl into a megalopolis overwhelmed efforts to shape and polish and keep pace with modernity. Buildings replaced woods, and pavement prairies; the drained lake filled with noxious air; industrial forestry for pulp crowded out reforestation, pushing out both campesinos and ciudadanos; and fire bans unhinged landscapes that had coexisted with flame for millennia. By the 1980s liberalism returned, and the cycle began anew.

Yet throughout, the valley remained the cultural and political nucleus of Mexico. Its story was the nation's. What it experienced, the rest of the country, allowing for local peculiarities, would also.

NATIONAL PARKS

The restoration of communal lands took many forms, but among the most striking were national parks. Initially concentrated around the Valley of Mexico—the first, the Desierto de los Leones, was declared by President Carranza in 1917—they then spilled out in ways that reflected the development of the country, with two in Coahuila and Oaxaca, and one each in Nuevo León and Chihuahua. The great outburst occurred during the Cárdenas regime. The project included 40 parks totaling 656,676 hectares, with sizes ranging from 8 hectares (El Sabinal) to 246,500 (Cumbres de Monterrey). When Cárdenas finished, Mexico had more national parks than any country in the world.[47]

Yet they were an unexpected result of a revolution that sought to grant practical meaning to political values. Cárdenas's first proclamation, declared on November 8, 1935, covered 25,679 hectares around the twin volcanoes, Iztaccíhuatl and Popocatépetl, to the southeast of the Valley of Mexico. The mountains broadcast values that both Mexico and the world could agree on: they were majestic, their forests needed protection, they had long defined the skyline of Mexico City. But they also represented a unique reconciliation between people and place, an exercise in "social landscaping," because the newly established parks had to promote traditional use as well as nature protection and tourism.

Though made of natural materials, even on a monumental scale, these were deeply cultural landscapes. They were places which people didn't just visit but in which they lived: they had to coexist with, and were often coextensive with, ejidos. Unlike parks in the United States and Canada, where Indigenous peoples and settlers were removed in an effort to deculturate the place and cleanse it of a permanent human presence, Mexico's parks were a way to promote both nature and people. They were, in Emily Wakild's words, "revolutionary parks," a "common cultural patrimony of nature," which embodied the goals of the Mexican Revolution. Here was Cárdenas's compromise: one between communal rights and federal stewardship, which made national parks and ejidos not rivals but complementary means to a shared goal. Mexico's parks were the outcome of a historical moment when revolutionary ambitions found bureaucratic means of expression. Ejidos emphasized the human dimension of social landscaping; national parks emphasized the landscaping itself.[48]

✵

Reserving land from intensive exploitation did not set it aside untouched by fire. Lightning was common, of course, which made the volcanic arc around the Valley of Mexico prime ground for natural fire and ensured that fire would occur without people present. But humans accounted for most ignitions—tourists set accidental fires, along with locals who still lived and worked in and around the reserved lands. Research at Ajusco suggests a 3-to-15-year fire return interval in pine, but detailed reconstructions from fire-scarred tree rings in and around the region suggest a lower range of values, 6 to 7 years under a "natural" fire regime, and still lower where people start most of the fires (outside the valley at Orizaba, as low as 2.1–3.5 years).[49]

Firewood gathering and charcoaling were relentless, and the politics were complicated when nominally local usage was being hauled to cities. Resin tapping was also abundant, and the collected resin and the pitchy scars on the ocote trunks were highly flammable; this made tapped trees and resin orchards vulnerable to surface fires, which argued for clearing around the base of the tree or lightly (and regularly) burning the grasses and forbs.

Where livestock was a part of village economies, pastoral burning was routine, both in grassy glades and within the pine forest. San Miguel and Santo Tomás Ajusco in Cumbres del Ajusco National Park exhibited a *sistema silvopastoril*. Campesinos burned in traditional ways ("an empirical knowledge inherited from earlier generations") to promote forage within the ocote pinelands. Mostly beneficial, but occasionally detrimental when used too often or with too-great intensity, the burning had effects that could persist for two years, especially if it occurred during the rainy season. The belief among pastoralists was nearly universal that the fires were both beneficial and essential.[50]

Official records speak little of the fires that were surely present. Yet the foresters who served as administrators argued that urban demand for fuelwood should be met by drawing from more distant forests, and that firewood and charcoal should be replaced with coal, oil, and electricity, lessening the burden on protected parks. They sought to modernize the resin industry, improving techniques and ultimately replacing resin with products derived from fossil biomass; they pleaded to harvest timber on a model that was less artisanal (like that practiced by locals) or industrial (like that typical of pulp mills) and rely on scientific silviculture; they promoted alternatives to burning for forage, which would eliminate the messy tangle of grasses, forbs, and shrubs that formed an encumbering understory; they wanted to educate campesinos away from mixed agroforestry economies and their reliance on fire; and, where possible and necessary, they demanded measures to actively suppress fire. Quevedo himself wrote Cárdenas that "fires destroy our natural forest riches" and that protected lands like parks required rapid detection and suppression of all fires.[51]

Quevedo elaborated on the situation regarding firewood. The customary practice, brought from Mediterranean Europe and the Pyrenees, was to harvest low branches, dead branches, and windfall from oak woodlands and coppice, with a return cycle of 5 to 10 years, a usage that could perpetuate indefinitely. But increased demand led to clearcuts that were destructive, and pushed gatherers to expand from oak to pine, fir, and other conifers that could not resprout. By such reckless and "irrational" means, the forest was being wrecked, with all the knock-on effects on climate, water, and soil such ruin brought. Other, more advanced techniques were possible. Even before the parks were declared, Quevedo, as president of the Sociedad Forestal Mexicana, had argued for a system of forest reserves to protect the woodlands from such abuses.[52]

It was less a case of removing the hand of humanity than of deciding whose hand should prevail. Foresters wanted not only to protect what existed but to restore what they believed was being lost. Quevedo, for example, contrasted the stewardship of the campesinos, with their former economy unhinged by urban demand for charcoal, to the industrial pulp mill at Rancho de la Venta, which planted a hundred trees for every one felled and so rehabilitated the land. In revolutionary Mexico it was not possible to remove the traditional users or to replace the native woods—the techniques generally preferred by global forestry. Rather, the primary goal of the state was social justice, so authorities sought to redirect traditional uses into a more scientifically informed set of practices.[53]

Among the many ironies, however, was the fact that there was no adequate science of either forestry or landscape fire, and that traditional methods had contained fire better than imported and industrial ones.

Pruning forests of low branches, thinning brush and reproduction, picking up windfall—all kept the forest floor free of fuels; the wood burned in stoves and hearths rather than on the land. Routinely burning that surface to improve grazing cleaned out more windfall, and in conifer forests, it peeled off layers of needles and litter, which further lessened the likelihood of blowup fires while also furnishing ideal seedbeds for regeneration. Cattle and sheep cropped off the grasses, which kept pastoral burning under wraps. But it was not all just about combustibles: routine burning, which was possible because of cleaning and prior burning, rejuvenated the biota. The burning regimen of the campesinos was neither irrational nor ruinous.[54]

What unbalanced the old order was economic and demographic pressure. City demands forced charcoaling to quicken its rates and expand into areas not previously cultivated for firewood; markets for meat could prod ranchers into overgrazing; industrial logging, often illegal, left forests of slash rather than replanted woods. As Paracelsus had famously lectured, the toxicity of a poison resides in the dosage, not its substance. Traditional fire was not an ecological problem; traditional fire in untraditional circumstances could be.

Mexico's parks established a political context that demanded that tradition and modernity, campesino and científico, reconcile, and this applied to fire as well as to land tenure. Campesinos had little reason not to burn, and científicos had little power to stop them. Landscape fire remained for the former a companion, tool, and inevitable presence, and for the latter an all-too-visible emblem of the country's stubborn backwardness. But, with regard to fire, despite occasional public protests, they learned to live with each other. What mattered was the visible display of social landscaping.

It was a particularly Mexican solution—how much it diverged from the United States' model became quickly apparent when, in 1935, the two countries discussed the prospects for a transnational park along the border. The United States and Canada had Waterton-Glacier International Peace Park; the Americans sought a comparable expression of goodwill with Mexico, preferably at the Big Bend region of Texas. But while the negotiators found plenty of goodwill, discussions broke down not only over location but because of, in Emily Wakild's words, "culturally distinct definitions of conservation and divergent purposes for a park's existence."

A national park proved not to be a transcendent category of nature, capable of transfer anywhere, but—like landscape itself—the result of interactions with people. U.S. parks tilted toward wilderness, a place presumably unmarred by human use and stripped of permanent human residents; Mexican parks slanted toward places that needed rehabilitation and that had to accept traditional inhabitants, celebrate local patrimony, and support the resident economy. The remoteness that made a place attractive to Americans made it less desirable for Mexicans. The differences proved insurmountable. Of four national parks, six wildlife refuges, and six reforestation projects proposed for the border, none made the transition from paper to actual project.[55]

Daniel Galicia, the lead Mexican negotiator, balked because the Big Bend region was too difficult of access for Mexican tourists, and because it had a thriving livestock economy for which tourism would be unlikely to substitute. The proposed park conformed more to U.S. ideals than to those of revolutionary Mexico, and it was more likely to serve U.S. visitors than Mexican. Amid the environmental largesse of the New Deal, the United States' National Park Service could readily provide the infrastructure for another park. A more strapped Mexico needed to put its available funds toward restoring damaged landscapes, and needed to reserve for national parks those places at imminent risk or worth rehabilitating.

The concept of a national park proved elastic and capacious, taking on protective coloration suitable to its circumstances as it moved outside the United States. Even in the concept's homeland, the administrating agency, the National Park Service, absorbed historic sites, recreational areas, urban parklands, and legal wilderness. So, also, national parks proved to be only one of many varieties of nature protection. For Mexico they also proved transitional.

Mexico found in the park idea an institution it could adapt to create a new kind of communal land that celebrated the national patrimony in a globally recognized way but in which people could be residents, not merely tourists. The fusion proved unstable—the great archipelago of parks proclaimed by Cárdenas was a historic anomaly that fractured as counterforces of city and industry increased in intensity. The national park served as a transitional device until later, more specialized categories of protected nature could be promulgated. When reserves were established to the south of Big Bend in the 1990s, they appeared as biosphere reserves, not parks, much less wilderness.

And fire? One of the founding goals of state-sponsored conservation was to shield forests "from fire and axe." In the United States this meant that even lightning fires in parks were suppressed until the 1970s, when select parks experimented with allowing lightning fires to burn. In Mexico, where the servicio forestal administered the parks, the forest mattered more than the park, and while halting logging and stopping fires were official goals, logging and fires continued in tamer, traditional forms as firewood gathering and routine burning, however offensive to the global principles of parks and forestry. Mexico's national parks had to answer first to Mexico.

FORESTS

Like the peoples who lived in Mexico, Mexico's fires were varied, endemic, lambent, and shape-shifting, stirring together new and old. But megadiverse Mexico also displayed the freak and the feral.

Mormon colonies in Chihuahua reported fires started by meteorite fragments. In Aguascalientes a cigarette stub reportedly ignited a fire that destroyed 23 freight cars of the Mexican National Railway. Outside Tampico a field-clearing fire bolted into a row of oil tanks. An oil pipeline from Tampico to Mexico City erupted into flame at Azcapotzalco; it was possible that it was caused "by persons attempting to steal oil during the night and that a match or lighted cigarette started the fire." Fires broke out regularly in mines and in timber mills. Around San Luis Potosí, officials so worried over the "constant succession of forest fires" that they prohibited oil convoys by rail for fear of ignition. In Veracruz state a fire kindled during a hurricane wiped out the village of Jalapa de Díaz. Outside Amecameca near the base of Popocatépetl fires "raged," having reportedly been set by "Indians in revenge for not being allowed to have a religious procession from the chapel and hill of Sacro Monte during Holy Week." A volcanic eruption in Colima set off an earthquake (its epicenter in Michoacán) that shook small fires into "a roaring mass of flames" that blasted through outlying hamlets. In April 1928, *Excélsior* printed accounts of disastrous fires along the Sierre Madre Oriental that incinerated "mountain farms and huts" of Indigenous people and sent bears, panthers, deer, mountain lions, and anything else with four legs pouring down the mountains for the valleys, while clouds of birds searched forlornly for places to rest. Efforts to fight the fires were "futile." All creatures waited for the rains. Large fires in northern mountains were attributed to "Indians" intending to "drive out American prospectors." During the U.S. incursion into Mexico in 1916, supply lines were stretched to the limit to bring sufficient fodder for horses because "at numerous intervals along the line of communication the sky is lighted every night with the blaze from great fires sweeping over the prairie, and fields on which horses were finding meager grazing a week ago are now black"; army officers "regarded as highly suspicious that the fires should have sprung up so nearly at the same time along a line more than 300 miles in length." A few years later fire destroyed the international footbridge that connected the United States and Mexico at Laredo. Where possible, fires were fought by locals, army troops, the forestry corps, and workers drafted from oil, railroad, and timber companies, along with organized agricultural laborers.[56]

U.S. interests, especially in northern Mexico, frequently made reports about vast grass fires in the plains, sprawling forest fires in the mountains, and anything that burned around El Paso and Monterrey. Still, no Mexican fire of record caused anything like the immense damages left by the conflagrations that stormed around the Great Lakes and North Woods in the United States and Canada in the later nineteenth and early twentieth centuries, so breathless reports by U.S. newspapers about Mexican fires deserve a healthy dose of skepticism. For Americans to take interest, there had to be an American hook, preferably a celebrity. In 1935 smoke from fires in "central Mexico" was reported in U.S. newspapers because it grounded Amelia Earhart Putnam's flight from Mexico City to New York.[57]

The loss of Mexico's forests under liberalism and the Porfiriato was real, and it was convenient to demonize fire as a visible token of that loss and waste. Importing U.S.-style logging, it seemed, had brought in U.S.-style conflagrations. (Ironically, in denying rural people access to slash for firewood and charcoaling—literally domesticating a fire problem—authorities prevented a Mexican-based solution to a U.S.-catalyzed problem.) "Stop the bleeding" is the first rule of first aid. For Mexican forestry, that meant stopping the hemorrhaging caused by fire. Without fire protection there was no point in reforesting, for the new forests, like the old, would dissolve into smoke.

The Quevedo Years. If Mexico's land occupied the minds of its politicians and revolutionaries, its forests expressed the environmental ills that conservationists most sought to heal. From lowland rainforest in the south to forest-clad sky islands in the north to the wooded volcanoes that framed the center, Mexico's trees seemed to be where land, air, water, habitat, and natural resources intersected. Reckless deforestation had encouraged soil erosion, trashed watersheds, sparked fuel crises, and unhinged climate, leading to cycles of drought and deluge. Repairing the inequalities of land ownership could not solve Mexico's agrarian crisis if the land was damaged. The científicos of the Porfiriato might not view forests as much more than chips in the great game of political economics, but many of the científicos who came after the revolution made forest protection and restoration central to national success. For the revolutionaries, the forest problem was an agrarian problem. For the científicos, the agrarian problem was a forest problem.

A national forestry law promulgated in 1861, while well-intentioned, was narrow in scope and hobbled in execution. Leopoldo Río de la Loza published a blistering response that, among other proposals, demanded a total prohibition on fire (for all motives, though pastoral burning and charcoaling were especially obnoxious). Offenders should pay the cost of the damages their fires caused, along with six months to four years of public service. In 1892 the Díaz regime passed legislation (ineffective) to limit access to forests around Mexico City. In 1894 it enacted a forest law (premature) to establish guidelines for logging, create forest reserves, and appoint forest wardens who, among other tasks, would fight fire. Forest reserves had become a primary vehicle for a doctrine of state-sponsored conservation. The United States commenced a system of forest reserves in 1891. Four years later the Díaz regime proclaimed a reserve in Hidalgo.[58]

Mostly, the liberal reforms after independence and through the Porfiriato had sought to convert Mexico's forest heritage into marketable goods—treating Mexico's woods like its mines. Where protection was warranted, it was to ensure production. Like missionaries who appreciated that conversion could only follow a change in life circumstances, advocates for rapid industrialization could even regard deforestation as a means to remove the habitat that retarded the conversion of campesinos to modernity. A movement for the national conservation of forests, even arguing the case on economic grounds, had to wait for the last gasp of the Díaz regime and the revolution that followed. It had to wait for the public career of Miguel Ángel de Quevedo.

To an astonishing degree Quevedo not only was involved in nearly every aspect of the development of modern forestry in Mexico but was its instigator. Redact him out of the chronicle and there is little to read, and scant narrative to connect what remains. He was, and is remembered as, the Apostle of the Tree. Yet his entry into the politics of forests, like Mexico's, was indirect. In Quevedo's case the portal was the endless water crisis in the Valley of Mexico.[59]

Two accidents shaped his career. The first left him orphaned. His mother died when he was 10, his father when he was 17. An upper-class world of privilege and education in Guadalajara ended as his family sent him to an uncle, a canon in Bayonne, France. His full-immersion baptism in matters French redirected his education from a possible priesthood to engineering, his primary subject matter from Aztecs and astronomy to hydrology, and his enthusiasm about nature from mountains to forests. He took a science degree from the University of Bordeaux, then pursued his studies at the École Polytechnique, finally graduating with a degree in civil engineering from the École Nationale des Ponts et Chaussées. Along the way he was befriended by Gaston Planté, a member of the French Academy of Science, and he came under the influence of Alfred Durand-Claye, a hydraulic engineer who insisted on the role of forests in dampening cycles of floods and droughts, and Paul Laroche, who had worked on the Suez Canal and who taught him about maritime projects, especially modern ports. But forests, too, were important in stabilizing (and making productive) coastal wastelands dominated by dunes.

Behind these particulars was a tradition of French *dirigisme* that had bonded state and forest since the Code Colbert. Deforestation in the French mountains had led to notorious outbreaks of "torrents" downstream.

Massive afforestation with pines in Gascony's Landes had reclaimed a wasteland of sand and converted it into productive woods. An old belief, modernized, argued that removing trees in the countryside had destabilized the climate, leading to famines of fuel and food. The peasantry could not reform such ruin—had, in fact, been responsible for the crises. They cared little for the woods, which they converted to farms and pastures and which they invariably burned. Only the state had the necessary knowledge, resources, and will to rehabilitate damaged lands at the scale required. Hydrology and forestry were inextricably linked; below ground, on the surface, in the air. The French forestry agency was named Eaux et Forêts (Waters and Woods). To control one was to control the other.

When he returned to Mexico in 1888, while the Porfiriato was hitting its stride, Quevedo became supervisor of the Great Drainage Project (*el desagüe*), a scheme to drain the Valley of Mexico by means of a grand canal and a tunnel through the mountains to carry off excess waters. But his French training also alerted him to the forest condition of the surrounding mountains; as the waters of the valley had fallen, the clearing of its forests had risen, aggravating floods and so defeating the purpose of the desagüe. What some other Mexican observers thought were problems caused by the wholesale draining, such as desiccation and dust storms, Quevedo invariably attributed to deforestation. That the immense project, once completed, was overwhelmed by flooding convinced him that the source of the Mexican torrents had to be controlled.

By then he had been off the desagüe payroll for a dozen years, thanks to his second career accident. In 1889 he was thrown from a cart in the tunnel and nearly died. What might have been his life's work abruptly ended, and what might for a different man have led to lassitude prompted Quevedo, when he had recovered, to plunge into a gamut of the engineering works that were underpinning Mexico's modernization. An urban railroad for the Valley of Mexico. Earthquake-resistant structures. Hydroelectric dams. A makeover for the port of Veracruz focused on a great dike. A return to France—the model of modernity—for an international conference on urban hygiene. Everywhere he noted the changes, all malign, that resulted from feckless forest clearing. Flooding that overran the rails and raced through Mexico City and Veracruz. Soil erosion that damaged agricultural productivity and silted up reservoirs and harbors. Droughts everywhere, and desertification in arid zones. In cities, dust, foul air, and sunburnt barrios. Common to all these ills was the need for an aggressive program of reforestation. Mexico required a program in silvicultural engineering equivalent to the desagüe. It was, to Quevedo's growing sense of alarm, an "absolute necessity."[60]

In 1901, as flooding overwhelmed the desagüe, the showcase project of the Díaz regime, Quevedo returned to Mexico City in its public works department. This time, instead of forestry as a buttress to hardscape engineering, he emphasized forests as a core public-works program. Trees in the city could scrub air, shade walks and parks, and improve amenities. Trees on hillsides could hold soil and water and make marginal and abandoned lands productive again. Trees could replace an industrial economy's demands for wood. What a properly wooded watershed could mean for the Valley of Mexico, a reforestation program could mean for Mexico overall. Forestry had become Quevedo's passion. "It is to me as a religion."[61]

In December 1901 he gathered like-minded colleagues into the Junta Central de Bosques del Distrito Federal to lobby for further forest improvements, serving as the Junta's president. Meanwhile, from his post within the Ministry of Public Works, he advocated for urban parks—again on a European model. What had begun as 2 parks that claimed a scant 2 percent of the cityscape swelled to 34 parks and 16 percent. Success led to success. In 1904 the Junta Central was integrated into Mexico City's Ministry of Public Works. Another Paris conference allowed a refresher course on European forestry, especially planting efforts in Austria, Germany, and France; there Quevedo connected with the director of the French forest department, toured forestry schools, and then visited the export of the French model of forest reclamation to Algeria. Quevedo effectively renewed his French connection: If Paris was a model for Mexico City, France's Department of Waters and Woods became the model for Mexican forestry. If France could transplant afforestation from Gascony to the Algerian coast, Quevedo could carry it to Mexico.[62]

When he returned, he did so, arguing at Veracruz that the new forests could help drain wet lowlands, improve air, reduce disease, and replace problem sand dunes with productive woods. A promise extracted from the French foresters to help establish a forestry school became reality in 1908; the school echoed the classic curriculum and was staffed with French instructors. Surveys of forested land commenced, beginning with the

Valley of Mexico. The next year Mexico acquired its first forestry bureau when the Junta Central was reconstituted into the servicio forestal and dispatched Quevedo as its national representative to the North American Conference on the Conservation of Natural Resources, where he surprised the other participants with the degree of Mexico's engagement in forest conservation. He then presented an agenda of forest conservation to the Díaz regime, including expropriation of some public lands and regulation elsewhere, some of which was adopted in the form of more forest reserves around Hidalgo and a protective belt around the Valley of Mexico's mountain brim. The convulsions of the Mexican Revolution undid much of this work, and when he was told he might be a target for assassination, Quevedo self-exiled to France until Victoriano Huerta (not he) fell at the hands of an assassin. When he left, Quevedo was 62 years old.

The revolution was, in a sense, the third tragedy that threatened to derail his career. The forestry school shuttered, the nursery was abandoned, the Veracruz reclamation project was mauled. Upon his return, his wife died. Yet once again Quevedo renewed his mission. It was as though restoration through forestry had become a personal as well as a national need. In 1917 he persuaded President Carranza to transform the Desierto de los Leones into Mexico's first national park. In 1924 he helped reinvent the Junta Central de Bosques as the Mexican Forestry Society, complete with its own journal (*México Forestal*). In 1926 a forestry law was enacted that, at least on paper, included most of the major provisions Quevedo had campaigned for and pertained to all Mexican forests—private, public, and communal. The 1926 forest law gave Mexico's foresters legal sanction to control fire outside of urban settings, which meant most of Mexico.

Still, while the forest law granted legal authority, it did not come with many resources with which to apply it, certainly not at the scale Quevedo and his colleagues believed they needed. They could advise, they could publicize, they could plead, they could denounce. They could promote El Día del Árbol. They could produce plans for national parks and forest reserves. In 1928, on the U.S. model, they erected a demonstration fire tower on the summit of Rumurachi in the Sierra Madre in Chihuahua. But even in the Valley of Mexico fire protection was limited.[63]

In 1929 Quevedo seized on press publicity about a spate of fires in Río Frío around Iztaccíhuatl to argue for enforcement powers and a corps of forest guards. The fires were visible from the railroad between Mexico City and Puebla and so proved an occasion for endless commentary by travelers and journalists. What made the outbreak more than an item of public gossip, however, was that the fires burned on two adjacent lands. One was part of an ejido (Río Frío), the other part of an hacienda belonging to a bank (Caja de Préstamos).[64]

The fires had multiple causes and two dramatically different effects. Some were started by traditional pastoral burning, some by protest on the part of ejidatarios upset by bans on traditional forest use, others by travelers tossing cigarettes and matches along the main road, and still others by the railroad. The fires on the ejido burned freely. The fires on the hacienda crossed into land being developed for "orderly exploitation" under a plan "in accordance with the requirements of the Forest Law," overseen by a forester from the Mexican Forestry Society. Most of those woods were tapped for resin, and there was a nursery, but careful planning, along with the aid of a hundred forest workers from the hacienda, thc department of roads, and some officials of the servicio forestal, suppressed the flames before they could harm the woods. The lesson was clear: a corps of dedicated workers and a system of firebreaks could protect forests. Quevedo suggested that the National Roads Commission train its laborers to extinguish the fires that so frequently clustered along routes of travel.[65]

Such lessons were fundamental when Mexico held its first forest congress the next year. But particulars mattered more than generic exhortations. In 1932, for example, Quevedo lectured on the problems caused by changing methods of charcoaling in the Federal District. The traditional technique—coppicing of oak, first introduced by Basques—was ceding to clearcuts and extending to pines and firs, which, unlike oak, could not regrow their hacked branches. Pines could tolerate turpentining; they could not endure coppicing. Rail worsened the scene by increasing the range of commercial influence and allowing greater volumes of wood to be harvested. Add to the burden local firewood gathering and a fast-maturing pulp industry, and the region's woods were melting away. Quevedo urged a general ban on cutting throughout the Valley of Mexico and the creation of a system of protected reserves. Rare exceptions would be allowed only for sites (like the Rancho de la Venta) where modern forestry was under way, since such enterprises, if properly executed, could expand the domain of the woods through aggressive replanting. More fundamentally, he urged a deeper transition that would replace wood as a primary energy source with fossil fuels and electricity. Success lay in further technological development. Modernity, not traditional use, would save the woods.[66]

Other foresters like Antonio H. Sosa surveyed the panorama in the wider penumbra of the Valley of Mexico. Tlaxcala exhibited the usual suspects of deforestation and their typical consequences. But to these obvious sources must be added the "intentional fires" set by people "that year by year consume great extents of woods," which expanded the domain of farms and pastures, and in some cases expressed simple malevolence. What such fires did not burn directly, they helped ruin by enabling and interacting with other deleterious practices.[67]

Several years later Ricardo de la Vega declared that the primary motive behind a "growing deforestation" that "year by year" and with "greater frequency" affected the woods around the Valley of Mexico, Puebla, and Toluca was the bad behavior of the local people, or the competition between modern forestry and traditional agriculture. Fire was at once a means, a consequence, and a marker of that wretched process. Ejidatarios converted what were and should remain natural woods into inappropriate mountain milpas and pastos that lost not only their trees but their soils and the benefits of their natural endowment. In every "civilized nation" steps were taken to keep woods where the woods grew best and farms where farms best thrived. Mexico needed to do the same.[68]

Though the 1926 forest law did not create a state bureau that could exercise authority over the nation's diverse, diffused woods, it gave legal sanction to control fire outside of urban and agricultural settings. No other group—not agriculture, not anthropology, not economics—claimed fire: foresters in Mexico, like those nearly everywhere, became by default the engineers and oracles of landscape burning, a global corps akin to mining and civil engineers. A guild committed to extinguishing fire was charged with managing it, and given political legitimacy to do so. The shovel replaced the sword as a means of forced conversion.

Forestry's Mission Civilisatrice. Mexico's servicio forestal was not unique: similar institutions sprang up everywhere that came under the dual influence of the European imperium and fossil fuels. Everywhere forest reserves were an axiom of state-sponsored conservation. Everywhere foresters regarded fire control as their first, most fundamental task, the success of which determined everything else they might attempt. Everywhere they saw fire as the blood enemy of the tree, and everywhere they insisted that they spoke with the authority of academic science.[69]

In truth, Europe had no formal science of landscape fire. An enormous amount of traditional knowledge existed because fire was an axiomatic practice for most farming and herding, for turning heathy wastelands into rough pasture, and for converting woods to farms and fuel. Mediterranean Europe had the wet-dry rhythms that favored fire; boreal Europe had much longer cadences that from time to time allowed fire before green-up and after curing; but central Europe, which is to say temperate Europe, had no natural basis for fire and, without dry lightning, no natural source of ignition. It had fire aplenty—fire kindled and sustained by people. Fire's presence was a metric of human habitation: fire's ecology was an ecology of people. Folk burning was, as dismissed by Bernhard Fernow, the first professional forester in the United States (himself an émigré from Prussia), an expression "of bad habits and loose morals."[70]

For foresters fire was an existential threat. It was the means that permitted swiddeners to break woods into fields, pastoralists to replace tree regeneration with forage, hunters and foragers to promote shrubs, browse, and covert over closed forest. Agriculture grew at the expense of woodlands. When forests became important to the state as a source of naval stores, land reclamation, and watershed protection, foresters rose in influence, and they carried their fear and disdain of fire with them. By the eighteenth century Enlightenment elites also despaired of and condemned the peasantry's addiction to burning. A rational economy, a modern society, would find alternatives.

Germany excelled at silviculture, fire as a graft on the great rootstock of European agriculture. France specialized in bonding forestry to public works. Great Britain mastered the task of transforming foresters (originally from Germany) into agents of empire. An appeal to spare forests from "fire and axe" inspired vast forest reserves, which led to forestry as a guide to good practice and foresters as overseers. A temperate Europe removed from natural fire established a profession devoted to fire-immune woods and then dispatched its emissaries to places prone to burning—many overflowing with fire, for which seasonal flame was as much a presence as monsoons, dry winds, and a fire-thirsty biota. Britain made India (practically a dictionary definition of an intrinsically fire-prone land) the training ground for its imperial forestry corps. The shock could be intense.[71]

France offers an interesting case because its temperate north came to absorb a Mediterranean south, then Corsica, then swaths of North Africa, and then, crossing the Sahara, West African Sahel and savannas, and other places beyond. In ways not true for Britain, the Netherlands, or Germany, France had to deal with endemic fire within its national territory, not just in colonies. Converting peasants into Frenchmen equally meant taming folk fire habits. Suppression was not possible: quiet accommodations were made for the customary *petits feux* that stippled agricultural lands and formed a tag team with pastoralism in Provence. Corsica presented a fire insurgency. Algeria became a prolonged firefight that ultimately had to be ceded. Red Madagascar—red from its flames as well as its soils—had colonial administrators that publicly condemned burning but in practical reality had to overlook it. Unsurprisingly, perhaps, it was a Frenchman, Georges Kuhnholtz-Lordat, who in 1938 published the first quasi-global survey of fire.[72]

It was Mediterranean Europe, whose landscapes were rich with routine fire, that conquered Mexico, but it was to northern Europe, particularly France, that Mexico first looked for its models of fire and modernity. The United States soon offered another—geography and ties were too close to ignore. The two countries' forestry bureaus met at the North American Conservation Conference, then remained in informal contact. In 1928 Gilberto Fabila, an agricultural specialist attached to the Mexican legation, reported on the U.S. Forest Service, noting how its fire mission effectively began with the 1910 season, how it managed the federal forests, how the 1911 Weeks Act (expanded by the 1924 Clarke-McNary Act) established formal cooperation between the federal and state governments, and how private associations emerged for fire control, a model also common to France and Quebec. Inevitably, relations between the United States and Mexico could be testy and would not significantly improve until the two economies reconnected during World War II.[73]

The deeper problem was that there was no genuine science of landscape fire behind forestry's nominally enlightened project. Forestry examined fire because—like Sahagún documenting Nahua religion—it wished to know its enemy, and because wildfire and postburn scenes encrusted in black made a visually powerful statement of what conservation was about. Fire ecology did not acquire a name until the 1960s and did not merit its own journal until the early 2000s. Mexico's fire modernity was not premised on empirical fact so much as on the fraternal prejudices of forestry. Certainly not all of Mexico's fires were good; the issue was that there was no intellectual methodology for sorting the good from the bad. In Mexico, as in France and France's overseas possessions, landscape fire was accepted as a metric of both social and environmental order or chaos. The less free-burning fire a society had, the more progressive its ranking; the more landscape fire it displayed, the more slovenly and atavistic its character.

Yet the French example suited Quevedo, suited a version of Mexico that had been shaped by the Conquest, and suited elites that sought both to protect the Indigene and to civilize the country to the standards of Europe. Campesinos might work the land, even communally own the land, but they should follow the guidelines of the educated and progressive. In Quevedo's vision modern societies regarded forest conservation as a necessary "obligation of the state." Modern industry was rational and could reconcile with science. But rural populations lacked the training, the knowledge, and the "conscience" to manage forests in a way that could guarantee Mexico's future. This held even when forest cooperatives were organized the way missions might have been in an earlier age—the forester as Franciscan of modernity. Campesinos would have to be proselytized and converted to a proper understanding of fire.[74]

Yet there seemed to be no institutional means that could counter the vastness of the challenge. Even the Mexican Revolution with its reforms, by expanding the dominion of the campesinos, had broadened rather than contained low-tech agriculture. Then the revolution reached its climax in the administration of Lázaro Cárdenas, who, as governor of Michoacán, had become fascinated with forest issues.

Quevedo's Second Coming. President Cárdenas fashioned what he hoped was a compromise that would combine the social goals of the Mexican Revolution with the ambitions of modernity: the old mirage in which the directing minds of elites could uplift the lower classes came to Mexico's forests. Simultaneously, in 1934 the ejido program skyrocketed and an autonomous forest department was created. After repeated appeals Miguel Ángel de Quevedo, now 72, agreed to become its director. On the eve of the Cárdenas *sexenio* (the six-year presidential term) he had published in *México Forestal* a program for forest protection, which he regarded as "undoubtedly" of "great urgency." The Cárdenas administration gave him the chance to apply it.[75]

In scope, the servicio forestal was a department of natural resources that also had responsibility for wildlife and fish. In principle, the department had standing equivalent to the department of agriculture—forestry acquired the political clout its colleagues in Europe and the United States enjoyed. Its officials oversaw national forest reserves, liaised with forestry cooperatives, cultivated nurseries, reforested degraded lands, and by means of taxes and permits had oversight of logging programs, or more broadly a wood products industry. As many commentators have noted, this was for forestry what the ejido program was for agrarian reform (and what Pemex would be for oil after 1938). In principle, they were complementary projects, reclaiming a common national patrimony. In practice, the forest department remained weak and underfinanced, and forests competed with other land uses—that was, in fact, why so many were mauled.

In early 1937 Cárdenas instructed Quevedo to prepare an action plan to advance forestry's goals through the remainder of his term. Boldly, Quevedo proposed a full-gamut program that embraced national parks, national forest reserves, protected forests around cities and in critical watersheds, nurseries for reforestation, training schools for foresters and guards, a research institute, and even a museum. To meet its fire protection ambitions, the department established a division of fire control. That corps sought out suitable peaks for fire towers, placed firefighting tools in critical sites in national parks and protected forests, worked with railroads to lessen locomotive fires, promoted fire prevention, tried to rally state governors and city mayors to the cause (some responded, many ignored the call), and sought to craft a suitable workforce. Forestry called on the Department of Indigenous Culture to study points of mutual aid; on the Mexican Automobile Association and the Department of Roads and Railways to help with prevention and reporting along routes of travel;

and on public relations programs to broadcast mass publicity on fire prevention. Bureaus worked with bureaus.[76]

Extinguishing fires on the ground was another matter. The biggest check on fire spread was the sheer cultivation of the countryside—the more intensively cultivated, the better shielded it was against free-burning flame. But much of Mexico either was extensively farmed and grazed or seasonally relied on the monte. The department's own guards were laughably meager, best used to direct the efforts of others. For manpower the office relied on local authorities, on the national army, and especially on local laborers organized into fire brigades (*corporaciones de defensa contra incendios de montes*) that could be called out as necessary. By the end of 1936 the office had enrolled some 16,000 men into 955 companies. Their geographic distribution ranged from Baja California to Quintana Roo. When local resources failed, the department appealed to the army.[77]

On April 4, 1939, Quevedo prepared anticipatory plans for the succeeding sexenio. He praised Cárdenas for recognizing the problem, but despite "good will," implementation had fallen far short of what was needed. The forest issues had not been approached integrally, had not penetrated into the foundation; the forest department had only fussed with palliative measures and had left the concerns for the next regime. Quevedo insisted that the problem would require decades of work and enormous sums of money, that it would demand an undertaking both endless and continuous. The politics had proved complicated. While ejidos involved a publicly visible transfer from large landowners to small, forestry's schemes seemed to deny ejidatarios access to their woods. Quevedo was an ardent patriot, but he was also an apostle of the tree, so while he pondered the "agrarian problem," he fretted that ejidos might lead to a worsening of deforestation. The next administration would need to amplify the measures the servicio forestal had begun.[78]

Quevedo's observations about the two fire scenes at Río Frío—the competing responses of ejidatarios and a bank-owned hacienda—foreshadowed the attitudes of the Mexican forest service. In 1937 a national park was proposed for Oaxaca, but a good chunk of the lands belonged to a U.S. company (Cargill Lumber) that was willing to cede it to the federal government in return for equivalent land in Chihuahua. The servicio forestal dispatched Antonio H. Sosa to the Alta Sierra Tarahumara to investigate the prospect. The land was rugged, around the Barranca de Cobre; it was public, both *bosques nacionales* and *terrenos nacionales*; and it boasted a magnificent forest of mature ponderosa pine.[79]

A forester of his time, Sosa argued that good practice could exploit without destroying the woods, and in fact could sequentially improve the forest. Cargill Lumber, he concluded, wished to work the Chihuahuan woods legally, in a rational manner, adapted to all the rules and technical regulations that were appropriate. These were people the servicio forestal could literally do business with. There were prospects for lumber, resin, and an expansive grazing industry. What the land needed was access in the form of roads and rail. It also needed—this was "absolutely indispensable"—a system of forest protection.[80]

The reason was the Indigenous people. The Rarámuri were destroying the forest with their scattered milpas and endless fires; unless interrupted, their habits, which were nothing more than vandalism and "persistently" opposed the life of the forest, would lead to the "total ruin" of the Alta Sierra. Why did they girdle trees and burn the woods? It was, Sosa concluded, simply their instinct to destroy trees. They had never been incorporated into civilization. They lived in a semisavage state, not harming anyone else, but almost entirely lacking formal work and aspirations, tending their "most miserable" milpas and tiny flocks of goats, all the while more or less unconsciously destroying the woods. They could not be reasoned with.[81]

It was, Sosa confessed, an intractable conundrum. But the *problema indígena* was also enfolded into the Mexican Revolution, and Lázaro Cárdenas addressed the issue directly. Mexico, he affirmed, had no wish for the disappearance of the Indigenous people. Rather, a government of the Revolution, such as his, would find suitable ways to incorporate and improve them without seeking to replace or enculturate them. That was not a political calculation that came easily to the servicio forestal.[82]

More and more, Cárdenas received complaints about foresters interfering with customary practices, not supporting the spirit of the revolution, in effect seeking to replace rural mores with those of modern industry. The complaints mounted: either the agency didn't enforce laws fairly or its enforcement was too harsh. Frustrated, Quevedo wanted to expand logging bans, eliminate forest cooperatives, widen prohibitions on grazing, and enhance forest reserves. The goals of expanding small farmer holdings and those of enclosing forests—liberating campesinos to engage in traditional practices while demanding controls on hunting, burning, and firewood collection—were bound to clash, and when the political pressure seemed to require a choice between one or the other, ejidos won.[83]

Quevedo succeeded in Mexico City; he stumbled in the countryside. The power of forests—that they touched every other feature of Mexico's natural endowment—was also the strength of Quevedo's critics: they were so diverse and so many. In January 1940, a month after Quevedo had submitted proposals for the future administration, President Cárdenas demoted the agency and its agenda by subordinating it to the Secretariat of Agriculture, and dismissed Quevedo. Forestry mattered; farming mattered more. Seven years later the Apostle of the Tree died.[84]

Miguel Ángel de Quevedo had ample reason for dismay. Like Mexico's forests, his career, born of desperation, had been cultivated, cleared, replanted, and felled again and again. Yet the forest scene he left behind was significantly different from what he had first found. There were institutions, both governmental and civil, devoted to forest research, forest education, and forest management. There were journals, reports, and statistical surveys where before there had been random anecdotes. Mexico celebrated Arbor Day. Among científicos, forests and parks were regarded as something worth protecting, not simply commodities to shove into the maw of a developing market. Mexico had the trappings of global conservation. It had trees planted where they had previously been removed. And it had an apostle, its own patron saint, to grant Mexican forestry a face before the world.

Quevedo's was a unique career; there is none elsewhere in North America to compare it to. Quevedo's story is more or less the story of forestry in Mexico until 1940. If he failed to achieve all that he wished, or even much of what he regarded as the most elemental reforms, if he could not engineer politics as he could water and trees, he could console himself with the traditional solace of foresters: that what he planted today would grow for future generations.

Three Americans Look South. Three visitors from the United States, all with literary ambitions, toured Mexico before and at the onset of the Mexican economic "miracle" that began in the 1940s. Two were trained in forestry at Yale; the other was a self-taught ecologist who learned his craft on guano islands. All achieved international acclaim, and all looked askance at Mexico's fires.

In 1936 the U.S. prophet of modern environmentalism, Aldo Leopold, visited the Río Gavilán region of the Chihuahuan sierras and compared it to the north-of-the-border landscapes he had known as a Forest Service ranger years before. On both sides of the border, the "low country" had been "equally abused and spoiled." The Mexican sierras escaped because the Apaches had held off newcomers, and then Pancho Villa's troops "performed the same ecological function," followed by unstable land policies and the Depression, which collectively kept "the mountains green." The Chihuahuan mountains struck Leopold as being what the U.S. Southwest's mountains had been like "before the Juggernaut." Despite the presence of rock check dams "on almost every slope," he regarded the scene as wild, or as close to pristine as possible.[85]

The secret was the absence of a settler economy, and especially a modern one, accompanied by logging and grazing. "The Chihuahuan Sierras burn over every few years. There are no ill effects, except that the pines are a bit farther apart than ours, reproduction is scarcer, there is less juniper, and there is much less brush." But the critical watersheds were intact, "whereas our [U.S.] watersheds, sedulously protected from fire, but mercilessly grazed before the forests [reserves] were created, and much too hard since, are a wreck." Regular burning kept the mountains healthy; the intrusion of axe and hoof had sent them into a spiral of decay.[86]

Émigré Mormons from the United States had established a few colonies in the valleys of the Gavilán that included grazing, but the effects were microscopic compared with the vaster countryside. In fact, the influence was as likely to go from Mexico to the newcomers as from the transplanted communities to Mexico. In 1938 Leopold returned, this time with his 24-year-old son Starker. Starker was shocked to watch their Mormon guide, Floyd Johnson, toss lighted matches into the grass as they rode along in the late fall, and was later startled when Johnson casually burned a batch of witches'-broom in a pine. Yet nowhere had they seen a more magnificent grass-forest savanna. They discussed this curiosity, appropriately enough, "around a campfire."[87]

Many years later, Starker recalled that "it began to dawn on me that fire was a perfectly normal part of that sort of semi-arid country, and might even be an essential part of it." Aldo had a harder time grappling with what they saw. After all, he had helped set up the fire protection system of the national forests in the Southwest, had argued strenuously against light burning, and had dismissed constructive burning by American Indians ("It is of course absurd to assume that the Indians fired the forest with any idea of forest conservation in mind"). Still, the experience stayed with both father and son, forcing them to question inherited wisdom, even when this wisdom was putatively supported by formal science.[88]

Leopold the father worried about what the future of settlement might hold for the Mexican mountains. "Our own Southwest was pretty badly misused before the idea of conservation was born," he lamented; the consequence had been that the United States' "conservation program for the region has been in a sense a post-mortem cure." Grazing was particularly pernicious—Leopold doubted that the region could tolerate any grazing "without ultimate deterioration." Yet Mexico had enshrined moving small farmers and herders into the public lands as a national goal. With ejidos, especially with the relocation of rancherías into the mountains, would come roads, and grazing; and with grazing might well come both erosion and lethal blows to the ancient fire regime.[89]

That seems to have happened. In the summer of 1948 Starker returned, along with a couple of colleagues. To their distress they "found the loggers were moving in" with their trucks; they listened to the grating hum of a sawmill and noted that with the influx of settlers came "the livestock." It was a pattern Starker had witnessed "elsewhere in Mexico," so avid was the lust for land and so powerful was the thrust of the Mexican Revolution to grant it. Many of those relocated to the region had previously lived on ranchos and reestablished that pattern. First came the cattle and horses, and then the summering sheep; the concentrated grazing "slicked off" the bottomland grasses, then the "more accessible mesas," and finally the most remote places. With livestock the grasses receded, and without fuel the fires shrank.[90]

William Vogt was a skilled organizer and bureaucrat, a gifted writer, and a self-trained ecologist who learned about complex food chains and resource limits studying guano islands west of Peru. But his real understanding was what might be called political ecology: the unstable intersection of nature's bounty and humanity's appetites. Like others who studied islands, he understood limits, and could see continents and even Earth as islands. Unlike many, he understood constraints in highly personal terms: polio had left him partly lame and reliant on crutches. Every step he took reminded him that life came with limits.[91]

His talents converged when, after doing some "intelligence work" for the U.S. State Department, he was appointed head of the Conservation Section of the Pan-American Union in 1943. What the Chincha Islands, climate, and populations of seabirds had encapsulated, he expanded to embrace the Americas. Indefatigably he visited every nation in the Western Hemisphere, and everywhere he saw humanity's froward hand stripping away more than nature could replace. More people consuming ever more stuff under governments committed to giving them still more was a formula for ruin.

He spent 10 months in Mexico. It was, he bluntly informed the Mexican government, "a sick country." It was losing its soil, its forests, its ecological balance. Yet increased productivity also increased environmental loss. Much of the crisis was simply about numbers: there were too many people (and too much introduced livestock) for the carrying capacity of a mountainous land with an erratic climate. When Vogt thundered similar jeremiads to most of the countries he toured, nearly all were outraged, not just offended but angry. In Mexico, however, the secretary of public education enlisted him to write a brief, popular book on the subject. *El hombre y la tierra* was published in 1944.[92]

Forests were too important not to single out as critical to everything else, and fire was too visible not to target for special scorn. The woods had three "terrible enemies": fire, pastoralism, and bird hunters (Vogt was an ardent birder). If all the children of Mexico were burned to death each year, Mexico as a nation would vanish. But that, he argued (with considerable hyperbole), was what was happening with Mexico's forests. Farmers, pastoralists, charcoalers, even careless hunters, all added fires. Little by little, year after year, the fires killed the young trees, vaporized the organic soil, left exposed earth to be washed away in the rain, and unhinged the ecosystem. "Disgracefully," many of these fires were set by campesinos, the people who would most feel the ill effects and who literally held the future of the country in their feckless hands.[93]

In 1948 Vogt expanded his vision in an article in *Harper's*, "A Continent Slides to Ruin." (Among its charges it singled out Chile for lacking "a single fire warden or forester" although fires had recently rambled over a quarter of a million acres. Fire is always good publicity.) That piece only distilled what followed, a best-selling book, *Road to Survival*, which opens with a panorama of eight people from around the world and how they confronted the one thing they had in common: the environment they share and depend on. "One common denominator controls their lives: the ratio between human populations and the supply of natural resources, with which they live." Unsurprisingly, one of the representative eight was a Mexican, a woman named María Dolores from Michoacán. Her response to her burdens and the "sterile landscape about her" was a form of fatalism: "*Se aguanta*" (You accept it).[94]

William Vogt was writing at the dawn of a war-stimulated and postwar boom, what would become the Mexican Miracle. Yet he saw more being extracted than could be replenished—the first prophet of a long line of "limits to growthers." He proved too outspoken, too unyielding certainly, for a political appointment like his in the Organization of American States. He was allowed to resign, and he took his organizational skills to Planned Parenthood. In 1968 he tried to recoup his triumph with a successor, *People! Challenge to Survival*. The book failed to launch. By then it had too many rivals, not least Paul Ehrlich's *The Population Bomb*, published the same year.

Road had appeared at the false dawn of the Mexican Miracle, its warnings eroded by the boom. *People!* arrived as the Green Revolution rooted and agricultural productivity exploded. But too many people taking too much remained the core dilemma. To Vogt's mind a "*dementia economica*" prevailed, more concerned with symbols of national wealth than the realities of nature's well-being. Traditional Mexican agriculture could not feed the population, and productivity measures by themselves would only be overtaken by ever more mouths to feed. "Only a progressive and accelerating destruction of natural resources" allowed Latin American countries to feed and shelter their citizens. The future pointed toward "biological bankruptcy." Half of Mexicans, he was informed, did not get enough to eat.[95]

Like so many observers, Mexican and American, Vogt saw in Mexico's endless fires both a reason for retardation and an undeniable illustration of it. "Fire and water are the limiting factors" in milpa agriculture; fire was the "rampant ally" of the axe (Vogt quoted a Nicaraguan proverb: "One man in one day with one match can clear a hundred acres!"); charcoaling was a principal cause of flooding. Flying from Guatemala to Mexico City, he counted "seventeen forest fires, most of which were allowed to burn themselves out."[96]

Tom Gill had studied Mexico for decades. A linguist by instinct and education at the University of Pennsylvania, he took a master's at Yale in forestry, then worked for the U.S. Forest Service in Colorado and South Dakota, with time out for a tour during World War I in the Army Air Service. In 1922 he transferred to the Washington office, where he worked in the Branch of Public Relations. Three years later he became associate editor of *American Forests and Forest Life* for the American Forestry Association, and after another year he assumed the executive director post at the Charles Lathrop Pack Forestry Foundation, where he remained until 1960, the duration of the foundation's existence.

And he wrote. He wrote Westerns, a dozen novels and a blizzard of short stories, many of which landed in the leading magazines of the era, such as the *Saturday Evening Post* and *Cosmopolitan*, and a few of which were filmed as movies. But he also wrote, with Charles Pack, *Forests and Mankind* (1930) and the classic *Tropical Forests of the Caribbean* (1931), which first tethered him to Mexico and made him an international oracle. He was among the drafters of the UN Food and Agriculture Organization's (FAO) forestry division; he founded and served as president of the International Society of Tropical Foresters; he became executive director of the International Union of Societies of Foresters; he contributed, through the Pack Foundation, to the creation of the Mexican Institute for Renewable Natural Resources; and he was honored with special awards in the United States, Germany, France, and Mexico. Like others of his generation (including Leopold), his broad education had allowed him to engage with big questions, and his talent for expression allowed him to convey his understanding to a wide public. In 1951 he wrote *Land Hunger in Mexico*.

He brought to that inquiry a scientist's eye for landscape but a novelist's ear for story and the moral pathos of human tragedy that moved his text beyond the reports of a technocrat. Like Quevedo, whom he admired, Gill argued for the critical value of forests not only as timber or aesthetics but mostly in terms of what they meant for water and soil, which is to say agriculture, which was still the nuclear core of Mexico's revolution and remained the basis for Mexican life. In its European origins forestry had been pitted against agriculture, part of an ancient vendetta. In Mexico, they had to complement each other. It was unimaginable to have a forestry program that did not accommodate agrarian concerns, and it was equally impossible for agriculture to survive without the life support offered by forests.

More than Quevedo, Gill empathized with the conundrum of the campesino who needed more land when land alone, managed in the traditional ways, only worsened the underlying conditions of poverty. Instead of lifting up the poor, a programmatic transfer of land, especially marginal mountain land, only spread the immiseration. Prerevolutionary Mexico's attempt to modernize had alienated campesinos. Postrevolutionary Mexico honored campesinos' desire for land, but at an environmental price that could undercut campesino livelihood and future. Granting more marginal land to the agrarian poor slaked land hunger, but, because such lands quickly degraded, it set up a vicious cycle that demanded yet more land.

Land Hunger opens with a vision of pine forests between Morelia and Mexico City, while "the rising smoke of forest fires covers the countryside in the dry season." Together with the axe, that smoke was "the picture of a people destroying the very stuff their life depends on." "Tragic and, in a way, heroic figures, these campesinos of the high plateau and hinterlands, yet they are the most ruthlessly destructive predators that over the centuries have degraded and destroyed Mexico's soils, forests, and water sources." Like all tragedies, the outcome was "probably inevitable." Given Mexico's "climate and topography and the long-established customs of her people, there was scarcely any other road for her to tread." Yet while the fire and axe of roza-tumba-quema were bad, the burning by pastoralists was worse because Mexico was "essentially a stock-raising country," and pastoral fires constituted the bulk of the burning (Gill cites Mexican authorities who estimated the contribution of herder fires as being from 20 to 80 percent of all fires). At the inaugural convention on Mexican forestry, Rigoberto Vásquez attributed "99% of the forest fires to the primitive practice of burning" for crops and pasture.[97]

There is an element of tragic despair in Gill's analysis. Reform demanded a makeover of Mexican land and livelihood. The Mexican Revolution had begun transferring ownership of land, but had not fundamentally altered how it was used. "One can scarcely overestimate the difficulties," Gill wrote, since they would affect all aspects of Mexico's social composition. "Every effort is sure to be fought, resisted, and willfully distorted, for it will interfere with age-old patterns, and will run athwart of human cupidity and organized greed." Unlike Aldo Leopold, who saw the land without people, or Starker Leopold, who could come to see fire as an ecological benefactor, Gill viewed a land inextricably intermeshed with the people who lived on it, and saw fire as an enabler of their pauperism.[98]

As much as any outsider, Tom Gill understood Mexico's fatal conundrum. He spoke fluent Spanish, he studied Mexico's history, he connected Mexico to international forestry, he empathized. Upon his death, Enrique Beltrán, Quevedo's successor in Mexican conservation, paid homage to him as a "sincere friend of Mexico, of which there are innumerable proofs." He worked with reformers and understood the depth of the dilemmas they faced. Equally, he sympathized with the campesino. But he had little sympathy for the fires that made the land habitable for the campesino's way of life.[99]

The Old Fire as New Fire. New fires, old fires—they have long been a Mexican tradition.

In 1925, after the violent revolutionary fires had entered a quieter phase, Mexico City revived the Aztec fire ceremony "using reproductions of the costumes worn by the Aztecs." Advising the project was Manuel Gamio, second secretary of education, a student of Franz Boas at Columbia University, and later a founder of the *indigenista* movement. Selections of "minor priests and high priests [were] made from representatives of the purest Indian races left in Mexico."[100]

The event was restaged in 1953 amid a period of rapid growth, what might seem a new world for Mexico's political economy. The "Pageant of the Fifth Sun" was enacted before 50,000 people, this time not at a historic site but in a stadium ("a combination of ancient and modern design") in which "copies of the pyramids were reconstructed on its grounds." The authorities hoped to attract tourists and arouse awareness throughout Mexico of the history, culture, and "glory of its pre-Hispanic civilization."[101]

Still, it proved easier to celebrate ancient fire ceremonies than contemporary versions of ancient fire practices. It was okay to re-create a pageant in a stadium but not to tolerate enduring emblems of the ancients on the land. How to reconcile the seemingly irreconcilable, the ancient with the modern—to mix, fuse, segregate, hold in common suspension—had long been a Mexican challenge. In many respects, fire was no different from so many other features of land and life in Mexico.

Yet whether seen or not, fire remained, as the reenactments showed, symbolically if not practically, at the center of what connected Mexicans with their land.

MEXICAN MIRACLE ON FIRE

By time of the 1953 reenactment, Mexico was fast changing. The Mexican Revolution was institutionalized and tamed. With the stimulus provided by World War II, Mexico uncoiled from its revolutionary turmoil and leaped ahead. Commercial agriculture boomed. Dams, irrigation projects, and later the Green Revolution boosted production. Roads and mills felled remote forests and opened rangelands to new markets. Industry boomed, with mines and mills revived, riding the postwar surge that swept over the United States. Here was the breakneck modernization that liberal reformers had long yearned for.

Most of all, Mexico's population boomed, and began a migration to the cities and to the United States. In 1930 the population was 16,696,000, 80 percent rural; by 1950, it was 26,185,000, 71 percent rural. In 1980 Mexico had a population of 68,562,000, some 60 percent urban; by 2020, it was 99,000,000, with 75 percent in cities. In 1950 Mexico City's population was 3.4 million; in 1990 it was 15.6 million, which is to say a population almost as great as the entire country had had 50 years earlier. By 2000 over 9 million Mexicans had migrated north of the border—nearly 1 Mexican-born person in 10, or more than half of its 1930 population.[102]

The transition should have relieved the burdens on Mexico's lands and woods that both internal and foreign critics had identified. But the population growth was so steep that even with 60 percent of it in cities in 1980, a rural population remained that was equal to the country's entire population in 1950. The stresses remained, the revolutionary project continued to redistribute land, and, thanks to accelerated industrialization, population pressures increased in many landscapes. The effects, moreover, were not distributed equally; to avoid social conflict, too many ills were downloaded to Mexico's land, water, air, and woods. Mexico's environment continued to bend until, finally, it began to splinter. Some of the most dramatic outcomes were felt in soils and forests, and in keeping with tradition, reformers targeted that most visible of failings, fire.

FIRE PROTECTION INDUSTRIALIZES

In January 1941, a presidential decree created a program of forest protection by employing inspectors and guards. Soon afterward, the first National Convention on Forests was held at the Palacio de Bellas Artes. Some 198 papers were read, of which 6 focused specifically on fire—an astonishingly small number considering the voiced alarm about the damages fire wrought. A summary of the fire scene by Guillermo Gaitán was one of 25 presentations published with the proceedings.

Gaitán argued that fire control was of "primordial importance," and that without it fire would consume existing woods and overrun efforts to reforest. He documented the causes behind the promiscuous burning: agriculture (40%), pastoral burning (25%), illegal logging (8%), arson (5%), charcoaling (10%), campfires (5%), locomotives (4%), and cigarettes and matches (3%). The statistics showed that fire could in fact serve as a tracer of Mexico's slow industrialization of the countryside. The overwhelming source of burned area was traditional fire use; outside railways, these figures could probably represent the scene a century earlier. (What is striking, however, is the absence in these statistics of what was the oldest cause, lightning, and historically the largest cause: unknown.) Among Gaitán's 13 recommendations were amendments to the forest law, the creation of a national committee to integrate agencies involved in fire protection, the establishment of local committees to control pasture and roza-quema use, an increase in publicity, more fuelbreaks, and the dedication of forest guards exclusively to fire control during dry spells. Mexico's fire problem was another expression of its still-hobbled progress toward industrialization.[103]

A succession of forest laws, decrees, and amendments followed from 1942 to 1949, with the 1947 legislation recognized as the third forest law.

Of 9 published sections, 1 referred to fire. The 1949 amendments contained 19 sections, of which 1, again, addressed fire. Foresters struggled with verve and zeal on behalf of conservation. But their fire policies and programs were misinformed, misguided, and ultimately misplaced. While forestry had assumed responsibility for fire, and rural Mexico had fire aplenty, the issue was not fundamentally a *fire* problem: the flames were a manifestation, not a cause. They were no more a problem that could be solved with strict fire control than were the flames that in the late twentieth century flickered across the Amazonian rainforest or the peatlands of Indonesia. The problem was one of political economy. Forestry was not empowered to solve it, and trying only embarrassed its agencies and enfeebled its public standing.

Fire mattered, and it made for good propaganda, but in terms of commitment it mattered to the amount of maybe 5 to 12 percent of the forest problem. The wartime boom had favored rapid production over expanded protection, and in practice the power to declare logging bans was compromised by the temptation to issue exceptions, which created legal dissonance and tended to render the servicio forestal an inept annoyance to all parties. Even as a new generation of researchers scanned the fire scene—legally, statistically, practically—the official response was clearly less than what official alarms declared needed to be done. Apart from political will, the Mexican state lacked the resources to contain the flames it denounced. In 1952 it created a dedicated Department of Fire Protection (Departamento contra Incendios Forestales) organized under the Agriculture Secretariat. The agency fielded seven brigades, all in the Valley of Mexico.[104]

Mexico sought, in time-honored fashion, to promote growth, sustain its rural population, and assert government control, all at the same time. As Christopher Boyer concluded, "Mexican forests were shaped less by market forces, management policies, or population pressures than by the effects of political negotiation among the people and institutions that vied to determine how, and in whose benefit, they would be used." The forest codes that followed the 1917 constitution attempted a "Solomonic compromise" that allowed timber companies and rural communities to access the nation's forests so long as they did so under guidance and plans approved by the servicio forestal. To assist governance over a patchwork of forest ownerships, both companies and rural communities were organized into collective units subject to "expert scrutiny."[105]

The upshot was the fashioning of two coextensive realms for forests. One promoted a political economy informed by (nominally) scientific forestry and international capitalism; the other, traditional usage as adapted by rural residents. For the commercial economy this meant outsourcing to *unidades industriales* (logging companies), which functioned as parastatal agencies. As part of their concessions, the companies were required to staff fire brigades to substitute for an understaffed servicio forestal. But without effective oversight, these often fell by the wayside, like other nominal obligations (such as funding schools). For the rural economy, the arrangement meant that, while communities held forests, they did not have the right to use them as they saw fit but had to submit plans to a servicio forestal, again ill-equipped to meet those requirements. There was no official mechanism of fire control. Traditional burning continued, an informal economy of combustion.

The two realms worked by ignoring irreconcilable issues, by not policing unenforceable laws, and by engaging in corruption. The process left Mexico with some of the most magnificent forest and conservation laws anywhere, and some of the worst enforcement. The narrative reached a macabre climax in the 1970s as the Echeverría regime allotted record numbers of ejidos, targeted forestry as a means of rural development, and sponsored vast deforestation projects in the tropical regions to establish commercial plantations. It is estimated that Mexico lost half its forests over the course of the twentieth century.[106]

Critics dismissed the woods laws as "paper forestry." Even Gill disparaged the new code as "largely inoperative." It could restrain neither campesinos nor capitalists. The law required that nearly every forest enterprise have a formal plan approved by an official forester. The servicio forestal, however, had a hopeless budget, little political clout to enforce unpopular laws, and a small and often technically inept staff (in 1950 Mexico had a dozen forestry students). Of the servicio's 1,137 personnel, 30 percent were administrative and 13 percent technical. Elaborate laws without provisions for adequate enforcement meant they were "almost universally flouted, evaded, and disregarded," prompting cynicism, inefficiency, and corruption. Such laws were a legal fiction; revolutionary origins unhappily ended in reactionary outcomes. "In spite of the most stringent laws," Gill lamented, "the forests are being destroyed at a faster pace than ever before."[107]

In 1955 Abel López Caballero wrote a thesis at the National Agricultural School on forest fire control. So much had been written in recent years on the subject of fire by so many (the servicio forestal, the FAO, forest technicians, persons interested in forests and conservation), and so much of it was erroneous, that his role, López announced, was to record the "most important enemy" of Mexico's flora and to suggest sensible solutions. He thought most of the fires were the outcome of disputes over land tenure ("estados de litigio" and "tenencia de la tierra"), though that illustrated what had happened to a subject that should be simply technical. It was a theme that encouraged lots of words, but few practical deeds. His bibliography had eight entries, one of them the forest law and another a Mexican description of the U.S. Forest Service. Two years later Guillermo Moreno Arellano submitted a thesis on organizing for fire control that had seven references, only one Mexican. Perhaps the most remarkable feature of the theses is that they were written at all.[108]

Commerce had its say, and positioned itself on the side of economic development. In 1958 a self-appointed advisory committee organized by the paper industry surveyed the forest scene and like everyone else pronounced it "grave." Agriculture, pastoralism, forestry—all needed to intensify and adopt modern methods to avoid soil erosion and climatic disruption, but the story resembled a Greek tragedy in which the players are unable to halt their own self-destruction. Traditional farming and herding left the land exhausted, but unregulated logging was no less shiftless and damaging. Every enterprise had its signature fires, which branded the land and wrought "incalculable" damages, before again compelling the desperate, the feckless, and the greedy to move on. Mexico's population was simply too poor and growing too fast. Its "*tribus aborígenes*," malnourished and oppressed, were a "predatory force" that had over the centuries, without much reflection, destroyed the woods. Mexico could not recover its patrimony until it restored its soils, waters, and woods. For that it needed science, modern industry, and a supportive state. All these ideas were reinforced at the second forestry convention, held in 1959.[109]

In reality, extensive logging created explosive fuels in the form of slash and blew sparks with abandon. But flames made it easy to distract critics. A forest warden in Chihuahua in the 1920s had even proposed that the "culling of the forest would end 'the continual forest fires that occur in those distant lands.'" A later version argued that it was better to cut the woods than leave them to the mercy of aboriginal burning. (Timber companies in the United States' North Woods had argued that it was necessary to log ever faster because of the danger from fires gorging on the slash already strewn about the landscape.) More likely, rural people's habit of extending their wood gathering for charcoal and hearth fires into slash, like poor miners working through tailings, did more to reduce fire hazard.[110]

Still, the premise persisted: modernity would abolish burning. The sooner the economy fully industrialized, the sooner traditional fires would vanish. Industrial forestry would do what proselytizing foresters could not. A Green Revolution would abolish agriculture burning. There was truth to that argument, for even as landscape burning continued as both symbol and taunt, working fires were receding from homes, cities, factories, and urban consciousness, with unanticipated and often deeply paradoxical consequences. Fire suppression joined other industrializing enterprises in Mexico that strove to maximize productivity, while leaving pesticide pollution, soil salination, slashed forests, degraded landscapes, and disgruntled ruralites in their wake. Smoke thinned. The Mexican cornucopia steadily emptied.[111]

THE BELTRÁN BOOM

Perhaps more than anyone else, certainly internationally, Enrique Beltrán Castillo became the face and voice of Mexican environmentalism. He embraced conservation, a doctrine of rational use. He relocated Mexican science and forestry from France to the United States. Teacher, researcher, public administrator—Beltrán grew up in the Mexican Revolution but advanced his adult career during the boom years of the World War II and postwar epoch. His special contribution was to show that the revolution needed to promote not just social but environmental justice. In 1950 he lamented that Mexicans "are always talking about our turbulent, strife-torn history," and observed how much of that strife applied equally to Mexicans' relationship to their land and that "to continue the criminal abuse of the past would obviously be suicide."[112]

Beltrán trained as a marine biologist, but his academic career, given the state of the subject in Mexico, spanned all aspects of biology. Like

Quevedo he went into personal exile when his views conflicted with the Obregón regime. A Guggenheim Fellowship placed him in Columbia University, where he subsequently earned a PhD in zoology (1933). When he returned, he passed through an exhausting number of posts as a teacher and researcher, increasingly drawn to conservation issues. In 1952, with money from the Pack Foundation (through Tom Gill), he established the Mexican Institute of Renewable Natural Resources. In the postwar era he became the spokesman for Mexican conservation. From 1958 to 1964 he served as a subsecretary of forestry and wildlife in the Ministry of Agriculture and Grazing. As much as anyone he understood the complexity of the interlocking challenges that plagued Mexico's environment, and he appreciated the limits of resolving them. Rural backwardness was not simply the product of ignorance but of poverty. Poor forest conditions reflected laws that invested too much power in state bureaucracies that lacked the means to enforce those laws—a politics of anarchy.[113]

In 1960, on Arbor Day, with both President Adolfo López Mateos and agriculture secretary Julián Rodríguez Adame present, Beltrán seized the occasion of a new forest law to address what he regarded as "one of the principle causes of forest destruction and one of the more difficult to control." He meant forest fires. They affected all three of the charges Mexican forestry had been handed: the protection of the existing *patrimonio forestal*, the rational development of a forest industry that could be conserved into the future, and the regeneration and restoration of areas that had suffered forest loss. "Fire the destroyer" touched each. Its visible damages were a fraction of the invisible harms it wrought, because fire not only damaged the woods but degraded their capacity to regenerate.[114]

There was plenty of blame to go around. Industrial forestry engaged in "irrational exploitation" under an economic philosophy that seemed to argue, "Get a tree, trash it, and seek out another tree to trash as well." But fire, Beltrán argued, was worse. Not only did it destroy standing timber, it led to soil erosion and floods, and the routine burning that persisted destroyed any recovering forest, be it natural or planted. Fires inspired conditions that encouraged more fires, some of which would spread into surrounding lands. Fires set for pasture or milpas or cleaning, sparks cast by travelers or picnickers or arsonists—all Mexicans contributed, and all needed to rally to the national crusade to eliminate the menace.[115]

The 1960 forest law sought to correct the overly centralized character of previous forest codes, devolving some tasks to state councils and parastatal corporations. It also included special provisions for fire, among them requirements for official approval of any fire use and obligations for anyone using the forest to actively prevent and help suppress fires and to report fires to the servicio forestal. The law sought to regulate, not ban outright (an impossible task), agricultural use and hence fire, where land was not mountainous. An aspect of "great importance" was that the law prohibited salvage logging in burned areas, thus reducing the incentive to burn woods in order to evade approvals to harvest them. All physically able inhabitants were obliged to cooperate in firefighting. Plans would be drawn up for all the country, with five regions ranked by significance. The federal government would contribute toward a special fire corps supported by 5 helicopters, 48 engines, and 48 lookout towers. The organization of local workers into firefighting companies would be revived and augmented. A national fire prevention program of public education was projected. And in reserve stood the army, due special homage for its importance to the servicio forestal (though it lacked adequate firefighting tools). A realist, Beltrán argued that the program should concentrate in the Valley of Mexico, where most of the population and potential damages resided. He believed a 50 percent reduction in forest fires was possible.[116]

Two years later the Mexican Association of Professional Foresters systematically elaborated on Beltrán's comments and their crystallization into a five-year plan (1960–64). It identified as the principal threats agricultural burning, pastoral burning, and campfires left by charcoalers and travelers. The real total of fires and area burned was likely double the official count, but it is revealing that the plan isolated only rural Mexicans as causes. It noted that this was the "first time in its history" that the government of Mexico was addressing the problem of forest fires, that a subsecretary had made it a personal mission, and that the plan proceeded with the explicit approval of President López Mateos—the first time in Mexican history that the country's leader had personally involved himself with the issue. A commitment to controlling fire was an important broadcast to the world that Mexico was claiming its place among the industrial nations.[117]

The allocation of significant resources was virtually unprecedented: modern fire protection in Mexico dates from this commitment. While the amount was only "a tenth" of what was needed, that it existed at all was a demonstration to both Mexico and the "entire world" that Mexico did not "ignore" its natural estate and was committed to conserving it with the best contemporary means. In 1960, 8,262 fire brigades enrolled 82,830 campesinos; by 1962, there were 10,150 brigades with 101,500 men. On paper this was a huge firefighting force. In practice, many were never mobilized and had no more presence than the schools and social services that were promised by companies but never materialized. For most companies, timber concessions were all benefit and little cost.[118]

Still, the program needed specialists under the authority of the servicio forestal. All the other bureaucracies touched by fire in some way, from the secretariat of public works to those of education and even tourism, were official partners. A program of public education ranked high among projects. A new journal, *México y Sus Bosques*, carried the message to professionals. Mexico joined the United States and Canada in the FAO-sponsored North American Forestry Commission, which promptly created a fire working group. Not least, there was the army, which not only assisted with patrols during the height of fire season, but granted the program political stature and a certain moral standing. The presence of the army was a public statement that the program had the support of the government. All in all, the Mexican Miracle was devoting a fraction of its new wealth and power to fire control.[119]

At the end of his tenure, Beltrán reflected on what had been accomplished. In *La batalla forestal* he summarized what had been done, what had not been done, and what remained to do. It was of course a political book, but also one of a personal nature. He opened with his own history of commentaries on the forest crisis dating back to 1946 in order to compare achievements to aspirations. He knew he wouldn't like the criticisms that would result, but he claimed he welcomed them. That is how progress in science works, and it should apply to politics as well. Though he was not a forester, his passion had always been for renewable natural resources, and the woods were where many of his preferred themes came together. The forests synthesized all of Mexico's environmental crises.[120]

The obstacles to reform were formidable. Public opinion was confused, policies were often mistaken, the character of the servicio forestal was uncertain, and funding was inadequate. It was easy to identify pathologies and "not simple" to prescribe remedies. The great, ancient enemy was fire, and what was needed was a holistic approach ("sistema completo") of prevention and suppression. Symbolism aside, the actual allocation of resources against fire was minimal and "brought smiles to the skeptics" who believed nothing would be accomplished.[121]

But Beltrán countered that the proper response was not a choice between "all or nothing," but incremental progress, neither "magical" projects nor "speculative" (and delusional) results. Within Mexico's circumstances, progress in taming fire was happening. Even the United States, he pointed out, which dedicated "stratospheric" amounts of money to firefighting, had not eliminated wildfire. Ten years after his friend Tom Gill had wryly observed that the United States had not yet "conquered forest fires," Beltrán noted that Los Angeles was burning again.[122]

NORTH AMERICAN MEXICO

The allusion to the United States was hardly casual. Even Quevedo regarded the United States and Canada as models because of their discipline, perseverance, commitment, and progress. In 1914 Coert duBois, a Gifford Pinchot acolyte and head of California's national forests, argued that the United States' fire problem was unique, distinct from Europe's, and required its own study. In the postwar era, the United States became a standard reference for the world. Gill had helped author the forestry charter for the UN's Food and Agriculture Organization. Stuart Show, longtime head of the Forest Service's California region, had written the FAO manual on fire control. The first college textbooks for fire were written by Americans (1946), and revised by them (1953). Americans proclaimed that fire control was their contribution to global forestry.[123]

There was an embarrassing gap between what was taught in classrooms and what was needed in the field. Young foresters were given "only an introduction to forest fire control as an innovation peculiar to the United States. The effect of this has been that many young foresters have found themselves on a job of which four-fifths was protection of the forest from fire, but with their training in inverse ratio." Yet practice had more than compensated. The Depression-era Civilian Conservation Corps had allowed "systematic fire protection," as the concept was termed, to expand across the country almost overnight. The postwar era replaced the Corps' manpower with war-surplus mechanical muscle and special firefighters, notably hotshot crews and smokejumpers. The United States, it seemed, was engaged in a cold war with fire, that other "red menace." The paramilitary spectacle of B-17s dropping retardant, helicopters shuttling crews and water drops, bulldozers punching fire lines through woods—these became the ruling images of modern fire control.[124]

As the U.S. Forest Service dominated the U.S. fire scene, so the U.S. example did the world. Institutional alliances followed, most notably the North American Forest Commission, one of six such bodies worldwide. (Mexico also belonged to the Latin American and Caribbean Commission.) Soon afterward the NAFC organized a Fire Working Group, which first met in 1962 in Washington, D.C.[125]

Working groups came and went, but the fire group persisted and was, by consensus, regarded as a genuine success. It met annually, rotating hosts among three countries, reviewing themes in prevention, research, and control, and making field trips to sites of special interest. It sponsored occasional fire study tours: in 1968 (to Canada and the United States), in 1975 (to Mexico and the United States), and in 1980 (to the United States, for an introduction to prescribed burning). In 1968 it helped arrange for a mutual assistance agreement between the United States and Mexico. In 1974 it devised a logo for itself (created by Rudy Wendelin, a Forest Service artist who had shaped the modern image of Smokey Bear). In 1978 it published a glossary of fire terms in Spanish. It created eight international sign symbols for fire prevention. Beginning with a four-page "Forest Fire Equipment News" report, it published for the next 19 years *Forest Fire News*, a veritable news magazine that was widely regarded as one of the working group's "finest projects." Beginning in 1978 *News* was published twice a year.[126]

The cross-border flow of information led to exchanges of people and equipment, demonstration tours, and airings of policy concepts. The Fire Management Working Group made it possible for the forestry bureaus of three countries to work with one another. In 1983 the U.S. Forest Service and Mexico arranged a memorandum of understanding to help strengthen all aspects of Mexico's fire program. In time, familiarity and confidence segued into outright mutual assistance treaties and practical support through, for example, the U.S. Agency for International Development (USAID). The Fire Management Working Group had no funds to disburse, but it gathered senior staff and offered a formal mechanism by which soft welds of personal contact and exploratory concepts could bind a North American fire community that was otherwise confined by national borders.

Hardware was easier to transfer than software. Helibuckets could go on loan from Georgia to Mexico, engine operators from the Los Padres National Forest in California could be sent to the Federal District, and train-the-trainer programs from the National Advanced Resource and Training Center in Arizona could be deployed in Chihuahua. Still, fire prevention posters and proposed fire film festivals, while avidly sought by Mexico, had to pass through force fields of culture. At the 1973 meeting in Guadalajara it was proposed "that 'Smokey Bear' be established as the common symbol" for fire prevention by all three of the member nations, though "each country will search for and use prevention symbols believed appropriate." Each went its own way.[127]

Forest Fire News ceased in 1989. Or rather it went global, succeeded by *International Forest Fire News* under the auspices of the FAO, which also sponsored the Global Fire Monitoring Center in Freiburg, Germany, which in turn established a worldwide consortium of fire networks. Increasingly fire was treated as part of a suite of natural disasters, along with earthquakes and floods, addressed by the UN's International Decade for Natural Disaster Reduction.

More broadly, satellite imagery revealed a planet dividing along two kinds of fires, one that burned living landscapes and another that burned lithic landscapes in the form of fossil fuels. In 1988 the United States experienced a major drought and outbreak of fires, concentrated—to endless media fascination—in Yellowstone National Park, while Mexico suffered from Hurricane Gilbert's rampage through the Yucatán Peninsula, slashing forests that would subsequently dry out and burn. Behind both sets of events lurked warnings about planetary warming and climate change: thanks to the globalizing quality of the atmosphere, the burning of fossil fuels shifted from competing with landscape fires to colluding with them. Climate and land use saw Amazonia burning, Indonesian peat smoking in much of Southeast Asia, abandoned land along the northern Mediterranean catching fire and threatening villages, and flames moving into suburbs and exurbs that had pushed into former rural lands in developed countries like the United States. Mexico found itself linked not only to North America but to Earth.

Its position was awkward, always developing, or rather, half developed and half not. The fire group within the NAFC succeeded in part because it had to absorb only three countries and two languages. Two of those countries were among the Big Four of Earth's firepowers. The gradient between Canada and the United States was real but mild; between both those countries and Mexico it was so steep that surely the benefits would flow southward. In truth the fire fundamentals formed a sharp border along Mexico's northern frontier, not because of technological backwardness or policy lapses but because of the character of the land being protected. Canada and the United States developed fire institutions to control fire in public wildlands. Thanks to its demographic history, confirmed violently by the Mexican Revolution, Mexico did not have wildlands, only cultural landscapes. The strategies for fire protection so spectacularly displayed by the North Americans could not simply transfer to Mexico. At best they might be translated.[128]

The North Americans were actually the odd ones out. Modern fire protection had emerged out of a global project of conservation developed by Europeans as they tramped and trammeled around the world. Conservation was a doctrine to contain the wreckage and breakdown wrought by imperialism and the onslaught of footloose capital. Everywhere, Europeans found themselves in a chronic firefight, not just against open insurrection but against an insurgency of traditional burning. The British in India, Myanmar, south and east Africa, and Belize; the French in Corsica, Algeria, Madagascar, and north and west Africa; the Dutch in Indonesia—all faced routine fire, some because of political protest, most as the assertion of traditional usage in the face of a global economy that wanted to turn the woods into marketable commodities. To the authorities, fire control not only shielded reserves from damages but asserted control over the population.[129]

The experiment generally failed, or limped along with major (if silent) accommodations in which the resident population continued to live in and around the reserves and use them. In India British foresters adopted "early burning" around the protected forests to insulate them against the landscape fires that annually flared around the reserves, and then began to burn interior firebreaks, so that in the end they probably burned as much as would have burned in their absence; they just did it in another pattern. In Myanmar the Conservator of Forests withdrew fire protection altogether. The British wanted teak, and their German-inspired program of fire control prevented teak regeneration, while indigenous fire practices encouraged it. Later, the nominal authorities fashioned a taungya system of agroforestry that allowed them to save face while permitting the local peoples to persist in traditional ways. The French in Madagascar denounced and criminalized burning but had to ignore it in practice. That kind of subrosa acquiescence would sound familiar to Mexican foresters.[130]

North Americans' apparent success at dampening fire rested on a political condition that resulted from a history of colonization subsequently stiffened with a philosophy of conservation. Canada and the United States were populated in their formative years by northern Europeans and immigration, and their national state was fashioned by Enlightenment ideals; they had extensive backcountry stripped of people by war, disease, and law. People visited those parks and forest reserves, they built roads and felled forests, they set fires by accident and occasionally arson, but they did not live on that land or require it for their daily sustenance. In principle, they could be strictly regulated or banned outright. What in principle the authorities could not control was lightning (though, incredibly, they did experiment with weather modification projects in an effort to suppress strikes). While fire agencies had to answer to a general public, they did not have to deal with residents who stubbornly insisted on routine access to burning and had a revolution to back up their demands. And both the United States and Canada rapidly evolved into fossil-fuel civilizations, leaving their populations ever more removed from flame in quotidian life.

Mexico was constitutionally different. It had little immigration, had Enlightenment ideas of governance imposed as a retrofit, and had almost no places that were not inhabited as working landscapes. Instead of global forestry's paradigmatic ideal—vacated land, subject only to the workings of officially sanctioned foresters—Mexico had a messy overlay of protected reserves and local residents either within or around them, not so much regulated as finessed. Thanks to the revolution, Mexico stabilized that messiness in law: it displayed an unassimilated mixture of both fire realms, with neither in a position to erase the other. This meant that pumps and helicopters could cross the international border, while notions of how and where to use them might not. Mexico's fire scene more resembled India's or Madagascar's than Canada's, except that the colonial power was its own ruling class. This meant the fire exclusion project was doomed to stumble if not fail outright.

Strikingly, even as Mexico looked to the North Americans for inspiration, the United States was engaging in a fire revolution of its own. What had started as a fire crisis caused by too much bad fire was morphing, by the late twentieth century, into a crisis of too little good fire. Ecological blowback and an incessant accumulation of woodland fuels caused federal programs to reverse policy in order to promote the reintroduction of good fire. However troubled that project turned out to be—and there were plenty of interests vested in fire's suppression, including a private sector that critics dubbed a fire-industrial complex—U.S. fire agencies dealt with public lands empty of permanent residents who lived by using fire to farm, herd, forage, and recreate, or who burned the woods in their hearths and stoves. Mexico and the United States had fundamentally different political realities to reconcile with fire in the countryside.[131]

The assumption behind the original Fire Working Group was that knowledge and technology would flow from the more advanced nations to the less, and that policy would follow. But 1962, when the NAFC launched, also witnessed the opening salvo of what became for the United States an extraordinary reformation of how it engaged with landscape fire. In 1971, to reflect more progressive thinking, the fire group replaced its founding emphasis on "fire control" with "fire management" and inserted "management" into its formal title. By 1978 the United States had committed to a national policy that matched—and for many partisans sought to replace—air tankers with drip torches. Fire on the land, it seemed, was not as easily coded into technology as fire in cities or furnaces.

BANKING THE COALS OF REVOLUTIONARY FIRE

The bold burst of interest and investment that culminated in the 1960s—what became known as the Mexican Miracle—could not be sustained.

Throughout the 1970s population growth continued while economic growth slowed. Political unrest was one response; emigration and environmental deterioration were others. What was true generally held also for fire programs. In retrospect, the Mexican fire community considered those early years to be their moonshot, an Apollo program for fire. Then it stalled. Other visions arose; the environmental knock-ons of industrialization and pollution clamored with louder political voices than fire, which could seem a relic of an earlier, unlamented economy. A few decades later, however, the very premises behind the hard-welded alliance between forestry and fire control were called into question.

Through the 1970s the early boom became codified and institutionalized. The servicio forestal restructured the program of forest-industry fire brigades, which by 1969 numbered 12,164 *corporaciones* with 121,640 members. To complement that field force it published guides and texts. Handbooks for firefighting, mostly based on U.S. models (adapting even illustrations), were revised. In 1968 José Verduzco Gutiérrez and Guillermo Gaitán Medina updated their 1955 edition of *Combate de incendios forestales*. In 1970 Alfonso Gutiérrez Palacio published *Texto guía forestal*, revised again in 1977 in collaboration with Verduzco Gutiérrez. In 1976 Verduzco Gutiérrez expanded the scope of treatment in *Protección forestal*, which, revealingly, was published through the university at Chapingo. Both books spoke to themes and maladies beyond fire, but they gave fire an academic standing it had not had before in Mexico, much as the reforms in the early 1960s gave it an institutional legitimacy.[132]

Yet fire remained an ancillary program that could seem not so much a token of modernity as something that should have been resolved as a precondition for modernity. The fire that mattered most was the invisible combustion powering trucks and dynamos, not the cleansing flames in a countryside that resisted novelty. What underwrote Mexico's boom was its reservoir of fossil fuels that could be processed into energy, petrochemicals, fertilizers, pesticides, and of course government revenue. Fire control was a means to promote a forest products industry, which was part of an expansive project to transform Mexico into a developed country. Still, if the understandings that figures like Gutiérrez Palacio and Verduzco Gutiérrez conveyed regarding fire were those habitual to forestry, they gave the topic a Mexican accent. The books went through regular revisions and expanded authorship.

The national fire program meanwhile experienced serial name changes and shufflings among bureaus. Fire had no autonomous standing within the institutions of government (but then, the only fire department at universities worldwide was the one that sent emergency vehicles when an alarm sounded). Fire was something that programs that were dedicated to other purposes such as agriculture and forestry assumed as a condition of their mission. It was a seasonal annoyance that came and went. Much as allergenic pollen or the flu was best handled through broader public health agencies rather than a department solely dedicated to it, so fire could be addressed through agriculture and forestry. (The U.S. Forest Service had been moved to the Department of Agriculture in 1905, a move mostly regretted ever since.)

So even as population and environmental problems multiplied in the 1970s in the wake of the latest boom, even as the number of fires rose and burned area increased, institutional support struggled to hold constant. Still, such consistency was an advance compared with the havoc and uncertainties of revolution and the unequal competition between forestry and agrarian reform that had followed. If fire did not have its own bureaucracy, it at least now existed as a bureaucratic category. And with a fleet of helicopters it had showcase technology.

Mexico's approach to fire was not unique—was in truth closer to the response found in many other countries on Earth than the approach characteristic of a handful of firepowers. Mexicans had many perspectives on fire in their landscapes and what to do about it. But the official position was clear: the less fire the better. Fire protection was as much an obligatory duty of the state in public forests as it was in cities. Money and resources were scant, however; claims on them were many; and a political chasm widened between official decrees and actual behavior. It was possible, with little cognitive dissonance, to denounce fire in principle while simultaneously overlooking its everyday presence and the role of campesinos in promulgating it. In this Mexico was not alone.

The dissonance was practical and could be personal. Ing. Antonio H. Sosa was a prominent member of Quevedo's forestry corps. He studied the Desierto de los Leones, he wrote about the French influence on Mexican forestry, he took a party line on fire. In 1937, as the servicio forestal became a bureau, he scouted the Sierra Tarahumara as a prospect for forestry. His conclusion was much like that of Quevedo and the rest of his generation: the primitive economy and fire practices of the indigenous Rarámuri were destroying a magnificent forest. Turning the land over to modern commerce would replace superstition with reasoned regulation. It would remake the sierra; the forest would survive in perpetuity.[133]

In 1965, with the Mexican Miracle still aboom and the Green Revolution taking off, he returned and found that the land had in fact been refashioned. It boasted "caravans" of logging trucks, indiscriminate felling and high-grading of timber, phony plans approved by pliant officials, abuses of agrarian reforms, encroachments beyond legal allowances that bred eroded rangelands. The technologically primitive Rarámuri had tended a majestic forest. The modern Mexican official, engaged in converting the Sierra Tarahumara into marketable commodities and enabled by scientific technologies, was avidly trashing it. Sosa still believed in the principles of scientific forestry. Its practice was another matter.[134]

In his own way, Antonio Sosa was another revolutionary who looked back on his life's work with some dismay and perhaps regret. If the Mexican Revolution had failed to fully enact its founding vision of agrarian reform, so modernity had failed to promote its promised forest of the future. Such unease could not be passed off as paradox, or pushed aside with irony. It exposed a contradiction that, unresolved, yielded the worst of both worlds. The failure of colonial forestry was not unique to Mexico. It just found unique ways to express itself.

THE AMERICANS RETURN

Wartime demand and U.S. markets sparked the Miracle, but as Mexican goods went north, U.S. observers, variously versed in fire, went south with contrasting visions of what it meant. Starker Leopold's return helped convert him to a belief that fire not only could be tolerated but ought to be encouraged. Ed Abbey's travels confirmed his sense that people everywhere damaged what they should leave alone.

With his father Aldo, Starker had first camped in the Río Gavilán region of northern Chihuahua in 1938. In 1948, shortly after his father died (fighting a grass fire), Starker returned and recorded the effects of agrarian reform and the postwar boom.

A new forest law had been promulgated, roads now punched into the mountains, ejidos had formed, and livestock and loggers seeped through the passes and valleys. The party had to relocate from their original site to one that had "too little timber to attract the mills." Small mills dotted the forest, littering the landscape with flammable slash and mounds of sawdust that found its way into streams. The slash increased fire hazard, the intensive grazing temporarily reduced fires, and together they began restructuring the landscape. The formerly abundant game was scarce. Even the Mormon colonists who served as guides, born with dual citizenship, were preparing to decamp to the United States.[135]

In 1958, after another 10 years had passed, Starker Leopold published *Wildlife of Mexico*, an encyclopedic survey of game birds and mammals that came with a flattering foreword by Enrique Beltrán. The book mentions fire as an inevitable accompaniment to agriculture that worsens the damages done by unwise practices ("Fires occur with monotonous regularity, and as a result the forests do not recover and revegetate the mountains"). Yet by 1963 the view of fire that he had joked about at the Gavilán—that it did no harm, and might do good—had replaced the conventional wisdom that had informed his career. It was not destabilizing fires that degraded landscapes, but destabilized landscapes that degraded fire regimes.[136]

In 1962–63 he chaired a commission on wildlife in the United States' national parks whose report pleaded for a restoration of naturalness, for parks to be "vignettes of Primitive America." Strikingly, the report concluded that much of what had gone ecologically wrong was the result of fire's removal, not its chronic presence. The forests of the Sierra Nevada, once open and abloom with biodiverse patches, had become a dog-hair thicket paved with pine needles as impenetrable as peat. What had once resembled the premodern Río Gavilán had degenerated into a firetrap that could mutate surface fires into conflagrations. If the United States wanted a primordial world, it would have to relearn primordial practices. It would have to reinstate fire. The next year Congress passed the Wilderness Act. In 1968 the Park Service adopted Leopold's conclusions as the basis for policy. The wild and much that was associated with it, from wolves and grizzly bears to mistletoe and free-ranging fire, were deemed prime features in the national heritage, rightly in need of preservation, and in some cases of re-creation.[137]

Something had changed besides the Gavilán landscape. Behind Starker's flipped perspective lay Carl Sauer—a historical geographer, Starker's dissertation supervisor, and a longtime student of landscape fire and Mexico—and Harold Biswell, a professor of range management, himself a prophet of restored burning and a colleague at the University of California with whom Starker had co-taught seminars. His experiences on the Río Gavilán had also marinated with wider wanderings, both geographic and intellectual. Still, it was one thing to accept fire as a neutral presence that might be tolerated; it was another to advocate for its active application as a restorative tool. In the early 1960s this was a stance beyond the pale of professional forestry. It was a leap that even his famous father had never made.[138]

Ten years after the Park Service overthrew its fire policy, so did the U.S. Forest Service. No such reformation occurred in Mexico.

If William Vogt and Tom Gill spoke for classic conservation, and the Leopolds for restoration, Ed Abbey voiced the language and vision of an unyielding preservation. In 1975, then a writer best known for *Desert Solitaire*, soon to be celebrated as an environmental monkeywrencher, and a sometime fire lookout for the National Park Service, he took that sensibility to the Sierra Madre Occidental in Chihuahua.

The sky, he noted, was "extremely hazy" from windblown dust and "the smoke from forest fires to the east." Abbey at first felt alarmed, then was placated. Later in his travels, as evening highlighted the flames, he became "aware of fire burning around us, miles away in all directions, on hillside and ridgetop, enormous bonfires," as "potential forest" went up in "gaseous waste." These were not natural fires, but ones mostly set by the Native peoples of the region in what he somewhat whimsically speculated was the "intentional extermination of the plant cover of an entire region." Maybe, he mused, "they *want* to eliminate the forest and wash the topsoil down into the dammed-up rivers of Sonora."[139]

That Cactus Ed, the famously antiestablishment, endlessly provocative spokesman for wild nature, who published *The Monkey Wrench Gang* the same year (1975) he toured the Sierra Madre Occidental, could not see the ecological value of such burning speaks volumes for the stubborn hostility among the literate classes on both sides of the border toward landscape fire. During his years as a Park Service lookout, Ed Abbey had observed through his tower windows the much-altered forest that overzealous fire protection had wrought and the early attempts to repair those damages. But he saw those scenes through the binoculars of a preservationist philosophy that not only challenged the industrialization of previously unroaded areas but questioned inherited doctrines of conservation.

For Abbey the primary problem was that these were human-caused fires, not natural, and thus a symbol of humanity's malevolent presence. Those flames were the spray cans of graffiti vandals, not much different than the chainsaws of industrial logging. What he most valued was not so much the primitive as the primeval. He could imagine—had written some of his best prose about—landscapes empty of people. For his Mexican counterparts such imaginings were not possible. Their objection to the fires was that they were both symbols and means of a primitivism that kept the rural and especially the Indigenous populations mired in poverty, that degraded soil and tainted waters, and that made traditional practice an economic millstone around the neck of a fast-modernizing Mexico. In Mexico there was no landscape without people.

PYROGEOGRAPHIES OF THE MIRACLE

In 1958, at the crest of the Miracle, two events highlight the consequences for landscape fire, one from the perspective of the movement to modernize and the other from the vantage point of traditional agrarian reform.

That year La Michoacana de Occidente acquired a 1.25-million-hectare logging concession in the Tepalcatepec watershed. All in all, Michoacán offered an interesting cameo of the country's national fire economy. Whipsawed by changing laws that privatized, nationalized, and redistributed land among Nahuatl and Purépecha speakers, mestizo rancheros, and haciendas and companies, fire practices were a palimpsest of lightning-kindled blazes, milpa burning, pastoral firing, and assorted hunting, foraging, accidental, and revenge burns that creeped and occasionally flared amid a lumpy landscape of patchy fuels thinned by relentless firewood gathering. Foresters considered the simmering and occasional "boiling" of the land to be "stupid custom." But Michoacán was where, during his tenure as governor, Lázaro Cárdenas became aware of forestry.[140]

La Michoacana was mostly interested in sawtimber pine, with some attraction to oak and resin. Across 20 years the company managed to haul out massive amounts of lumber and modest quantities of resin but it failed to meet concession requirements for local hiring, infrastructure construction, social justice, and even reforestation. Environmental and social conditions were too complex for a company, however large, to function as a surrogate government. Land tenure became further complicated by the creation of two ejidos within the concession. What La Michoacana could do was to cut. Rather than improve, it accommodated. Local customs continued, adapted to the company's desire to protect existing commercial woodlands.

So fire persisted. Locals resisted attempts to restrict traditional burning and compelled the company to dispense burn permits to nearly all who asked. Soon the company was instructing locals in how to burn in ways that did not (to the company's mind) threaten the surrounding woods. It organized fire brigades, again offering training. Rather than enforce official policy, forest rangers found themselves working with local people to nudge burning habits in ways that would not harm the concession. However much their formal education told rangers that fire was atavistic and damaging, their working reality invited them to overlook theory and live with fire, which might or might not even be reported. Still, while they reluctantly tolerated its harms, they were not yet ready to appreciate its ecological benefits.

Criticism arose regarding the company's performance; even Cárdenas fumed that La Michoacana had worsened both the social and forested scenes. When the company's profits sagged, it was sold to San Rafael y Anexas, and in 1978 it ended operations just before its concession was canceled and as the Miracle was running out of steam. Along with its unfulfilled promises, the experience left a tangled landscape of fire.

Meanwhile, the same year La Michoacana was granted its concession, Starker Leopold published his magisterial *Wildlife in Mexico*. His travels had given the observant Leopold personal experience with how the ancient cycle of burning and farming had metamorphosed. The process began with roads that, once opened, provided entry for everything else. Leopold reported, almost as an aside, how the process worked in the Sierra Madre del Sur. Loggers led. "Itinerant farmers" followed the roads after being "forced to abandon their former lands because of erosion." They converted the cutover forest to small fields ("which later are abandoned in turn when the topsoil is lost"), and livestock were grazed throughout the year, while roads allowed the hauling of "charcoal and firewood to the communities in the valleys." Fires followed, "with monotonous regularity, and as a result the forests do not recover and revegetate the mountains." In place of logging as a prime mover, the ejido program could start the cycle by relocating people to unfamiliar environments or moving too many people to marginal land.[141]

Leopold didn't blame either the loggers or the campesinos themselves—they were doing what they felt they had to do. The villain was overpopulation and a shortage of arable land. But the results, for both groups, were suicidal. The logging industry was mining the woods and would expire "when all the virgin stands of pine have been exploited." Traditional agriculture (and firewood gathering) would fade once overuse led to soil erosion, complemented by wanton burning that retarded regeneration. Once they had exhausted a site, both axe and torch would migrate to other locales and restart the vicious cycle.

Mexico's two castes, two economies, and two realms of fire were interacting in ways that could upend the expected outcomes. Mexico was betting that it could industrialize fast enough to remake economy and society before the dire consequences of their perverse interactions rotted away the foundations of Mexico's environment. Fire tracked that wager.

CENTRAL MEXICO

Formal fire protection had first manifested itself in the Valley of Mexico and its periphery. Here the arguments for diminishing and, if possible, abolishing fire were sharpest. By protecting watersheds, fire control bolstered the drainage projects that were remaking the physical geography of the valley. By suppressing flame and smoke, it protected life and property, enhancing the health and habitability of the city. By stabilizing forests, it seemingly assured the survival of a critical natural resource. If only indirectly, state-sponsored fire protection forced reform on rural residents.

The Valley of Mexico furnished the model that the rest of Mexico would emulate. Here was where Quevedo created his prototype for forest conservation. The national parks—the crown jewels of conservation under Cárdenas—ringed the region like a green necklace. But neither forestry nor parks could exclude fires because they could not deny locals access for their traditional economy. Milpas, fuelwood gathering, and pastoral burning that passed through the woods like the summer rains all continued, with minor shifts in timing or location and with effects, good and bad, that varied accordingly. Ill effects stemmed from overburning, overharvesting, and overgrazing. What happened around the valley was visible to elites and politicians in ways that what happened in Quintana Roo or Baja California was not.[142]

The servicio forestal strove to make the valley into a demonstration site for fire control. It focused its efforts on lands under operational control of timber companies and along routes of travel overseen by railroads or the ministry of highways (especially the road from Mexico City to Puebla), and so avoided disputes with local people. Where the conversion to a modern economy had progressed, traditional fires receded, sometimes replaced by modern fire avatars started by locomotives and fed by logging slash. Outside those zones, the historic pattern persisted. Granted its feeble resources, the agency had few other options.

As the Miracle matured, the fire scene attracted more attention and money, building on the 1952 reforms that established the Department of Fire Protection under the Agriculture Secretariat. In 1959, at the second national forestry convention, Antonio Sierra Pineda and Guillermo Gaitán revisited the reasons for fire protection and summarized what was required to install an acceptable system in the valley and its neighboring lands. The project needed vehicles, firefighting apparatus and tools, lookout towers, and fire guards—ideally, a corps of firefighters similar to what existed in cities. Since 1956 the region had recorded 5,960 fires that burned 730,624 hectares.[143]

Those resources and more arrived during the administration of López Mateos. A plan existed, the 1960 forest law bolstered protection, Enrique Beltrán served as subsecretary in charge of natural resources, and President López Mateos spoke publicly for the project. The result was a five-year Program for the Prevention and Combat of Fires. Roberto Garduño García assumed the role of director general of the fire program;

Sierra Pineda and others contributed. Before the program, the Valley of Mexico had 3 fire caches, 3 lookout towers, 10 vehicles, and 75 guards organized into 3 patrols. After the program, there were 48 brigades, ample equipment, 48 tool caches and lookouts—some with radios—and even 5 helicopters. The Mexican Miracle had come to fire.[144]

Although the program came with unprecedented funding and display of political interest, Roberto Garduño was careful when presenting its lessons to the Third National Convention on Forestry in 1966 to emphasize that the fundamentals of forest destruction were socioeconomic and that, just as soil, water, and forests were "indivisible," so a proper response required all the resources of the government, not just the servicio forestal. Mexico had to find the "correct relationship between the agents of destruction and those of control." The wholesale conversion to a fossil-fuel society was reconfiguring the parameters of that relationship.[145]

Studies in the more remote pine forests within the eastern fringe of the volcanic belt track the consequences. At Cofre de Perote National Park, fire-scarred trunks inscribe a chronicle that traces back to 1461, with fires returning roughly every four to six years, almost always in the spring (95 percent), just before the rainy season. The cadence of fires continues through the Conquest, through the wars of independence, through the Mexican Revolution. Then 500 years of routine burning end in 1973. The break in the chronicle of fire corresponds to one of several major development programs (in this case, forest productivity) that involved roads, logging, and fire prevention. At the Pico de Orizaba researchers in 1975 pondered a thinning and retreating timberline and attributed the recession to "deliberate burning . . . by herdsmen at the end of the dry season." A follow-up study 35 years later, however, concluded that there was no evidence of change. From the late nineteenth century through the twentieth, fires had been continuous, leaving scars in 90 out of 100 years. Most burns were small and "asynchronous," likely the product of lightning, or patchy stringers from lower-elevation starts. This time, however, a decrease was evident in recent decades, likely the result of a telescope constructed on the Sierra Negra, which meant a road and a concern about fire protection for the site. Add Chignahuapan in northern Puebla. Here suitably scarred trees span from the Porfiriato to 2011. The average return interval (5–5.6 years) is close to the average for the region. More extensive fires broke out when wet years toggled into dry ones, the ancient fugue of anthropogenic fire and Mexican climate. Then came 1988, and drought, and the onset of a new order.[146]

Whatever new policy or strategy the servicio forestal adopted, it would premiere in central Mexico. The national story was the Valley of Mexico's writ large. At a symposium sponsored by the fire working group of the North American Forestry Commission, held in Denver in 1972, Jesús Vásquez Soto lamented that fire ecology studies in Mexico were "very scarce." Modern fire science, like industrial forestry or machine-based fire suppression, was a product of developed countries. But already Mexico's latest boom was collapsing into a bust. "The problems of a well-planned forest management are overwhelming and appear to be aggravated by the great increase in human population." Mexicans were proliferating faster than Mexico's ability to absorb them. Good fire was morphing into bad, or disappearing.[147]

If too much traditional burning was a problem in undeveloped areas, too little surrogate burning—prescribed fire, in technical terms—was a problem in developed areas. After a century of alarm over fire's presence as a metric of environmental deterioration, a few observers began to fret over its absence and what that meant for ecological degeneration and worsening fires. Modernity, too, had its discontents. The deeper threats to the Valley of Mexico were the transition to a megacity and the ever-heavier weight of the human presence. Ground subsided, the once famously champagne air became notoriously foul, lands overgrew with streets and slums, and much of the metropolitan sprawl became marginally inhabitable. Fires in living landscapes receded, fires from lithic landscapes overran everything. Flames and scorched trees were visible, however, in ways that the consequences of fossil-fuel combustion were not.

The reformers were correct. What happened in the Valley of Mexico would spread throughout the country. Eventually, the pressure of industrial combustion reworked fire regimes nearly everywhere.

SIERRA MADRE OCCIDENTAL

Chihuahua felt with special intensity the logging boom catalyzed by the war. It was closer to U.S. markets, and U.S. companies had in earlier decades probed and penetrated its pine forests. So savage had the cutting been that a general ban was proclaimed in 1949, which extended into Sonora, Sinaloa, and Durango, before exceptions made a mockery of the gesture. But between them, industrial logging and ejido creation transformed the fire regimes of the Sierra Madre Occidental.

Lightning and Rarámuri had largely kept the mountains simmering with surface fire. The transfer of land to ejidos and logging concessions brought in newcomers. Roads were punched into the mountains to open them up for herding and logging—the one took the grass, and the other the big trees. Together, they unmoored mountain ecology. Official animus, however, tended to be directed toward the Indigenous people with their fires and goats—the evil twins that had historically haunted European forestry. Locals left campfires unattended, they fired patches for grazing and milpas, they refused to join fire brigades (unless paid), and likely, as colonized people around the world did, they also set fires in protest. Four municipalities concentrated the fire problem of Chihuahua; three were inhabited by Rarámuri, whose "social integration" into Mexican life was "incomplete." Yet foresters had to accommodate them, and overlooking their fire practices was a simple solution. Indigenous burning was useful for propaganda, but otherwise was ignored.[148]

Once again, fire-scarred trees testify to the transformation. At the Tutuaca reserve, a pine-oak forest in southwest Chihuahua established to protect the thick-billed parrot and a relict spruce forest, the chronicle repeats the standard narrative: regular spring fires, recurring roughly every four to five years; surface burning, most expansively in dry years following wet ones. Until 1955. Actually, the cadence expires around 1945, with a single widespread burn in 1955, after which fires vanish. The 48-year hiatus between burns is two to five times longer in duration than any previous fire gap since 1702. The break corresponds with the widespread arrival of goats, cattle, and sheep.[149]

From the sky islands along the Sonora-Arizona border to the southern border of Durango, a virtually identical scenario plays out across the Sierra Madre Occidental. Fires recurred at 3-to-6-year intervals, with widespread regional burning on a rhythm of 6.9 to 9 years—a minimum value. It would seem that ignition was essentially constant and the land burned whenever there was sufficient fuel to carry flame, with combustibles riding on the waves of dry and wet spells. In only a few niches was there interest in preventing fire or the capability to suppress it.[150]

Then the fire lights go out as the reorganizing pincers of agrarian land reform and industry arrive. At some sites fires expired during the 1930s; many succumbed during the war and postwar boom in the 1940s and 1950s; remote landscapes like the Sierra de los Ajos did not shed routine fire until the 1970s. Still, fire didn't vanish altogether. Terrain and land use preserved a few fire refugia, and some communities continued to use fire, though in less expansive forms. Even in sites where routine fire appeared to vanish, it could eventually rebound in less frequent, more intense forms, powered by the fuels stockpiled during the phases of fire's exclusion. At Tutuaca big fires broke out in 1987 and 1995. At Charcos, Mezquital, in Durango, routine fires ceased in 1976, and more severe outbreaks followed in 1988, 1996, 2009, and 2012.[151]

The process differs in details, as one would expect. For some sites the provocation is a road for logging, grazing, or mining that cracks open woods and pasture; for others, a relocation of peoples that changes how they live on the land, as they shift from seasonal walkabouts to more sedentary habitation. Almost everywhere, it involves an intrusion of livestock. Most of the Sierra Madre Occidental had known patches of herding since missionary times, but the livestock existed as familial animals like milk goats and transport donkeys, or as small herds, not as export commodities. What had played across the land like dust devils mutated into haboobs. As in the southwestern United States, the onslaught arrived as a shock wave that rattled fire into new patterns. In mimicking U.S. practices (often to supply a U.S. market), Mexico acquired U.S.-style fire problems.

Most of the tree-ring research was done by Americans, or by Mexican-U.S. partnerships, and focused on what most engaged Americans eager to restore fire to their own fire-starved wildlands: protected sites that could serve as benchmarks for a more "natural" state. In Mexico designation as a park, biosphere reserve, or local preserve did not mean the land was uninhabited. So while Mexicans learned a technique, those from the United States learned to see landscapes differently. They came to Mexico, like Aldo Leopold, to understand primeval places. They left with an appreciation of the fact that landscapes were cultural, and that anthropogenic fire could disrupt regimes otherwise governed mostly by climate.

Two places in southern Durango show the fine-grained factors that fire regimes integrate. The La Michilía Biosphere Reserve experienced regime change in the 1930s, with widespread fire exclusion after 1945. A livestock-laden hacienda emerged in 1932. By 1940 the nearby municipio of Suchil tallied 9,771 cattle; 14,946 horses, mules, and donkeys; 3,337 sheep; 2,866 goats; and 4,877 swine. The rich pastures slow-combusted in their rumens rather than fast-combusting in flames. Eventually, low-grade fire suppression by local landowners with hand tools matured, assisted when needed by army troops. Only the core of the reserve evaded the otherwise general abolition of burning.[152]

The Las Bayas forest reserve, also in Durango, had a less dramatic shake-up in land use and a less pronounced shift in fire regimes. A transition occurred in the 1950s when ejidos arrived and traditional semi-migratory burning by Indigenous people receded. Interest in logging—which brought roads and some measure of fire protection—and a more field-centric pattern of burning promoted less-extensive fires, and perhaps more intense ones that occurred when conditions were most favorable. Small, mild fires, as frequent as the woods allowed, disappeared; large, severe ones, possible under extreme conditions, replaced them.[153]

Change came at different times with varied tempos. But come it did. One of the most striking discoveries occurred at Pino Gordo, a Rarámuri settlement in the Sierra Tarahumara. In 1961 a railroad spanned the range from the inland region to the Sea of Cortez, and the tremors spread rapidly outward. Fifty years later some 99 percent of the Madrean range had been logged, accounting for 97 percent of Chihuahua's timber production. But Pino Gordo evaded the shock. Its chronicle of surface fires shares the signature rhythms apparent elsewhere, if slightly more frequent (2.4–3.3 years). What is remarkable is that the fires simmer along through an industrially unharvested, ungrazed forest, flush with big trees and snags, with a preference for pine over oak. The Rarámuri maintained traditional ways—burning in cleared patches of oak, harvesting firewood, timing burns with the onset of the rainy season. Not least, they did not extinguish fires kindled by lightning.[154]

As much as might be hoped for, Pino Gordo holds the reference fire regime against which to measure the immense changes spurred by the rush for economic development that, with dappled intensities, rippled across the Sierra Madre Occidental. That the most "natural" fire regime should be one maintained by an Indigenous people speaks volumes about the character of fire in Mexico.

SIERRA MADRE ORIENTAL

The eastern Sierra Madre tells a more familiar story. Surface fires played across the mountains, though a bit smaller and less frequent than in its western complement. Still, the years that had fires splash across the Sierra Madre Occidental had them in the Sierra Madre Oriental as well. But the interruptions in the rhythm of fire came earlier to the east, with some in the 1920s, prior to the major thrust of agrarian reform and postwar industrial logging. Afterward, the fires became both less frequent and more severe.[155]

In the pine-oak forest of Peña Nevada in southern Nuevo León, the earliest fire scar dates to 1541, roughly coincidental with Spanish arrival; a sufficient density of scars to establish a credible chronicle appears in the seventeenth century. The record tells of a consistent pulse of burning on a roughly 9-year interval (and for individual trees, "a conservative estimate" of 17 years), with some 18 percent of fire-scarred years extending across all of the three peaks sampled. Then the chronicle abruptly ends. Two sites experienced one fire between 1929 and 1998, and the third had none. The fires vanish, and so does the correlation between ENSO cycles and fires. Climate and fire bond again in 1983, and monstrously in 1998, the second-driest year since 1508. The returned flames mutated from a burn that skimmed across the surface into a stand-consuming inferno.[156]

All in all, it's a similar history to that in the Sierra Madre Occidental, though not quite interchangeable. The reasons why are not yet understood. The sampled pine is different (ocote rather than ponderosa), at higher elevations, and amid more rugged terrain, all of which divides and hobbles the spread of fire. The regularity of the western fires speaks to a consistent and ubiquitous ignition, beyond what lightning alone might muster. The peaks of southern Nuevo León may not have had a comparable character of human occupation. The fire annals of the Peña Nevada close with the establishment of the Ejido La Encantada in 1937. To what extent people boosted fire's historic presence is unclear; that they ended it seems incontestable.[157]

In the northern peaks flanking Monterrey, along three parallel ranges, the first scar appears in 1422 and the last amid the general conflagrations of 1998. The mean fire interval for all fires runs from 7.8 to 16.4 years, with some sites as low as 1 to 4 years and others as long as 19 to 79. Likely, terrain has mattered more than climate. The three mountain ranges shed fire at different times. La Viga had its last multiple-scar year in 1917. Rancho Nuevo saw its last in 1909 or 1952, depending on when the old rhythm is considered to end and the new one to begin. San Antonio had three anticlimaxes—1887, 1918, and 1962.[158]

Just as the steeply corrugated character of the Sierra Madre here broke up climate's impact on fire, so it seems to have fragmented human impact. Ownership was a "patchwork," with "some areas under ejido management and other areas privately owned." As always, fire integrated its surroundings, as varied as the push and tug of wind, woods, bunchgrass, ridge and road, lightning and torch. The fading of the old economic order and the arrival of the new one did not happen suddenly or through an overwhelming force of arms, politics, or commerce, but piecemeal, threaded through the valleys and ridges. Yet happen, eventually, it did.

MOUNTAIN AND MESETA

Since the Conquest, Mexico had become a land of livestock, and during the Miracle ranching revived on a grand scale. Each of the major pastoral regions—the tierra templada, the tierra caliente, the northern mountains and meseta—had its characteristic version of livestock husbandry, each had its own regimen of fire, and each had its peculiar fire history.

In the tierra templada, the introduction of livestock invited regular, even annual, burning to refresh the fescue, muhly, bent, and other pasture grasses. The long litany of complaints about burning registered around the Valley of Mexico mostly referred to pastoral fires, which swept the highlands and mountains of central Mexico much as brooms did houses.

In tierra caliente, burning was routine, even annual, because only fire kept a revanchist flora from reclaiming land cleared for planting or pasture. What mattered overall was scale, which depended on the number of livestock and how they complemented (or competed with) other forms of agriculture. Historically, pastures were often small because livestock were few; pastures were one tile in a tumbling mosaic of land use. Around the Sierra de Santa Marta in southern Veracruz, until 1950, the mostly Indigenous people planted maize, beans, squash, tomatoes, papayas, cane, and plantains. For livestock a typical family might have five heifers and a horse. The landscape was parceled into milpas, pastures, second-growth woods, and a scattering of old-growth trees, fruit, and miscellaneous fields. The milpa cycle was five to six years. Fire applied and fire withheld shifted the biotic tiles, like ice floes blown, fused, and broken apart by wind.[159]

As agrarian reforms moved newcomers, mostly mestizo campesinos, into the region, they favored ranchos and expanded burning according to a choreography of fire, rain, and grass. This conversion to livestock could shatter the old mosaic, replacing it with a monoculture of pasture. Eventually agricultural advisors promoted productivity through such means as introduced grasses and fertilizer. But those additives required money, while fire could do all those tasks in a single sweep—less efficient for any isolated factor but cheaper and more robust overall.[160]

The northern region of mountain and meseta is vaster in area than the others (Mexico's largest physiographic terrain), more singular in its commitment to ranching, and especially revealing because its fire-fauna history not only contrasts with the rest of Mexico but also compares with the United States. The defining flora is *matorral*, a lumping of agaves, cacti, and drought-toughened shrubs; the deserts, plains, and foothills often mix the shrubs with grass, and occasionally with woodlands—all of it well suited for grazing. In general, the desert flora respond poorly to fire, the grasses well, and the shrubs, reflecting their more mottled composition, with mixed results. (Still, in desert areas the right burns can enhance grasses and make certain cacti available as emergency fodder, by burning off spines.) A 1973 survey mapped matorral at 47 percent and chaparral at 8 percent of the national biota.[161]

The complicating factor, of course, is the livestock. Livestock compete with fire for grass, and the more intensive the ranching, the less fire is possible. The ancient solution is to move the herds and flocks to seasonally available forage, which requires large estates or the ability to drive animals over long distances. The 1880s resolved the issues that had previously boxed in ranching. Politics was stable, the Indigenous people (especially the Apaches) were silenced, rail connected to U.S. markets, technologies devised for the Great Plains such as barbed wire and windmills inched south, and foreign capital, both U.S. and British, was abundant. The big haciendas got bigger, yet the vast size of the holdings meant profits were possible without cropping grass and matorral to mineral soil. The Hearst spread had a million acres; the Luis Terrazas hacienda had 6.7 million (which made it the largest private landholding in North America). By 1910 the latter was reportedly branding 140,000 calves a year.[162]

The revolution shattered northern ranching. What the combatants didn't eat, desperate ranchers shipped to the United States. Then came drought. Then the Spanish flu "finished off the *vaquero*," noted Donald Brand, "who had no cows to care for anyway." Livestock continued in small rancherías, but as farm stock, not as a commercial enterprise. Agrarian reform broke up some haciendas, though many (especially in Chihuahua) survived until the postwar surge in land transfers or were parceled into mid-sized ranches. But mostly the northern rangelands were emptied. Much of the land had time and space to recover. Texas and the southwestern United States had no such pause—drought and overgrazing crashed most of the grasslands, and production accelerated during World War I.[163]

Eventually ranching revived. Agrarian reform redirected smaller herds and flocks into the mountains, now opened by logging roads. The big ranches restocked, this time with better cattle, kept close by fencing with barbed wire; they installed windmills and stock tanks; they catered to a U.S. market. As Brand again notes, by the 1960s Chihuahua and Coahuila more resembled "southern New Mexico and Arizona than Durango and Zacatecas." That included fire regimes. Fire became rarer, and when it occurred, it could explode, both larger and more intense.[164]

Maderas del Carmen is a protected area within the Sierra del Carmen, a sky island where the continuity of the Sierra Madre Oriental breaks down in its northerly trend across Coahuila. Now a protected area, the forested Maderas offers a useful contrast across the international border to Big Bend National Park and the Davis Mountains, both in Texas. The Texas mountains differ in their fire history from most of the U.S. Southwest, and Maderas del Carmen differs from both of those areas. The pivotal factor is not just livestock, but a shift from cattle to sheep.[165]

Throughout the borderlands formal fire protection only came after hordes of livestock had gutted the grasslands and stripped away fire. In the southwestern United States, the newly constituted Forest Service, established in 1905, made fire control core to its mission and kept fires from returning after the swarm had passed. Because Texas never surrendered its public lands upon admission to the United States, the state retained responsibility, which meant that only late in the twentieth century did the apparatus of modern fire suppression arrive. Northern Mexico resembled western Texas: the northern Sierra Madre Oriental was too distant, access too troubled, and money too lean for active fire control. Instead, the mountains on both sides of the Rio Grande were shaped by hoof rather than shovel.

Cattle came first, an eddy from the great cattle drives that spilled out of south and west Texas after the Civil War (the Chisos Mountains held a small band of feral longhorns, which prompted the state to create a park, which it later ceded to the federal government). But cattle weren't suited for transhumance and couldn't clean out the mountains as they could the lowlands; sheep and goats could. In the 1920s and 1930s the herds withdrew and the flocks swarmed in. Fire frequency lengthened, then essentially flatlined. A similar scenario played out in the Guadalupe Mountains farther west, where a common return interval of four years thinned during the nineteenth century before vanishing after 1921.[166]

Compared to the U.S. Southwest, fire exclusion had come to Maderas del Carmen late. Compared to most of Mexico, it arrived more or less in sync with the agrarian reforms that extinguished the other, second-order fires of revolution. It arrived with the creation of ejidos in the 1930s and 1940s. Ganado menor rather than ganado mayor moved into Maderas del Carmen, along with much of Coahuila (the state has long boasted the highest numbers of sheep and goats in all Mexico). Not until Maderas became a protected area in 1997 did the numbers fall. In the meantime, some selective logging and a network of roads further carved up the domain available for free-ranging fire. Some small fires appeared but failed to spread.

The Sierra San Luis lies south and a tad east of the international four corners region where Arizona, New Mexico, Sonora, and Chihuahua meet. As often as conditions permit, the land burns. The earliest scar dates to 1654. The regimen of fires that followed is typical: frequent surface burns whose extent rises and falls with climatic pulses. Then, when roads, logging, and grazing arrive, the fires fall off. So far, so expected. But of course, in keeping with the endemic diversity of Mexico, the Sierra San Luis has to be different. The usual pattern in which wide-ranging fires occur in dry years that follow wet ones seems to have reversed prior to 1900 before edging into greater conformity with historic norms. The reasons are unclear ("an active and unresolved area of study"). So, too, the exchange of an Indigenous fire regime for a modern one has had its quirks and unexpected detours.[167]

Unlike the Rarámuri, the Apaches appear not to have routinely burned the land, or at least not on the scale of other Indigenes. The abundance of naturally ignited fires seemed enough for their purposes, given an economy at least partly based on raiding villages (the raiding, and later outright warfare, produced its own characteristic fires). The Apaches' effect was indirect: they didn't extinguish the fires lightning set, and they inhibited significant settlement and livestock grazing, which further gave those ignitions room to roam. North of the international border the Apache wars ended in 1886. Almost immediately, combined with railways, the removal of the Apaches to reservations opened up the region to a surge of mining and grazing that overwhelmed valley and mountain both and snuffed out the ancient rhythm of burning.

In the Sierra San Luis, however, penetration was slower and, thanks to complex terrain, incomplete. Instead, the process begins with agrarian reform in the 1930s. Roads and logging came later and were codependent: logging required roads and roads encouraged logging. The grasses that had fed fire since forever now fed livestock. Where the stock appeared, the fires left. But cattle in particular did not habitually trek up the higher slopes, so remote sites and peaks were spared. The upper mountains held their grass; they kept fire.

Farther south, the Mesa de las Guacamayas experienced a disruption of routine surface burns in 1945 at two sampled sites and in 1953 at a third. Severe fires hammered all three sites during the 1990s–2000s. Another site continued its ancestral burning unimpeded and experienced no eruptive conflagrations. The contrast to the Chiricahua Mountains across the border is striking. There regular fires ceased in 1886, with a final modest-sized burn in 1892. Grazing on a vaster scale than occurred in Mexico was followed by organized fire control after 1905—a one-two punch that restructured the regime for almost a century. Not until a century later, after policy reforms to restore fire took root, did landscape-scale burning return.[168]

The economics of commercial ranching has always been unstable. Climatic cycles, business cycles, ecological cycles—rarely do they align favorably. To the wet-dry cycles of climate were added the boom-bust cycles of global commerce. The stock traded on Wall Street mattered as much as the stock mustered in the northern meseta. Inflation mattered as much as forage, access to credit as much as stock tanks. The complexity is not additive but exponential.

Fire's status was further complicated because it required that cattle and sheep be removed from the range long enough that there be ample fuel available for burning. This was pastoralism's fallow phase. Fire's capacity to improve the amount and palatability of forage, drive back encroaching brush, and upgrade biodiversity comes with a short-term cost that can overlap with drought and depression and seduce ranchers into foregoing, for one more year, that catalytic burn. Each missed beat leaves the brush more deeply rooted, the forage more tired, and ecological health drained.[169]

The standard advice north of the border—get big or get out—clashed with the precepts of the Mexican Revolution, though many of the grand estates tended to retain their scale. That left the alternative precepts of the Green Revolution, which emphasized selective breeding and inputs of fertilizers and other petrochemicals (not to mention capital). Agronomists experimented with introduced species such as lovegrass, buffelgrass, and varieties of vetch that promised enriched fodder. In the north, however, lovegrass and buffelgrass, both originally from Africa, and bromes, from Central Asia, proved highly combustible, capable of overturning biomes. A common North American solution was to quit bringing animals to forage and instead carry the forage to animals, ultimately in feedlots.[170]

Fire became a problem—in some places a question of too much, in most a question of too little. Fires that had been light, patchy, and rare ceded the land to unburned biomes or fires more intense, widespread, and frequent. Yet these fires were not visible or emblematic to elites in the way burning around the Valley of Mexico or the southern highlands was. The missing fires were not routinely cited in catalogs of Mexico's environmental ills—overgrazing, not overburning, was the concern. Fire ecology's three-body problem morphed into a game of rock-scissors-paper. Until fires blew over the Coahuila mountains in 2011, the northern fire regimes remained outside the national narrative.

BAJA CALIFORNIA

Baja California offered a miniature mosaic of Mexico's history. The indigenous Kiliwas endured, "one of the last native cultures that still practice hunting and gathering in North America." The Jesuits established missions beginning in 1697, and after their expulsion from Latin America, the Franciscans assumed responsibility for Alta California and the Dominicans for Baja. Again, missionizing faltered and then collapsed as Native populations imploded under the lashes of serial epidemics. Secular explorers crossed this mini-Mexico in the eighteenth century, most famously the 1768–70 expedition of Joaquín Velázquez de León to measure the transit of Venus, the trek of José Longinos Martínez in 1792, and the traverse by José Joaquín Arrillaga in 1796. Gold rushes blew over the region in 1872, 1874, and 1889, though in the form of placer mines, and with one exception these were easily worked and readily abandoned (a few revived after the revolution). A land colonization project during the Porfiriato folded. A national forestry reserve in the sierras existed, though on paper. Several ejidos were established, mostly for grazing. An attempt to pry open the interior mountains for logging inspired some spare roads, but the ambition, like so many before it, wilted away. Not until the latter twentieth century did roads create wedges into the interior, a cross-border industry of maquiladoras flourish, and tourism colonize coastal locales like La Paz and Cabo San Lucas.[171]

As in much of Mexico, livestock replaced people. Even the San Pedro Mártir Mission emphasized ranching rather than farming. The herds included the usual complement of cattle, sheep, goats, swine, horses, donkeys, and mules, with numbers ranked in that order. Local rancherías took up the practice, though with much-reduced numbers. The mines stimulated a market, which like the missions then faded. Eventually an economy of mostly cattle emerged, with winter grazing in the valleys and summer grazing in the mountains—the Sierra de Juárez and the Sierra San Pedro Mártir—that form the spine of the peninsula, with mixed impacts. Some mountain meadows, heavily used, reflected that grazing and trampling; much of the mountains felt the impact in more diffuse forms. By the mid-nineteenth century cattle had increased "substantially," along with sheep organized into long drives by foreign financiers and Basque shepherds. In the early twentieth century the process swelled through the loss of mountain grazing in Alta California due to the creation of national forests. The sheep moved south. Efforts to control the stock drives failed. Denying access to sheep when a national park was established only encouraged a revival of cattle; a tax on cattle to prevent overgrazing was ignored. More significant was lowland development that denied livestock their winter pastures. As late as 1975, estimates put the number of cattle in the Sierra San Pedro Mártir at 4,000–5,000, and that of sheep at 10,000.[172]

In 1967, as though echoing the 1768–70 expedition of Joaquín Velázquez de León to measure the transit of Venus, the National Astronomical Observatory, having been moved from too-polluted Mexico City to Puebla, was again forced to relocate to the Sierra San Pedro Mártir, which had some of the clearest skies in the world. A road to the facility opened the mountains in 1968, which birthed other roads. In 1973 the transpeninsular Highway 1 spanned the length of Baja California. By then growth along the border was propagating southward and concerned observers sought to provide legal protection in the form of two national parks and a biosphere reserve for what, comparatively, they regarded as a near-pristine landscape.

Paradoxically, their isolation made Baja California's mountains attractive. Even as the U.S. economy crowded along the border and spread like a fungus south, U.S. scientists were drawn to these mountains as potential reference points, the ecological equivalent of a geodetic marker by which to measure the changes that had overturned so much of what became the United States. The U.S. Biological Survey conducted detailed reconnaissances in 1906 and 1907, and as early as 1927, two U.S. geographers, Carl Sauer and Peveril Meigs, noted that the "ruined mission" at San Fernando was followed "by a more complete return to wilderness conditions than this region has known." A preindustrial landscape was a landscape attractive to postindustrial scholars. The upshot is one of the most comprehensive surveys of fire history available in Mexico.[173]

Lower California had the same general Mediterranean climate as coastal Upper California, though it tended to be drier, and the mountains ran north-south without the big east-west kink that turned the Peninsular Range into a Transverse Range around the Los Angeles basin. Yet over the course of the twentieth century, fire history north and south of the international border diverged dramatically. The north had fewer fires, but the area that did burn clustered into ever larger blocks. The south had more fires, a finer-grained texture, and fewer monster burns.

Why? An assumption was that, as in Mediterranean climates generally, lightning was a negligible source of ignition, and that fires, which are frequent, resulted from humans. The installation of an automated lightning detection system north of the border challenged that notion. It showed abundant lightning in the Sierra San Pedro Mártir, plenty to account for the recorded fire regime. The problem, researchers concluded, was not ignition but fuel. Vigorous surface fires cleaned out the understory of mixed-conifer forests, and replenishment could take decades. Slow accumulation of combustibles, rugged terrain that tends to fragment spread, fires that can linger to creep and sweep across several months—these seem to define the regime. Previous fires determine future fires. Many fires start, few spread.[174]

An inflection in regimes seems to occur around 1800, which corresponds to the short-lived San Pedro Mártir Mission in 1794. Speculation over cause pivots around ruptures in climate, livestock, and Indigenous habitation. Livestock did not explode until later in the nineteenth century; the Indigenous population collapsed, but how much burning they contributed is unclear. The prevailing evidence suggests firing mostly in the lowlands, among coastal sage and annual grasses (and later, invasives), and some in the matorral, with a fraction due to escapes, but little if any deliberate firing in the mountains, which they visited in summers for hunting. The land may have had all the fire it could produce or that the Indigenous economy could accommodate without some other factor altering the basics. The best guess to explain the pivot is a tweak in the climatic ensemble, further nudged by livestock.[175]

Everywhere in Mexico ranching reshuffled fire regimes. Herders' burning added to starts; their herds' cropping subtracted from combustibles. Until later in the nineteenth century, rancherías were few, stippling the land like milpa plots. Still, ranchers burned the lowlands (the fires were made more potent when invasive grasses arrived), likely burned portions of the matorral, and patch-burned the uplands, concentrating around major meadows; some of these fires would creep and leap into their surroundings. It would be odd if, adapting Iberian transhumance from the Pyrenees to the Peninsular Range, shepherds did not also burn in the fall, before the winter rains, as cattle and sheep trekked off the mountains; but if they did, chroniclers did not record it. Herding made its major impact by altering the fuels. Small fires speckled the mountains before grazing arrived, and they continued after transhumance swelled into migrating flocks of thousands. Grazing further limited, in select places, the capacity of spot ignitions to spread.[176]

While formal fire protection was a policy from the time of the national forest reserve, it was not announced publicly until 1960, and never seriously applied. There is scant evidence over the centuries that anyone tried to extinguish fires; the question is how much people supplemented natural fire starts, and how much they influenced available combustibles. The power of fire depends on its capacity to propagate. The prevailing view—based on the conclusion that fuels determine the overall character of the regime—is that people influenced fire only marginally, that they added little to ignition and subtracted little from fuels. But that conclusion projects a relationship between fire and people typical of modern times into the past. Premodern peoples tend to merge more smoothly into the fire environment, acting like a pyric flywheel that smooths out the atonal rhythms of ignition and the patchiness of fuel.[177]

Is the Sierra San Pedro Mártir as close to a precontact fire regime as any place in Mexico? Before the 1970s that was not a question anyone cared about. When it did arouse scientific inquiry, it coincided with policy reforms in the United States; the research was conducted by Americans with some Mexican colleagues, and the inquiries sought to compare Baja California not with other sites in Mexico but with the southwestern United States and especially California.

Every conclusion, every identified factor, every proclaimed driver in the fire scene south of the border—fuel, ignition, climate—was answering a question deemed vital north of the border. The landscape burn scars in Baja California contrasted with those of Southern California and so highlighted different strategies of fire management. The burn scars in Jeffrey pine in the Sierra San Pedro Mártir suggested powerfully what the presettlement fire regime of the Sierra Nevada of California might have looked like.[178]

While a final reckoning of precontemporary fire history remains open to dispute, what is uncontestable is the curious inversion of perspectives that peered across the border. Even as Mexicans were looking to California for guidance in how to create a modern fire protection system, Californians were journeying to Mexico to better understand how their vaunted system had gone so wrong.

YUCATÁN PENINSULA

Southeast Mexico inverted the physical geography of northwest Mexico. Instead of a narrow peninsula with a central spine of mountains bordered by coasts, the mountains thinned and stayed west, while a great sprawl of lowlands curled east and north. Instead of sparse and scattered Indigenous habitation, the Native population had recovered and continued traditional land use and customary fires. Where central Mexico bonded closely to the north, with its mines and missions, and where it had fought a disastrous war with the United States, the southeast held no mines, boasted no quick wealth of plantations and livestock-thronged haciendas, and had a border with Guatemala and British Honduras that was not delineated until the 1890s. While the Conquest had created sagas for the center and north, Cortés's wretched entrada south left only miserable memories and trampled any prospect of wealth or glory.

Beginning with Hispaniola, the American tropics had always lured and disappointed. Soils could be tricky, diseases were rife, and access was poor. Yet Yucatán did have a forest of commercial timber like mahogany, patches of pasture, and prospects for commercial plantations of henequen that could entice modern capital if access was available. The Porfiriato's ministry of development applied, or attempted to apply, the same formula to the south that appeared to work in the north. It enacted land laws to furnish legal access, and opened rivers, erected ports, and sponsored railroads to provide physical access to foreign markets. Telegraph lines paralleled rail.[179]

Within a few decades millions of hectares of public and communal land slid into private hands, and valuable timber was high-graded on a vast scale. Prior to 1902 some 2 million hectares of Yucatán and Quintana Roo, and between 1902 and 1905 some 3.5 million hectares, were transferred to private control. The same process followed rivers and rails from Tabasco to Campeche and Chiapas. Traditional usage was overrun with pastures, plantations, and sawmills. Local populations were displaced at a scale comparable to the exported timber. What Herman Konrad concluded for the eastern peninsula came to characterize the whole: "In effect, the entire territory was now under control of private interests for the purpose of natural resource exploitation." Observers with an occupational interest in fire were few (in 1912 the servicio forestal had 17 subinspectors and 31 forest guards); and transients like travelers, scientists, and scouts for commerce generally ignored fire except where it cooked their food and lit evening conversations. Ultimately, unrealistic hope yielded to despair, and instead of installing a modern economy, the program helped inspire a revolution.[180]

So self-evidently ruinous did a fire-based agriculture appear to most observers, including archaeologists, that it was frequently invoked to explain the demise of Mayan civilization. As population increased, researchers speculated, milpas had expanded and cycles shortened, and the nutrient capital of the soil, never rich, degraded to the point that the empire rotted from within. The massive stone temples and urban clusters rested on a thin crust of fertile soil that fire ate away. What critics witnessed during Mexico's postwar boom—population explosion, agrarian expansion, traditional practices intensified—seemed to be reenacting the demise of ancient Mayan Mexico.

The exceptions were the odd naturalist and the stray anthropologist who took traditional agriculture and its fire practices seriously. Starker Leopold observed that, in Yucatán, the "Mayan system of rotating milpas has been in effect for centuries," and noted the cumulative effect was "less devastating" here than in the mountains because the land was flat and erosion negligible. Around Xocempich, where he camped for two weeks, the cycle involved slashing and burning, with two years of maize planting ("with little or no cultivation"), followed by encroachment by vines and brush during the third and fourth years, reducing yield and leading to abandonment in the fifth year. The cycle would renew 25 years later. In the mountains Leopold regarded the milpa economy as "disastrous"; in the limestone flats of Yucatán it could be perpetuated indefinitely, though the second-growth forests were "poor in vertebrate life and yield no timbers."[181]

More influential was the parallel emergence of *indigenismo,* which encouraged some scholars to reexamine traditional ways with fresh eyes and to understand customary practices within their own context. In the same year that Leopold published his observations in *Wildlife,* Efraím Hernández Xolocotzi published a lengthy inquiry into the still-prevailing agriculture of the Yucatán (*agricultura yucateca*), or what he elaborated into roza-tumba-quema-siembra (*rotación nómada*). Within the peninsula, he observed wryly, the technique had been little studied and much criticized.

Hernández Xolocotzi was a long-serving professor at the Colegio de Postgraduados at Chapingo, with time also at the Universidad Autónoma Chapingo (UACh), who together with Faustino Miranda helped found ethnobotany and the indigenista movement. In his hands Mayan agricultural techniques for farming, gardening, and selective silviculture were presented as versatile instruments, as subtle as an artist's brush or sculptor's chisel, by which to shape habitats. Writing in the 1950s as Mexico planned a wholesale conversion of tropical forest to commerce, he pushed back against blanket condemnation of the Indigenous system with detailed analyses of its abundance, durability, and subtle logic. Modern agronomy confronted the same challenges of climate, soil, and land tenure, and it had yet to contrive a more durable solution. For the near future this meant any agricultural expansion would depend on roza-tumba-quema. Hernández Xolocotzi asked that foresters isolate their commercial woods better and that fuelbreaks be installed to prevent escape fires. Later, he and like-minded colleagues could celebrate the Mayan milpa as a model of sustainability.[182]

Still, officials, commerce, and the Green Revolution looked covetously on the expansive lands of the peninsula. Surely the same techniques that had boosted productivity elsewhere in Mexico could be adapted to Yucatán and replace an antiquated agriculture with modern alternatives. But Hernández Xolocotzi argued for caution: Yucatán's environment was unique within Mexico. Equally, he insisted, swidden makes sense if one considers the peculiar limestone caprock and soils, the uncertainties over the rains (and in more recent times, the unstable markets for commodities like henequen), and the deep interconnection between maize and Mayan culture. In fact, roza-tumba-quema was and would continue to be the most stable economic basis for social well-being in the peninsula. There were places that might exchange it for something else, and places where it might expand (or not), but what is noticeably lacking in Hernández's analysis is a ritual condemnation of roza-tumba-quema as inherently slovenly and unsustainable. On the contrary, no other practice had survived so long and with such tenacity.[183]

The intrusion—in a way, an economic conquest—was real, alarming, and disruptive. But the peninsula absorbed and dampened the worst excesses. What was lost were the practices of traditional management. Transient companies and scattered outposts of forest guards were no substitute for farmers who resided on the site. Much of what regrew was no longer cultivated but feral. Still, the peninsula was accustomed to chronic, low-severity manipulation, punctuated by occasional blowouts from hurricanes and subsequent fires. The best guess about what evolved is that some places saw their prior landscapes upended, some continued as before, and some devised hybrid accommodations (such as intercultivation with henequen, agave, and maize). If the shakeout caused by sudden industrialization schemes was environmentally harmful, the mix of government institutions, commercial companies, Indigenous peoples, and mestizo campesinos seeking land for corn and cows was politically combustible. The new flames proved to be catalysts for change rather than recyclers.[184]

The postwar boom rekindled the process. Agrarian reform dispersed migrants from throughout Mexico into the tropical forest, with little choice but to emphasize livestock as a pivot into more robust farms or (as happened around Veracruz) a coarser pastoralism. Roads became corridors of change, including state-sponsored land clearing. Mechanized agriculture was attempted for citrus, sugar, coconuts, and other tropical commodities, and, in the mountains, for coffee. Similarly, artificial pastures with introduced forage expanded and redefined the character of grasslands. The Green Revolution added fertilizers, pesticides, and herbicides—all derived from and typically applied by machines powered by fossil biomass.

Burning for milpas and pastos had long argued for fuelbreaks around the fired patch. There was no comparable speculation about fuelbreaks that would halt the accidental escape of industrial combustion across Mexico's landed estate.

MEXICO'S PYRIC TRANSITION

What has been called the pyric transition—the replacement of fire in living landscapes by the combustion of lithic landscapes—affected Mexico as it did every other place on Earth. Though the transition had to stutter-step through revolution and agrarian reform, and though it splintered as it tried to move through the fracturing landscape of Mexico, by the latter half of the twentieth century the torches of the new fire were spreading everywhere, like a welcome contagion, to replace the fires of the former sun.

Pyric transition and political revolution made modern Mexico. For a while it seemed that a fossil-fueled economy might resolve what revolution could not. Pemex funded more and more of public spending, almost a state within a state. The Miracle industrialized an economy increasingly powered by fossil-fuel energy and the collateral effects it enabled. The Green Revolution was a fossil-biomass makeover of Mexican agriculture. So, too, the revolution in fire protection that would follow was a fossil-fueled phase change in the management of fire on Mexican landscapes. What began as supplements, additives to enhance performance, became the primary mechanisms of improvement. The fundamental fires were no longer those that grew milpas and fed sheep, cattle, and braziers, but those that ran trucks, bulldozers, chainsaws, and power plants.

The road to modernity was paved and powered by fossil biomass. The new firepower broke the constraints imposed by living landscapes. That had been its attraction: its capacity to smash the shackles that had hindered Mexico's ability to support itself and left so much of its population mired in poverty. But in remaking Mexican society, fossil biomass remade land, water, air, and fire in ways and at a tempo that left no place able to absorb the effluents. They became pollutants, not just overlying land, water, and air but overloading them with indigestible toxins or removing processes that had previously sustained ecological goods and services. The sources of contaminants expanded far faster than the sinks available to hold them.

The old environmental troubles lodged in soil, water, and woods were extensive, as with most land use after the Conquest; Mexico's fire history evolved with those changes in human land use. But there were always limits. Too much fire, too-frequent or too-intense burning, destroyed the basis for a fire's regime. Fossil fuels had no such fetters, and they expanded unbounded until, globally, their use and effluent overwhelmed air, waters, and life, and so altered the conditions within which fire in living landscapes operated. Fossil fuels could burn day and night, winter and summer, through wet and dry, and they had a bottomless reservoir of combustible fallow to draw from. But while the excess burning did not affect further industrial combustion, it did change the conditions for living landscapes and the people who lived in them. What had been extensive environmental damage became intensive.

The instinct was to stave off collapse by injecting more fossil biomass. The original use of steam engines in Britain had been to drain coal mines to produce more coal; the original use of coal in Mexico had been to run railroads, in part to haul coal. Likewise, an oil-based economy tried to mitigate damages by relying yet more heavily on the energy and processed fossil biomass made possible by oil. If fossil-fueled agriculture stumbled, the reaction was to unleash more fertilizer and biocides. Such beliefs were not unique to Mexico—they characterized all developed-world economies. Faced with consuming less or producing more, nations chose to produce more, and the fossil wealth of past geologic epochs made that possible. Excess led to excess, however, especially when that excess was not equitably distributed.

The downsides appeared later. The belief that converting from wood to coal and then oil, for example, would spare Mexico's forests often encouraged extensive logging that far exceeded the previous consumption of fuelwood, which had mostly concentrated on dead-and-down material, coppices, and select felling. More logging stimulated a demand for more wood products, which prompted more cutting. Air polluted by woodsmoke over the Valley of Mexico ceded airshed to industrial combustion, much of it by automobiles. Instead of replacing old irritants, industrialization overwhelmed systems with new ones. Mexico, it seemed, needed fewer fuelbreaks around milpas and more around power stations and petrochemical plants.[185]

In fire history, as in so many features, Mexico's experience was unique only in its particulars. All places have biodiversity; Mexico has megabiodiversity. All developing countries must contend with how modern science and technology interact with tradition; Mexico formally embraced modernity, yet could not abolish its premodern places and customary practices. Its recurring solution since the Conquest had been to exercise some degree of paternalism over Indigenous people, and later the rural campesino. All modernizing places, colonial or not, as they pass through the pyric transition, have to reconcile old fire with new, usually by condemning and trying to eliminate open burning through either substitution or outright suppression. Mexico publicly denounced traditional burning but in practice ignored its informal economy of fire (a literal black market). Like Xiuhtecuhtli, the ancient Aztec fire god, anthropogenic fire had multiple sides fused into a single presence.

Typically, as the pyric transition unsettles traditional lands and their fire regimes, it kindles a wave train of big burns. New sparks and fuels are added to the old, and the population of fires explodes, much as the human population does during industrialization due to increased births and declining deaths. (The exception are arid lands subject to grazing in which boosted production strips away grasses, and routine fire with it.) Eventually both populations, people and fire, shrink below replacement value, which in landscapes means fire can no longer do the ecological work expected.

Yet Mexico did not explode into the megafires that the rush to industrialize inspired elsewhere in North America or in Australia. (Or if it did, no one recorded the outcome.) Instead, there is a chronic background rumble, like pyric tremors, of rural fires burning around or into villages and driving a score or even hundreds of refugees. But the slash-powered conflagrations that swept around the Great Lakes, the Coast Ranges, even the Adirondacks and Appalachians, had no apparent analogue in Mexico. Terrain is one reason; the specific character of forests, another. But it's tempting to argue that traditional fire practices were also at play, sparing Mexico from the worst outbreaks by peeling away excess combustibles. Instead of big fires, Mexico had big fire seasons, mostly synchronized by climate. In northern Mexico, as chronicled by scarred trees, fires were present from Baja California to Tamaulipas in 1798, 1820, 1838, 1851, 1860, 1874, 1877, 1894, 1899, 1902, 1909, 1916, and 1917, and there were an additional 55 years when four of every five states experienced fires.[186]

That ancient coupling of fire with climate broke as roads and agrarian reform displaced the former inhabitants and began their awkward linkage with a modern economy. Instead of climate, fires realigned more closely with human behavior—the dual revolutions (political and pyric) that swept over Mexico. The ejido program disrupted the ancient rhythms by reconfiguring land use with its relocated communities, roads, and livestock. Then, as lithic landscapes overlay and interpenetrated those altered living landscapes, the at first feeble, eventually irresistible power of industrial combustion made itself felt.

Steadily, industrial combustion was applied to remove, not just replace, traditional fire. Open flame disappeared from homes and offices, from intensive agriculture, and, with very mixed outcomes, from the countryside. For the first time at scale in Mexican history, climate and fire decoupled. Then the new fire disrupted even the climate, and climate history teetered into the status of a subnarrative of fire history. It was no longer enough to live with fire. Like the rest of Earth, Mexico had to live with a fire age.

FIFTH SUN

FIRE AND AIR

NEW FIRE

LA CATASTRÓFICA TEMPORADA

THE FIRES *were violent, overwhelming, and all-pervasive. Fire season morphed into a* temporada catastrófica.

In 1998 Mexico found itself caught between top-down global forces and bottom-up local practices, neither of which it had much control over. Two hurricanes slammed into the Sierra Madre del Sur and the mountains of Chiapas, encouraging fuels, then ENSO flipped into the most severe La Niña phase of the century and the worst drought since 1508 (a year after the last Aztec New Fire Ceremony), which blasted much of the country with the hottest temperatures on record and one of the driest years—all this atop shrinking political attention to fire protection. Decades of fire's diminution, if not its removal, added fuels. Traditional fire practices, with ignitions peaking from May 12 to 16, met untraditional conditions. Normally benign fires turned malignant. Fires bolted free from pastos and milpas.[1]

Every Mexican state burned. The outbreak ran the gamut of Mexico, from Chiapas to Chihuahua, from Los Chimalapas, site of protected cloud forests, to Cerro San Martín in Baja California, site of the national observatory. The most severely affected lands were in Durango and Chihuahua, Nuevo León and San Luis Potosí, the Sierra Madre del Sur and most of Chiapas, and, with special vehemence, the Valley of Mexico. Officially,

14,445 fires were reported, which collectively burned 849,632 hectares, or three times the national average; some 26 percent burned woods, 32 percent grassland, and 42 percent brush. Fields burned, slash burned, forests burned, grass and shrublands burned. Areas unburned for decades and lands cleared for recent colonization burned. At Peña Nevada in southern Nuevo León, the outbreak destroyed forests that had survived repeated surface burns since 1829. Protected areas could not be protected. El Chico in Hidalgo; Chipinque and El Cerro del Potosí in Nuevo León; La Michilía in Durango; Sian Ka'an in Quintana Roo; even Desierto de los Leones in the Federal District—all endured serious burns. In Chiapas fires spread into protected sites such as the Lacandón forest and Selva El Ocote, and in Oaxaca into the Chimalapas cloud forest; both states declared environmental emergencies. Authorities and urban elites could not ignore the conflagrations that swept in leaps and bounds throughout the Valley of Mexico. No less than Villahermosa and San Cristóbal de las Casas, Mexico City suffered a plague of exceptionally bad air that led to mandated restrictions on automobile use. The smoke from the limitless burning spread north in a vast gyre, recorded in air quality measurements from California to Wisconsin and causing public health concerns from Texas to Florida.[2]

Even more shocking was the human cost. Communities in Coahuila, Guanajuato, Hidalgo, Tamaulipas, Oaxaca, Nuevo León, and Puebla were evacuated. Seventy-one people died fighting the fires, almost all campesinos. Outside Texocuixpan, in the municipality of Ixtacamaxtitlán in Puebla, 19 were killed and 11 injured when a small grass fire blew up during a wind shift. At Lagunas de Zempoala in Morelos a major tragedy was averted when a water drop from a CL-415 opened an escape route for 70 campesinos. Ten firefighting soldiers died in a helicopter crash in Tamaulipas. Nationally, no estimate was possible for the injuries and the additional deaths due to smoke. Fire had become, as one survivor noted, an "unforeseen enemy." The outbreak represented, as so many Mexican commentators insisted, a parteaguas *(parting of the waters) in national fire history.*[3]

Mexico mobilized for fire as it had never mobilized before. Secretaría de Medio Ambiente, Recurson Naturales y Pesca (SEMARNAP) created an "emergency intersectional" group to oversee the national response. An estimated 6,000 firefighters consisting of federal and state employees, along with volunteers and 139,000 soldiers, racked up 823,971 man-hours fighting the fires. In the years leading to the outbreak SEMARNAP typically fielded 3 planes and 6 helicopters. For 1998 it rallied 57 planes, 25 helicopters, 3 Skycranes, and a CL-415 (loaned by Canada). Twice President Ernesto Zedillo authorized extra federal funds, which eventually reached 290 million pesos. Texas senators railed against the smoke pall as though the plume were a column of undocumented migrants storming the border. Through the U.S. Forest Service, the U.S. Agency for International Development provided advisors, equipment, five helicopters, and an infrared-mapping aircraft, as secretary of agriculture Dan Glickman proclaimed the outbreak a "transnational disaster." As even woods long thought immune to fire burned, observers despaired that Mexico confronted an "unwanted enemy."[4]

After leading ceremonies to commemorate Cinco de Mayo—a reenactment complete with 3,000 soldiers outfitted in period costumes and Texas governor George Bush as guest of honor—President Ernesto Zedillo visited the tragedy at Ixtacamaxtitlán. There he informed the villagers that the army would henceforth fight the invading fires.[5]

Mexico had not asked for a visible expression of its malaise, but the fires gave it one. The flames illuminated hidden issues, like an MRI scan exposing tumors beneath the skin. The 1998 season blasted the presumption that conversion to a fossil-fuel society would eliminate fire as oil had replaced firewood. That logic could apply in built environments. It didn't apply in the countryside. The scale and ferocity of the outbreak stunned the servicio forestal, which had steadily improved its firefighting capabilities, only to find that those proud upgrades still lagged behind the pace of worsening conditions.

Of course, sparks don't by themselves create fires. They have to find kindling, and a setting conducive to spread; for disasters and conflagrations, that environment must include societies, institutions, and ideas, as well as wind, fuel, and terrain. The final surge of agrarian reform in the 1970s had moved many people to the tropics, increasing the vulnerability of communities; the creation of nature reserves during the 1980s and 1990s presented hazards not previously recognized, and the creation of novel institutions like the Comisión Nacional para el Conocimiento y Uso de la Biodiversidad (CONABIO) meant that new agencies had responsibilities they had not anticipated. A Mexican state weakened by *narcotraficantes*, rural unrest, neoliberal policies, and yet another fiscal crisis (in 1997) found itself at odds with Mexican nature. For a decade those points of ignition had individually flared and smoldered, and from time to time merged. Then in the summer of 1998 they found the conditions that could turn a rebellious notion into revolution.

Boom had segued into bust: night descended over Mexico. The lights of the Mexican Miracle seemed to go out, one after another, as though readying for the arrival of a new epoch. It came in 1998, violently, with a shock comparable to that of the financial, political, and environmental crises that in a sense it integrated.

The fires seemed to be everywhere; they were certainly everywhere in Mexican media. For a while they filled the imagination, as their smoke did the sky. They even burned over the Cerro de la Estrella. For Mexico's fire history, the *catastrófica temporada* announced a Fifth Sun.

LOST DECADE

It was, throughout Latin America, *la década perdida*—the Lost Decade. Certainly, it felt lost in Mexico. The Revolutionary state unraveled. The country's natural estate of land, water, air, and biotas deteriorated. Agrarian reforms that had long symbolized the Mexican Revolution stalled. The PRI, the institutional flywheel of Mexican politics, faced growing disenchantment, and rival political parties emerged. Multinational corporations and banks challenged control over the economy; instead of parastatal enterprises to do the work of the government, Mexico found itself overwhelmed with dark equivalents in the form of narcotraficantes and their associated collaborators, who challenged the state's monopoly on violence and undercut or replaced state authority (or wormed their way into the establishment). State legitimacy as well as state power tottered. Mexico had long had a flourishing informal economy; now it seemed to have an informal state.

Society and fire reorganized. For nearly five centuries, two peoples had tried to find a workable accommodation. The newcomers had pushed for conversion, whether to Christianity, capitalism, or modernity, while the Indigenous people (or later, rooted ruralites) had resisted in favor of costumbres. Each surge had a characteristic wealth that propelled it, then collapsed: silver, land, foreign capital, oil. Each reorganization resulted in a characteristic suite of fires and a dialectic between two fire realms that, for millennia, had structured fire's regimes. In the twentieth century Mexico added a third fire. The lithic biomass that had for eons underlain living landscapes now overlaid the country in patches, plumes, and palls—enough to disrupt biogeography, acidify the oceans, and unmoor climate. Human firepower was sufficient to create the fire-informed equivalent of an ice age. The three-fire problem resembled a three-body problem that had no simple or exact solution. Worse, the three fires no longer competed but began to collude.

All this is a lot to lay at the feet of fossil fuels. But few of the breakdowns could have occurred without the energy, the matter, the economics, the machines, the politics, and the ecology that oil had made possible. Even the establishment of fire protection in the early 1960s had relied on roads built by modern machines, on crews carried by trucks, and, as much symbol as tool, on aircraft. A century after the Porfiriato jolted the industrialization of Mexico, the country was wealthier, more populated, more intensively farmed, and more outfitted with the emblems and institutions of modernity, but also more burdened by the effluent of that boom and still rent by social chasms that the state was unable to close.

MIRACLE MELTDOWN

Reforms in place by 1937 meant growth was everywhere by the 1940s. The Mexican Miracle rode the crest of a surging economy. War production with a U.S. market, and then a Green Revolution in agriculture, drove GDP up at a rate of 6.5 percent per year. Commercial farming matched industry, increasing fourfold between 1940 and 1965, with most high-value crops destined for export. Ties to the postwar economy of the United States assured a solid run of economic development. A middle class began to form.

What the Mexican Revolution had failed to do by land distribution alone, a Green Revolution promised to restart. Mexico's natural estate made it a world hot spot for biodiversity, and its pre-Columbian civilizations made it a major hearth for cultivars. As it domesticated its political revolution, it hoped to repeat that agricultural achievement through state-sponsored science. The project began under the Ávila Camacho administration, with later assistance from the U.S. Department of Agriculture, the UN Food and Agriculture Organization, and, the major sponsor, the Rockefeller Foundation. In 1943 the International Maize and Wheat Improvement Center was founded near Chapingo, with major test sites in Sonora. Its success made it a model for rice and other critical staples in the Philippines, India, China, and elsewhere.[6]

But against those trends, others rose in tandem, and these were not so benevolent. Debt rose, an old affliction now revived. Inflation picked up, for a while below GDP, then above it. Agricultural production fell behind population growth to the point that by the end of the 1970s Mexico was again importing food. Much of the wealth Mexico generated left the country, along with exported manufactured goods and crops. (Between 1940 and 1970, there were only seven years when more money stayed in Mexico than left it.) As money flowed in and out, corruption went from a daily nuisance to an economic drag that, like illegal lines siphoning off electricity or gas, transferred money from a formal to an informal economy, which also meant from public to private purposes. What the PRI had glued together began delaminating. Meanwhile, births skyrocketed, which meant that industrialization would stumble much as agriculture had; national wealth did not translate into per capita wealth. In 1950 Mexico's population was 28 million; in 1994 it was 90 million. Rising expectations met rising resistances.[7]

Five centuries after Cortés sacked Tenochtitlan, Mexico's economy ranked fifteenth in the world by GDP and eleventh by purchasing power, yet it also ranked high in inequality, corruption, and environmental damage. The period from the assassination of Madero to the signing of NAFTA effectively defined the span of the Mexican Revolution and its successor state. The Russian Revolution had lasted from 1917 to 1991; the Mexican Revolution, from 1910 (or the 1917 constitution) to 1992.

LITHIC LANDSCAPES: THIRD-FIRE, THIRD NATURE

In Mexico, as in every part of Earth touched directly or indirectly by the pyric transition to fossil fuels, third-fire helped create a third nature. A lithic landscape, excavated from deep time, overlaid a living one in the present. From 1891 to 1982 it is estimated that Mexico's carbon dioxide emissions had increased at 6.5 percent per year; thanks to the magic of compounding interest, those numbers loomed large in the years of the boom. Mexico was doing what other developed countries did: it exhumed fossil biomass from the geologic past and exported effluents to the geologic future.[8]

Oil underwrote most aspects of the Mexican Miracle. Oil powered industry; oil made the Green Revolution possible; oil allowed the Mexican government to cover its costs, pay for some social services, and boast that, according to the World Bank, it was now among the developed countries. Oil made the Mexican Miracle more than a wartime fad—and oil lubricated its collapse as energy crises and new discoveries whipsawed Mexico between oil famines and oil orgies. What oil gave, oil took away. The black bullion pumped up Mexico's economy and government aspirations, much as silver had in earlier centuries, and, when global overproduction caused the market to crash, Mexico's economy sank like the treasure fleet of 1715 under the latest fiscal hurricane.

The transition redefined the relationship of fire to Mexicans and to the Mexican landscape as it reworked habitat after habitat. This was the story everywhere. Yet Mexico had its peculiarities. Lack of coal meant that wood had remained a fuel, even for generating steam, longer than in most places before the country pivoted to oil. The failure to swiftly convert exasperated elites, who saw the continuing use of fuelwood for cooking even in metropolitan Mexico City and the inexplicable reliance on fire for milpas as a stigma of backwardness and a token of how arduous and thankless would be the task of hauling Mexico into the status of a modern nation.

The transition required a national reconstruction. Oil suffused all aspects of Mexican life. It powered booms and left environmental busts. It remade landscapes. It carried a new fire throughout Earth.

Mostly, oil powered machines that did the heavy lifting. But the oil, cooked or raw, could burn on its own. Oil wells burned, and as drilling moved offshore, fires followed. In 1979 the Ixtoc I exploratory well in the Bay of Campeche blew out, burned, and remained uncapped for 290 days while 3.4 million barrels of oil leaked into the Gulf. Mexico's oil industry had begun with the largest terrestrial oil spill (Dos Bocas), and as the industry migrated offshore, Mexico experienced the largest marine oil spill to date. Efforts to contain the leakage mimicked those of 71 years earlier. Booms failed to halt the spread of oil, equipment broke down, dispersion chemicals delivered by aircraft proved ineffective, and relief wells were inadequate until finally enough mud was pumped through them to halt the outrush and allow the drill hole to be capped with concrete. Environmental damage was extensive. Pemex spent $100 million on cleanup operations.[9]

Explosions and fires occurred at platforms like Abkatún A in the Gulf, at gas plants like the one in Reynosa, Tamaulipas, in 2021, in oil tankers such as the *Burgos* in Veracruz, along pipelines through the countryside, and in villages like Tlahuelilpan, Hidalgo, in 2019. It was fuel because it was combustible, so wherever burning gas had unwanted contact with other fuels it could start a wildfire.

The major influence of third-fire, however, came indirectly through how it infiltrated all aspects of Mexican life and economy and how it interacted with first- and second-fire. It offered a technological substitute for open flame in hearths, furnaces, candles, and fields, and it powered machines designed to extinguish free-burning fire. In cities this was a critical improvement. In the countryside, it allowed a policy of fire suppression to advance, with very mixed, frequently unexpected, and often unpleasant outcomes.

The usual symbol of the Green Revolution is clever plant breeding. But those seeds needed other inputs in the form of water, fertilizers, herbicides, and pesticides, all of which required means of delivering them to the field and carrying the harvest to market. At every stage—from seed selection to market—this meant fossil biomass, whether burned or processed (chemically burned, as it were) into petrochemicals. The broad-purpose ecological catalysis that fire had performed for millennia was broken down by science and capital into its constituent parts, selectively maximized, and applied to promote growth.[10]

This revolution, too, failed to deliver to the rural classes. Wealth flowed to wealth because the other ingredient—the one necessary to access the other stimulants for the Green Revolution—was money. The production of oil unbalanced the economy, even siphoning funds away from agriculture.

Inequality between lands and peoples sharpened. Worse, landscapes lost fire, and they ceased to be managed to accept fire. For those trained in modern agronomy, this was exactly what was supposed to happen: fire was inefficient, fire was dangerous, fire was primitive. Second-fire demanded fallow, which took a sizable chunk of potentially arable land out of production; by contrast, third-fire extracted its fallow from fossil sources, which allowed farmland to produce without pause. The reformation wrought by oil did exactly what modernizers hoped. In this case the inequalities that resulted were not just between those people who got benefits and those who were left out, but in the character of the land itself, as it segregated between modernized and still-traditional places.[11]

Traditional agriculture, galvanized by customary burning, had accounted for much of the biodiversity, the ecological goods and services, and the resilience that the toughened cornucopia that was Mexico's natural estate boasted. Its multiyear fallow fashioned the environmental mosaic that sustained much of Mexico's diversity of flora and fauna. Living landscapes were elastic; humans could modify them, though only within limits. Anthropogenic fire was loosely shackled by those constraints because it fed on living biomass. Biotic sinks and sources were linked, though they more resembled the atonal cadence of predator-prey relationships than the steady flows of engineered pipelines.[12]

Third-fire had no such fetters. It could burn day and night, winter and summer, through wet or dry. It could overload ecosystems with its effluent, leaving layers and caches of pollutants, some of which (like CO_2) were effectively transferred into the future. The skies over the Valley of Mexico that had seasonally filled with the small smokes of cooking fires and milpas now flooded with smog. Previously, fire-released nutrients had been absorbed by fields that resprouted in postburn exuberance; now, those landscapes were moribund, and the chemicals released by third-fire overwhelmed ecological sinks and turned toxic. Fertilizers, pesticides, runoff—human and ecological health both suffered. Even a doctrine of fire exclusion demanded a counterforce to match fire's, and—for a time—petrol-powered machines made this possible.

Eventually, the appreciation began to dawn that fire was not just a tool, as the Enlightenment tended to imagine it. It was a general spark for ecological enrichment that third-fire surrogates failed to replicate. Its removal unwound traditional landscapes without a comparable replacement.

Eventually nature reserves had to assume the ecological role of fallow, which also meant they would have to burn on something like the pattern that had prevailed previously. So deep was the perception of fire as archaic that its revival was a hard lesson for authorities and intellectuals to accept.

The problem with the pyric transition was not only the often unprecedented damages it inflicted, but the extinction of a process that was vital to Mexico's landscapes. Still, Mexico's ruling classes regarded the control of landscape fire as a precept of economic necessity and cultural modernity. Third-fire made the experiment possible.

LIVING LANDSCAPES: COLLAPSE AND CATALYSIS

Until 1960 fire protection had enjoyed little improvement from the Miracle. Then, under the reforms introduced by Enrique Beltrán, some of the economic slop-over spilled into fire protection—not as much as concerned parties wished, but there were more crews, more lookout towers, a rough radio system, and the proud visibility of helicopters in paramilitary firefights.[13]

The reforms under Beltrán represented a jump in character as much as quantity. The specifics fell to Manuel Verduzco Tapia, who in 1964 brought vision and institutional weight to the Five-Year Program for Fire Protection. With patience, persistence, and professionalism, undeterred by limited funds and personnel as measured against the challenge, Mexico steadily improved its institutional capacity. From its hearth in the Federal District, the program expanded throughout the Valley of Mexico. In 1974 it began staffing on weekends and holidays. By 1975 the project had added another 60 brigades. In 1978 Víctor Martínez Rodríguez replaced Verduzco. Two years later the fire program was able to acquire additional helicopters, a Bell 212 and three Bell 206s.[14]

Critics carped that prevention programs still failed to stop the ceaseless kindling throughout rural Mexico, that industry crews loomed larger on paper than in practice, that scientifically trained fire experts were thin on the ground, and that the Miracle had moved Mexico farther up the league tables of industrialization than the ranks of global firepowers. The lapses were many. There was virtually no fire research. Fire protection was lodged in a forestry department that was endlessly shuffled among larger bureaus. But fire-related disasters were few. Many other aspects of Mexican society had a higher claim on political attention.

Still, tasks were added to the traditional duties, and as land use changed, so new occasions arose for inquiring into the proper place of fire. Fire protection was extended, for example, to Parque Chipinque along the border of Monterrey. In 1983 the city contracted for an air tanker that became by default a trial for a possible national expansion. The most stellar experiments were, unsurprisingly, in the Valley of Mexico. Under the inspiration of Ing. Antonio Sierra Pineda, Mexico City developed a program of aggressive reforestation (COCODA; later COCODER, "in the eternal churn of name changes caused by new governments"), which demanded better fire control to allow the planted trees to survive, which in turn led to fuel inventories and a fire behavior prediction model. The EXTINGE project became an exemplar not only of fire protection but of the application of science to a field that had mostly relied on oral tradition and informal analysis.[15]

Then the economy tanked. State-sponsored forestry suffered, and fire protection—already something of a problem child—was pushed farther to the margins. The sexenio of Miguel de la Madrid Hurtado began, as was customary, with a sweep of high-level administrators, but that purge also allowed for some fresh initiatives. Here and there—following a snowstorm that trashed forests in Durango; at a university field station in the Sierra de Manantlán; amid reforestation projects in the Valley of Mexico; at the Parque Chipinque—the shards of a new order appeared, which during the subsequent decade fused into a distinctive Mexican fire mosaic.

Surprisingly, perhaps, the spark came from the United States. The role France had served for Quevedo, the United States now provided for a generation that came of age during the 1980s and reshaped the Mexican fire establishment in the decades to follow. It didn't hurt that fire joined earthquakes and hurricanes on the list of disasters that demanded national attention.

FIRE'S NAFTA

The Fire Management Working Group of the North American Forestry Commission was one vehicle for disseminating information, and it gave Mexico both institutional and personal ties to Canada and the United States at a high level. But it was a series of fire training programs that sparked what Roberto Martínez, a long-serving official with Mexico's fire program, called "a radical change" and initiated the drive toward genuine professionalism and the capacity to manage fire suppression at a national level.[16]

Like so much of Mexico's modern fire story, the connection was serendipitous. In February 1983, the Dominican Republic asked for assistance fighting a problem fire, for which the USFS dispatched an advisor. Postfire reviews suggested that what the country needed was a training program to build in-house capacity. Within six months, with funding from the U.S. Agency for International Development (USAID), the U.S. Forest Service created a three-week course, all in Spanish, that it staged at the United States' national fire training center (the oddly named National Advanced Resources and Technology Center), located on a decommissioned air base at Marana, Arizona. Instructors included representatives from the National Park Service and the National Weather Service. The audience was filled with Latin Americans—eight Mexicans among them—along with a representative from Spain, Ricardo Vélez Muñoz. The project repeated in 1984 with 13 Mexicans, one returning from the previous year, this time as an instructor. Meanwhile, the U.S. and Mexican secretaries of agriculture signed a memorandum of understanding to promote scientific and technical cooperation. A 1985 supplement emphasized particular programs, fire management prominently among them. Thus began what might be regarded as fire's NAFTA, a North American fire training alliance.[17]

Much of the future corps of Mexico's fire suppression organization attended, or would attend, as the course was subsequently disseminated throughout Mexico. In 1985 the course migrated to Chile, then to Argentina. Ricardo Vélez went on to become director of Spain's national fire program, to author a text on the subject, and to host a similar course in Spain for the Ibero-American world. A follow-up FAO-sponsored project translated technical fire terms from English to Spanish. Thereafter, having trained the trainers, Mexico absorbed the course and began delivering it in various regions over the coming decade, for a while with U.S. instructors and again with USAID assistance. The "Marana" experience offered an unheard-of level of training for someone involved with Mexican fire, but just as valuable as the knowledge accrued was the companionship among those who attended. "Marana" became a rite of passage, a code word for a fire fraternity whose members would reconvene for decades. The course continued uninterrupted for 18 years.[18]

Meanwhile, under the auspices of the NAFC Fire Management Working Group, the U.S. Forest Service and Mexico's Department of Agriculture and Water (SARH) signed a memorandum of understanding to expand joint programs. The Marana courses were one such program; demonstrations and trials of aerial firefighting were another; there were select exchanges of personnel, and even hand tools like the Pulaski and the McLeod moved south. There was a workshop held at El Paso that accommodated regional representatives of the Forest Service and technicians from SARH. After the disastrous 1985 Mexico City earthquake, a National System for Civil Protection had been created to better cope with emergency response; the program included fires, which the next year led to a central command center for dispatching varied firefighting resources, mostly in the Federal District. U.S. advisors urged Mexico to adopt the incident command system (ICS) first developed for wildfire in Southern California and then abstracted to embrace all emergencies. In October 1986, at the invitation of SARH, two U.S. researchers, David Sandberg and Armando González Cabán, toured Puebla, the Federal District, the State of Mexico, Coahuila, Chihuahua, and Chiapas in a reconnaissance to identify points of mutual national interest, especially with regard to fire research.[19]

Essentially, they concluded, Mexico's fire program was underpowered, both in operations and in the knowledge needed to create a modern agency. (Mexico had five fire researchers, all in the Instituto Nacional de Investigaciones Forestales and all part-time.) Like Mexico's economic condition, in which productivity struggled to exceed inflation and population growth, the fire program was expanding, but not in proportion to needs, and it was concentrated in central Mexico. Mexico's fire project, like its economy generally, needed to transition to a more information-intensive order if the country was to modernize. Mexico needed its own corps of fire scientists. Usefully, it had Earth's largest concentration of fire science across its leaky northern border.

EDUCATING FIRE

The new order began with education. The Marana courses baptized fire officers into state-of-the-art fire suppression. But a thriving program required a corps of científicos as well as bomberos. It needed access to institutions that could offer models to emulate, conduct research, and provide training.

From its origins in central Europe, forestry had been regarded as a subdiscipline of agronomy. Under the direction of Marte R. Gómez, the National Agricultural School was conveniently located in Texcoco, where the International Maize and Wheat Improvement Center had launched the Green Revolution. In 1974 those developments prodded the Escuela Nacional de Agricultura to upgrade into the Universidad Autónoma Chapingo (UACh). Reforms included living quarters for professors and workers, laboratories, and other research facilities. Gómez even invited Diego Rivera to install a suite of original murals in the principal building on campus (which subsequently became known as the Capilla Riveriana). What Mexico's farms needed, its forests also required; fire protection was to Mexico's woods and forest plantations what pesticides and herbicides were to its row crops. In 1979 a Colegio de Postgraduados in forestry was established. Together those reformed institutions trained a new generation of agronomists and foresters and laid the foundation for a Red Revolution in fire. Of the six generally recognized "fathers" of modern fire management, five were *chapingueros*.[20]

Other institutions, notably the Instituto Nacional de Investigaciones Forestales, Agrícolas y Pecuarias (INIFAP), and eventually other universities, especially the University of Guadalajara and, later, the Universidad Nacional Autónoma de México (UNAM), established research groups. Novel institutions for land management, like the National Commission for the Understanding and Use of Biodiversity (CONABIO), which oversaw Mexico's constellation of biosphere reserves; CONANP, which administered protected natural areas; the Fondo Mexicano para la Conservación de la Naturaleza (FMCN); and NGOs concerned with environmental protection, like PRONATURA and the Nature Conservancy, demanded a higher grade of science to make informed decisions and monitor outcomes. A National Council on Science and Technology contributed some research grants.

Still, the creation of a fire research program was a minor Mexican miracle. The U.S. Forest Service had established a formal, if small, agenda in 1916, which had blossomed after World War II, partly with military and civil defense funding, and by the 1980s was pivoting from an emphasis on fire behavior to a more robust understanding of fire's ecology, as the agency shuffled from simple suppression to more pluralistic policies. By then only five nations had serious fire research programs—the United States, Canada, Australia, South Africa, and the USSR.

Mexico had none, and fire-related publications were rarely more than one or two briefings at national forestry conferences, followed by tables of ignition sources and damages. José Verduzco Gutiérrez's text on forest protection was published in 1976 (at UACh), but it addressed fire in only one small chapter on methods of control. Jerzy Rzedowski's *Vegetación de México*, published in 1978, mentioned fire in a handful of sentences, mostly as another (baleful) human influence. No formal course of study existed in universities, no tradition of fire research had rooted in bureaus, no dedicated line item in an agency budget targeted fire. Fire was a nuisance that festered as an embarrassment to reformers' pretentions of progress and which sometimes mutated into disasters. Mexico's fire science establishment was to the United States' as a milpa was to an industrial corn farm.[21]

All of the founders came into the subject of fire sideways. Ernesto Alvarado met fire accidentally through the study of forest diseases; Germán Flores, through fuels mapping, statistics, and computer modeling; Enrique Jardel, through an interest in biodiversity and a university field station; Robert Martínez Domínguez, through a bureaucratic shuffle; Alfredo Nolasco, through a Marana course at Cancún; Dante Rodríguez Trejo, through a reforestation program outside Mexico City. Their curiosity was piqued through personal, if unanticipated, encounters with fire in their jobs. Each discovered, as Alfredo Nolasco put it, that with respect to fire, theory didn't match reality. Yet within 20 years they had advanced Mexico to fire's equivalent of the Group of 20, and within 40 years they could proudly boast, as Dante Rodríguez Trejo did, that Mexico could stand within the top 10 countries committed to the research and management of landscape fire.

Yet what Mexico thought it was getting when exchanges began—better capability to extinguish fire in the countryside—was not all that its North American counterparts sent. Americans did dispatch better methods of suppression and demonstrate more powerful technologies. Mexicans, in response, welcomed training, appreciated air tankers, established a central command office in the Federal District, and devised a fire-behavior prediction system (EXTINGE), based on the Rothermel model developed by the USFS, to assist dispatching. Mexicans were receptive to technology transfer, and what most field personnel and the public accepted as an exemplar was the paramilitary model that had emerged to relentless media fanfare in the United States.[22]

But the U.S. fire community had, over the previous 20 years, undergone its own radical transformation in its understanding of and policy toward fire. In 1968 the National Park Service renounced the suppression-only goal in favor of an agenda of fire restoration. Ten years later the Forest Service followed. The evolving era orbited around a doctrine of fire by prescription, in which deliberate (prescribed) burning mediated between letting everything burn and letting nothing burn. Paradoxically, while Mexicans wanted access to the United States' air tankers, what Americans often wanted was access to Mexico's landscapes shaped by traditional fire practices.[23]

Joint research projects began as a trickle. In 1981 Mexico's principal forest research institution (INIFAP) signed an agreement with the U.S. Forest Service's Rocky Mountain Forest and Range Experiment Station to examine the effects of prescribed fire in the forested mountains of Chihuahua. The lead U.S. researcher, Jack Dieterich, had previously made a tour with the Peace Corps in Chile and was curious how ponderosa pine forests compared across the border. He conducted some dendrochronology-based fire history in the Sierra de los Ajos in Sonora. The formal agreement projected the scheme into the sky islands of Chihuahua, with the notion of expanding to Mexico's northern border the prescribed burning that was beginning to appear in the U.S. Southwest.[24]

It took another decade for that remote demo to mature into a wide-ranging network of fire history surveys, collegial relations, mutual projects, and a comparison between centuries-long fire histories across the border. In the meantime, Mexico cultivated small points of positive infection, like the prescribed fire trials at Paraje Los Cerritos and the land-use and fire history inquiries at the Las Joyas field station of the University of Guadalajara, that could help achieve what the 1986 reconnaissance by U.S. observers called making Mexico "more self-sufficient in forest fire expertise and management."[25]

A fuller conversion demanded a crisis that could mobilize the federal government to consider fire an issue worth investing in. That moment came in 1989.

AÑO CRÍTICO

In 1988 Hurricane Gilbert churned across the Caribbean and, as a Category 5 storm, slammed into the Yucatán Peninsula on September 13. It weakened over land, then regrouped after reemerging into the Gulf of Mexico, strengthened to Category 3, and struck mainland Mexico on September 16. Damages ran into several billion dollars. Perhaps 300 Mexicans died. A million hectares of forest were slashed. When the dry season arrived, so did fires.

Hurricanes are nothing new to Yucatán; what made Gilbert significant was its intensity (the second-largest Atlantic storm on record), the scale of its damages, and an official capacity to record in some detail its environmental effects. The storm stripped away most of the forest foliage within its path and reduced tree density by a third. Some trees were uprooted, many were broken, all were stripped of leaves. The amount of woody debris left behind was substantial, an average increase of 22 to 36 percent, with some places reaching 50 percent more than the amount prior to the storm. After Gilbert passed, the NAFC Fire Management Working Group toured the wreckage and advised about preparations for the fires that were sure to come. But rebuilding Cancún was more critical than preparing for the anticipated fires, and no special resources were made available. The local authorities and the servicio forestal would have to manage on their own.[26]

What then happened is what had happened for millennia: that tangle of debris dried and burned. When the first fires broke out, northern Quintana Roo had 30 firefighters, 2 small bulldozers, and a Bell 206 helicopter. The fires made news; their smoke occasionally shut down the airport at Cancún, which was hosting the Miss Universe pageant. Mobilization scaled up. Ing. Manuel García Gil was charged with leading a group of specialists (veterans of Marana) to cope with the outbreak, as a "great movement" of technicians, rural fire brigades, soldiers, sailors, bulldozers, airplanes, and helicopters converged on the scene. The fires were vast ("inmensos"), the resources to fight them sparse ("insuficientes"). Two American advisors sent by the USFS on request introduced the incident command system. The struggle went on from May until August, when rains finally extinguished the flames. In the end, some 135,000 hectares of that slashed landscape burned. The fires had turned the blowdown into a blowup.[27]

Mop-up involved inquiries by several groups of biologists (the Smithsonian Institution even had a field station at Rancho San Felipe). For really the first time, the ecological consequences of nature's exercise in roza-tumba-quema were investigated. Researchers found that within two years some 12.6 percent of the affected trees had died. Yet most of the landscape recovered quickly: nutrients were released, latent sprouts were liberated, and new growth flourished. More trees died from fire than wind. There were even speculations by noted observers in the popular press about the inevitability of hurricanes and their likely ecological role, even their value. It proved harder to project that same empathy toward fire.[28]

The still-consolidating fire community struggled to comprehend the magnitude of the burns. SEMARNAP's subdirector for fire protection, Roberto Martínez, called the Yucatán fires the "gran emergencia," the first of a triptych of outbreaks that defined the contemporary era. Alfredo Nolasco, later a national fire director, at the time a young fire officer on the scene, called them a "catástrofe ecológica," perhaps not only because of their impact on a landscape unaccustomed to fires of this scale but because of the shock they delivered to the servicio forestal. Dante Rodríguez Trejo, the encyclopedist of Mexican fire, seized on an expression in SEMARNAP's annual report and labeled 1989 the "año crítico." The struggle to find a suitable name mimicked the struggle to devise a suitable institution to cope.[29]

The year before, a fire season in Mexico and two fire scenes outside it helped frame understanding of the post-Gilbert burns.

Mexico's 1988 season had more fires than 1989 (10,942 compared to 9,518), and more land burned (518,265 ha to 507,471 ha), but the 1988 season belonged to the previous presidential regime, so it was recorded separately, and its impact was more diffuse. For the fire community the post-Gilbert conflagration came as a knockout punch. One big season was an anecdote; two looked like a trend. While the 1988 fires ended one national campaign against fire and the 1989 fires started another, the servicio forestal preferred to compare averaged statistics across administrations. It was also eager to measure Mexico's standing against its two preferred standards, the United States and Spain. From 1989 to 1994, by nearly all metrics, Mexico compared well with the United States and topped Spain, not least in fallen firefighters (in 1994 the United States suffered 34 firefighter deaths, Spain 36, and Mexico 2). The 1989 outbreak was bad but not devastating.[30]

The international stories focused on Brazil and Yellowstone National Park. Land-clearing fires in Amazonia presented to many observers a world-class act of nature mutilation, while prolonged wildfires in Yellowstone National Park suggested both a threat to a flagship nature preserve and, equally, the vivid workings of a little-trammeled nature. Clearly, suppression had endured serial failures and probably increased the final size of the burns significantly. Yet the remorseless fires had allowed a lodgepole pine forest that regenerated through crown fires to renew itself as it had done throughout its evolutionary history. Testifying before the U.S. Congress that summer, NASA researcher James Hansen announced that the signature of global warming was now clearly visible.[31]

Both the Quintana Roo and Yellowstone fire complexes were extreme events that brushed aside all the mechanical muscle fire agencies could muster against them. But, paradoxically, they could also be characterized as extreme versions of natural events. Before they had become a global media meme, Yellowstone's fires had been welcomed by the National Park Service. Similarly, the Yucatán fires were the latest installment of a process that had gone on for as long as the peninsula had been forested. Gilbert had increased by 50 percent the amount of woody debris (and the stock of nutrients) on the ground; the fires did at a scale of thousands of hectares the biological work that milpas performed on one or two. A consensus emerged that "modern-day forest vegetation in the Yucatán Peninsula appears to be influenced primarily by historic patterns of land clearing and by the combined impacts of hurricanes and fire," that its forests were resistant to hurricanes and resilient to fires, and that "medium-scale to large-scale disturbances caused by hurricanes and fires" might be critical for species such as mahogany.[32]

The fires were also a concern to fire officers. The scale of the conflagrations exceeded what experts believed had informed those landscapes historically, and it was well beyond what agencies could pretend to control. Because the attention of global media was elsewhere, Mexico's fire crisis was felt mostly by its own fire community, which saw in the post-Gilbert wreckage a bitter demonstration of how far the country had yet to go in its reforms. Perhaps, as at Yellowstone, once the fires had begun, no buildup of suppression resources could have halted the flames' progress. But Mexico had too few resources to know where that limit might be.

The notion that fire could be valuable, even essential, was too exotic to translate into any operational meaning. Yet speaking to the Mexican Forest Congress that year, Oscar Cedeño, then head of policy for the servicio forestal, described experiments with *quemas controladas* (prescribed fire) conducted since 1985 in the Santo Tomás community at Ajusco in the southern mountains of the Valley of Mexico. It was an effort to find a way to have fire that rejuvenated grasses without destroying woods, or, more broadly, to contrive a reconciliation between the fires foresters feared and condemned and those campesinos used and welcomed. With the right fire, the forest would regenerate naturally. The horizon-sweeping Quintana Roo fires overwhelmed those quiet flames on the slopes of Ajusco. The next year the servicio forestal hosted a Marana course at Cancún, with instructors from the United States and Mexico, and joined North American colleagues in organizing the first International Wildland Fire Conference, hosted in Boston with official recognition as an activity of the UN International Decade for Natural Disaster Reduction. Fire was going global, and Mexico with it.[33]

FOUND DECADE

Mexico reorganized. It further restructured its economy to accommodate neoliberalism: it joined the General Agreement on Tariffs and Trade, signed the North American Free Trade Agreement, and amended Articles 17 and 27 of the 1917 constitution in ways that permitted ejidos and Pemex to partly privatize. Yet even as it dissolved some communal lands, it continued to create others in the form of public preserves. Land use bifurcated, sharpening the divide between city and country, between an export economy and traditional use, and between private ownership and communal (or hybrid-tenure) lands. Everywhere, the state tried, with mixed results, to meld oversight with local (or, in the case of international relations, Mexican) autonomy.

Between Mexico and the United States political hard welds proved, in many respects, less transformative than cultural soft ones. NAFTA was less radical than its advocates promised or its critics feared, while remittances contributed to the Mexican economy an estimated $3.7 billion in 1992 and $39 billion in 2019 (foreign direct investments, by contrast, were $34 billion in 2019). Popular culture and foods seeped both ways across the border, with tacos moving north and Coca-Cola south. Many of Mexico's political class had studied in the United States (the primary architect of neoliberal policies, Carlos Salinas de Gortari, had graduated with a master's and a PhD from Harvard). So large was the Mexican diaspora that Mexican officials discussed how to allow them to vote in Mexican elections (in 2019 it was estimated that 10 percent of native Mexicans had moved north).[34]

All this Mexico's fire scene integrated. Instead of a common suite of fire practices, applied across the country with local adaptations, there were places that no longer had routine fires, and places that replaced customary tame fires with wild ones. There were places that retained a core of traditional burning, places that had transitioned to industrial fire, places that argued for hybrid models. In response, Mexico's fledgling fire community—a gaggle of barely linked entities and personalities, never really a self-conscious agency—strengthened into a meaningful presence.

Unsurprisingly, Mexico's fire economy echoed its political economy. Bilateral agreements with the United States blossomed into joint research projects, training and personal exchanges, and eventually a mutual-aid treaty to cope with fire across the border. (There was even a Mexican fire crew, Los Diablos, that assisted with fire suppression across the Rio Grande, especially in Big Bend National Park.) Global fire conferences brought Mexicans into an international context that introduced UN programs for fire management and nature protection, and the concept of integrated fire management. Canada's connection through the fire working group added another option for training. Throughout, Mexico retained and redefined its historic role as a nexus between North and Latin America. The United States and Mexico, with mutual enthusiasm, shared an enlarged Brotherhood of Prometheus (*hermandad de Prometeo*).[35]

LAND AND LIFE, AGAIN

As ever, the process centered on land. Again and again, from the Conquest to Articles 17 and 27 of the revolutionary constitution, national politics had hinged on control over mineral wealth and land tenure. Agrarian reform and the nationalization of oil had symbolized the success, however compromised, of the revolution and had made Lázaro Cárdenas its face. But there were too many Mexicans, not enough traditional resources to sustain them, and too few mechanisms to more uniformly distribute the national wealth, and those gestures gradually metamorphosed into symbols of the ambition's shortcomings. The makeover that resulted from the meltdown of the 1980s attempted to revive the economy by redefining the relationship between public and private.[36]

Agrarian Reform 2.0. Reigniting development meant, for the Salinas regime, that Mexico needed to move more public or communal goods into fuller production, which pointed to privatization. Oil was too symbolically sensitive to allow the government to reconstruct Pemex—partial privatization had to wait until 2013. But changes in land tenure were possible—what might be called agrarian reform 2.0—and, it was argued, were essential to prepare Mexico for NAFTA. In 1992 a new agrarian law sought to move more ejido land—"half of Mexico's arable lands held by 3.1 million persons in 28,000 ejidos"—into a market economy. Simultaneously, a new forest law sought to break up the timber parastatals, which had led to monopsony, and so open the forest sector to greater market influence. A third innovation promoted a different conception of nature reserves. Together, these reforms to tenure, forestry, and nature protection restructured Mexico's natural estate. Where land went, fire would follow.[37]

Most officials regarded the ejido system as effectively defunct by 1986, but it appeared sacrosanct and politically untouchable. In 1991 the Salinas regime proposed amending Article 27 of the constitution such that private ownership within ejidos was not only allowed but encouraged: whatever the political value of communal tenure, too much agriculture was too unproductive to allow Mexico to feed itself; U.S. agribusiness would overwhelm ejido farming. The legislation made ejidos, in David Barton Bray's term, a "corporate communal property." (The United States was doing something similar with Alaska Native corporations, having earlier tried privatizing by individual allotments.) Privatization and certificates for individual ownership still required ejido approval, so the project was transitional, and only a small fraction of lands made that transition (perhaps 8 percent). But a general shake-up was under way.[38]

Forests constituted a special category. Ejidos were not allowed to subdivide forests as they could arable and pasture, and if an ejido voted to legally end as a political entity, its forest would revert to the nation. As always, there was compromise. Mexico was prepared neither to abolish national control over major natural resources nor to deny local control. How both interests could coexist was, as it had been for centuries, the core of Mexican land politics. One outcome of the bureaucratic dissonance was a rash of illegal logging; another was the emergence of community-based forestry.

Community Forestry. State-sponsored conservation had been a hallmark of Europe's colonial era—led by Europe's second wave of imperialism, the one that arose in temperate Europe and included forestry. That outburst of colonial rule collaborated with a footloose capitalism that plundered landscape after landscape, especially traditional commons, and ran roughshod over local practices. Too often money and power had unhinged local politics: small communities couldn't stand against them, no more in the colonies than they had in Europe. (The Highland clearances might be counted a dress rehearsal for an empire committed to transferring public goods to privatized markets.) Eventually the state appreciated that it had to intervene: to stand between a global economy and local communities. A secularized paternalism, a scientifically informed agenda that presumed to replace tradition with Enlightenment, large-scale forest reserves—all helped justify European rule. Forest reserves proposed a network of national commons to replace those smaller habitats lost during settlement.

Sir William Schlich was a titan of colonial forestry, one of a triumvirate of Germans who helped set up Britain's Indian Forest Department. He was the author of a multivolume text, *Manual of Forestry*, that became the bible of global forestry; the editor of the *Indian Forester*, for decades the premier journal of forestry; and eventually a professor of forestry at Edinburgh. Like his peers, he saw fires as a human-prompted threat and a measure of social disorder. The best forest reserves had no human residents—they could be managed on strictly rational principles, as

German forestry understood them. When colonial foresters complained that residents within reserves caused complications, especially fires, Schlich retorted that the solution was to remove the people. That was the point of reserves. Another proconsul of Britain's empire, Sir D. E. Hutchins, insisted that a forestry department "*must have undivided autocratic control over all the operations*" under its dominion.[39]

Colonial foresters tried but failed to implement that ideal. It was more possible in settler societies where disease and war had decimated Indigenous populations and left relatively unoccupied lands ready to be set aside as parks and public forests. Decades later, authorities realized that much of what they valued in those first-encountered forests had been, in truth, a cultural landscape, and that banning the human hand (and torch) could be profoundly unsettling.

Now, as decolonization swept across the globe, those institutions either reformed or dissolved. Forestry bureaus shape-shifted and mutated into one arrangement after another; only the United States and the USSR kept their old agencies, though the U.S. Forest Service, like a heritage building, retained its outward structure while massively remodeling its interior. The old model of conservation, or its even more restrictive cousin, preservation, also demanded upgrades in the face of a planetary makeover. Mexico was "one of the ten countries in the world with the largest area of primary forest," yet "between 1990 and 2010 it was also one of the five countries with the largest net annual loss of those forests." Community-based conservation seemed a way both to satisfy decolonizing impulses and to halt environmental wreckage. Like state-sponsored conservation before it, community-based conservation soon became a global program.[40]

Mexico was well positioned for such a transition. From the beginning, colonial Spain had failed to replace lost Indigenous populations with immigrants from Europe and instead relied on Native labor; thus it had to agree to costumbres and local governance, and every attempt to modernize land use had tried to square the circle of national goals with local autonomy. Nor had Mexico possessed a deeply rooted forestry bureau, just an underpowered and understaffed one, rudely modeled on European and U.S. templates. Mexico's modernizers were aware of the shifts, which aligned nicely with the political economics of neoliberalism and privatization. The 1992 forest law tried to do for communal forests what the latest agrarian law had done for ejidos. The Salinas regime was more interested in markets than management, however, so another forest law was passed in 1997 to strengthen community governance and to shift some of the economic burden from natural forests to forest plantations. The Mexican solution—community-based forestry—granted local control, but with national oversight to ensure bad practices were avoided.[41]

The core purpose was local control. The goal was not primarily conservation, though that was a hoped-for collateral benefit. The purpose was to transfer decision-making power to local communities. Local boards would replace parastatal corporations. Put another way, communities would learn industrial forestry. It was assumed that more economic benefits would remain at the local level, and that communities' sensitivity to their habitat would also lead to more conscientious attention to the potential environmental consequences of a modernized economy.[42]

Yet even where community-based forestry was firmly embedded, studies seemed to show little difference in biodiversity or other measures of ecological health. What mattered was the character of the harvest, not which group made the decision to harvest. The problem was industrial forestry done at the scale a modern market required. A community-based strategy might improve social consequences and better distribute wealth, but it did not inherently improve environmental outcomes. It was the transition to an industrial model that wreaked havoc, not the particular hand that pulled the levers. The exception might be comunidades where traditional ways persisted and a community-based governance structure could ensure that those values informed economic choices and that traditional methods of conservation could endure. Still, community forestry was not necessarily community conservation. That left an institutional vacuum.[43]

Biosphere Reserves. In 1992, the year of NAFTA and the new agrarian and forest laws, inspired by the UN Rio Conference, Mexico floated the idea of a special fund by which to funnel money, both internally sourced and flowing from outside the country, to promote conservation. The Fondo Mexicano para la Conservación de la Naturaleza (FMCN) became reality in 1994. By 2020 protected sites amounted to 11.5 percent of Mexico's landed estate, overseen by the National Commission for Protected Natural Areas (CONANP), created in 2000. Some sites were historical and archaeological, but by area most were dedicated to nature protection. These included national parks (67), biosphere reserves (44), flora and fauna protection areas (40), nature sanctuaries (18), natural resource protection areas (9), and natural monuments (5).[44]

The most far-ranging arrangement was the biosphere reserve. The concept had emerged out of the Man and the Biosphere Program established in 1970 under UNESCO. The general design featured a core area devoted to research and high-level protection, surrounded by a buffer zone in which education, tourism, and research were typical activities, and a transition zone in which human influence was stronger, grading into the surrounding countryside. Governance, in principle, would involve local inhabitants. Only UNESCO, not countries acting unilaterally, could grant reserve status. By 2020 the program boasted 727 reserves in 131 countries.[45]

It was an institution that seemed preadapted to Mexico. Instead of banning people or relocating them to other sites, biosphere reserves reorganized land around existing peoples and their livelihoods. Instead of a binary all-or-nothing strategy—uninhabited wilderness versus third-nature landscapes—biosphere reserves proposed a degree of mestizaje, along gradients of settlement. What the ejido project was to agrarian reform, biosphere reserves were to nature protection. By 2020 Mexico had 44 (the United States had 48).[46]

POINTS OF IGNITION

The megadiversity of Mexico was meeting the inherent complexity of fire, a complexity squared when people were included, and a complexity cubed when official organs of governance attempted to control the interaction. Beginning in the 1990s, Mexico tried once again to square its peculiar fire triangle. The fire equivalent to the Green Revolution was years away, but the sparks to initiate it were finding appropriate kindling.

Old programs continued to evolve, sometimes helped and sometimes hindered by endless reorganizations. The servicio forestal was shuffled among larger agencies, downsized, and often, as its administrators lamented, overlooked; not since Beltrán's day had the servicio forestal enjoyed the funding it needed. SAG was succeeded by SARH, which yielded to SEMARNAP, which morphed into SEMARNAT. Still, the size of Mexico's fire establishment grew. The issue was whether, as with its economy, it could grow faster than the problems it confronted.

In 1990 forest development programs were rechartered, with the result that fire capacity was boosted at the state level, which had over the years become more significant than the federal. In 1992 agrarian and forest laws were amended. In 1993 SARH, which now housed the servicio forestal, saw an increase in funding and tallied 150 brigades, 133 lookout towers, 145 vehicles, 526 radios, and some 6,000 specialized hand tools. Adding in private programs from industrial forests, the national fire establishment could count on a further 380 brigades, 96 lookout towers, and 313 vehicles. State governments contributed still more. And in what was sure to delight officials, Mexico was renting helicopters from Canada and a DC-6 air tanker destined for Coahuila, the State of Mexico, and Quintana Roo. The next year it rented Bell 205s from Canada and a Bell 212 from the United States. Aircraft were visible and symbolically significant, but perhaps more helpful on the ground were modern firefighting hand tools like the Pulaski and the McLeod, invented early in the century in the United States and sent to replace traditional machetes and hoes.[47]

The administration of Ernesto Zedillo Ponce de León, begun in 1994, executed the ritual reorganization of the federal bureaucracy. SARH restructured into one massive Secretariat of the Environment, Natural Resources, and Fisheries (SEMARNAP), with Julia Carabias as secretary. Carabias had been educated in ecology, taught at UNAM, founded and directed an important NGO, the Programa para el Aprovechamiento Integral de Recursos Naturales (PAIR), and could, it was hoped, help clean up the environmental messes that were a legacy of the Miracle. The shake-up left forestry at a lower bureaucratic level, with fewer resources (a 20 percent reduction) and a fire staff often given split appointments with other duties, all with a more expansive mandate. It meant support for forest plantations, reregulation of industrial logging in natural forests, and promotion of community forestry, which sparked pilot programs with World Bank financing in Oaxaca, and later Guerrero, Michoacán, and Jalisco. All of this, of course, affected how fire was perceived and managed. Mexico's resident fire staff regarded the latest rearrangements as still insufficient for them to do their mandated job.[48]

The new director of forest protection, Oscar Cedeño, had long involvement with fire matters, including fire research with INIF; had participated with the NAFC fire group; and had served as a liaison with the Mexican military, mostly regarding aerial fire control. That theme continued, as Canadian air tankers were leased for fires in the State of Mexico and later transferred to big burns in the Yucatán, and in 1992 as British Columbia trained a corps of Mexicans in helirappelling. Both programs were expanded the next year, and rented helicopters were added from both Canadian and U.S. companies. Marana courses propagated into more Mexican states, from Coahuila to Chiapas. Thanks to the Yellowstone fires and their post-Gilbert counterpart, fires that in previous decades might have gone ignored were now attracting media interest.[49]

In 1994 Cedeño was called upon to lead a new initiative, the most ambitious since Beltrán's directorship, and importantly it had the support of President Zedillo. The National Program for Forest Fire Protection combined federal efforts with those of the 32 states to standardize programs, techniques, prevention activities, and reporting. The highlight was a National Center for Forest Fire Control housed in the offices of the fire protection division located in Viveros de Coyoacán. Its dispatch operations ran 24/7, 365 days a year. It became a resource not only for fire matters but for emergency response generally.[50]

Since Quevedo's day, fire protection had been regarded as vital for the protection of the forests Mexico had and those it needed to restore. That ambition had merged seamlessly with global campaigns for conservation and the creation of forestry bureaus. By the 1990s, both conservation and the agencies it spawned were in recession, part of an international rollback of colonial influences. Forestry was deemed to have too narrow a focus, forestry bureaus too anachronistic a structure and workforce, and forestry's fire programs too aberrant a policy. Forestry agencies created during the colonial era (or for countries like Mexico, during European-style industrialization) were downgraded, disestablished, or subjected to rechartering or at least renaming.

The shift had come earlier to the United States. Enthusiasm for environmental matters expanded policies beyond classic conservation to include wilderness and biodiversity. The Forest Service toppled from its public pedestal. In 1960 a famous study by Harold Kaufman had labeled the U.S. Forest Service a paragon of a public administration bureau. Fifty years later Francis Fukuyama and others considered the agency the epitome of dysfunctional democracy. Other agencies such as the National Park Service and the Fish and Wildlife Service claimed the mantle of environmental protection, and a rival agency, the Bureau of Land Management, swelled to nearly the size of the Forest Service. The commitment of public forestry to blanket fire suppression seemed at odds with progressive thinking. Between 1968 and 1978 the United States' federal agencies revised policy from trying to take all fires out to putting good fires back in. By 1994, as another collateral effect of the Rio Conference, the Forest Service adopted ecosystem management as a guiding principle. Not at first, but eventually, those changes in thinking and generations were reflected in bilateral relations to promote fire training and assistance between Mexico and the United States. What began as an exchange among members of a common forestry fraternity would diversify in often useful, occasionally unexpected ways.[51]

The tempo quickened and broadened. Marana courses were taught at the *escuela de ecoguardas*, fire contacts were established with Central America, Mexicans attended an Ibero-American course modeled roughly on Marana that was taught in Spain, and Mexican fire personnel traveled to British Columbia and California for training in helitack, later expanded to Arizona. The USFS and the servicio forestal arranged a review of their exchanges—what Mexico wanted, what the United States had available. The U.S. connection continued, partly with courses on fire control, helibase management, and the operation of remote automated weather stations, partly with personnel exchanges (Isidoro Solís of the USFS was a critical broker); even pilots in the Mexican air force were trained in fire tactics for helicopters. By 1997 a few of the helicopters and aircraft lost in the SEMARNAP shuffle were returned. Mexicans attended (and co-chaired) the second International Wildland Fire Conference, held in Vancouver in 1997, which also served as the venue for the NAFC Fire Management Working Group's annual meeting.[52]

All of Mexico's fire reforms focused on suppression. It was not alone. When the Soviet Union broke up in 1991 and fire exchanges were reestablished with the United States, what the Russians craved was better firefighting know-how; they had, they believed, fallen behind technologically and looked to the United States as the vanguard of firepower. So, too, with Europe's Mediterranean rim: when Spain, Portugal, and Greece were integrated into the European Union, their rural economies dissolved like sugar in rain, and feral fires raced through the countryside as it emptied of people and filled with fuels. Revealingly, the Marana course had 12 modules, just one of which addressed prescribed fire (it included slash burning under its capacious aegis), and none of which discussed policies aimed at restoring fire as a critical ecological process. The newer thinking about fire had to come from sources outside forestry and its legacy institutions. Still, the paradox held: Mexico was getting the latest means of fire suppression while the United States was redirecting fire management policy to other ends.

Research on Mexican fire remained paltry, and old attitudes evolved slowly. Industrial forestry hated fire, traditional use persisted, and nature reserves suffered if all fires were excluded. Gradually, reality and late-blooming science confirmed what people had known for tens of millennia: good fire can dampen or banish bad fire. All the actors needed to find some common ground. Fire sought its own version of the principle that conservationists since Beltrán had argued for natural resources: that they needed to be considered in a holistic way.

The point of convergence was the concept of integrated fire management (IFM). The term had two distinct understandings. For those interested in suppression, integrated fire management meant that a fire program had to include prevention, fuel mitigation and presuppression projects, and postburn rehabilitation, not just active firefighting. It meant that fire protection required a systemic approach, not just more helicopters and pumps. For those concerned about a broader, more pluralistic understanding of fire's management, IFM meant incorporating ecological, social, economic, and cultural considerations into fire programs. Both groups looked to prescribed fire as a useful concept, but again viewed it through distinct lenses. The fire-control group measured the value of prescribed fire in the amount of forest fuels it reduced, which made suppression easier, lessened overall costs and damages, and indirectly improved an ecosystem. The second group saw it as a broad-spectrum catalyst that recharged and renewed ecosystems (and traditional cultures), indirectly reducing fuels and easing the burden on suppression. The two perspectives came with distinctive cultures and were not easily reconciled.

Such reforms could shake up institutions and policies almost as radically as the 1998 blowups. Since the Conquest, Mexico had tolerated traditional fire use, and since the Porfiriato it had sought to replace that stigmatized activity with modern alternatives. But contemporary thinking was inverting those notions, identifying traditional fire use as a potential asset, not just a demonic driver of rural degradation, and it was seeking ways to promote landscape fires as well as traditional ecological knowledge to enhance conservation goals, notably biodiversity. What Mexico had long found backward and embarrassing was morphing, in an age of environmental concern and decolonization, into an object of interest in its own right and a paradoxical badge of advanced thinking. Those maligned fires might be necessary, and if lost, they would have to be replaced.

Colonial forestry had tended to overlook what it had been unable to abolish. Like Captain Louis Renault in *Casablanca*, who was "shocked, *shocked*" to find gambling going on in Rick's café, foresters tended to mouth the law and look the other way. (It was a typical ploy in colonized lands, common, for example, in India and Madagascar.) Even after legislation allowed burning under state guidance, the task of actual management overburdened most state foresters, who could barely cope with logging and wood gathering, much less with remote and often marginal landscapes that simmered with burns and were more valued for butterflies than for sawtimber. Now those fires might find a legal sanction, and in principle traditional restraints might be reasserted. A clear fire policy, however, could be difficult to reformulate and perhaps impossible to apply.[53]

REFORMING FORESTRY, READING THE TREES

Biosphere reserves, protected natural areas, community forests—classic forestry had little to say about managing such landscapes. Timber, reforestation, and fire control were its enduring ambitions. It was not prepared to consider fire, its blood enemy, as something not merely pervasive but valuable, even elemental in many biotas. Reform would have to come from beyond the bounds of forestry and forest bureaus.

Forestry's science focused on silviculture and the study of "forest influences" on watersheds, soils, and, curiously, climate stability. Ideally, forestry was a branch of agriculture; plantations were the model forest; diseases, pests, and fires were its ancient enemies. Natural woods needed to be used, and use demanded informed management, which required that science replace custom. Converting forests to, or intercalating them with, small-plot milpas and pastures was uneconomical and irrational. Delegating responsibility for mountain forests to traditional users ran counter to modernizing. Yet none of these ingrained attitudes and the methods devised to support them made sense for the new communal lands created by environmental reforms or for the trend toward enhancing natural ecological goods and services. Nature's economy needed a miracle, too. Such themes were better suited to ecologists, anthropologists, even political scientists.

Mexico was accustomed to having competing visions of fire, one based on European (later, North American) models and one rooted in traditional lore. Now it had to cope with the fact that the United States, the putative paragon of fire protection, was offering two versions as well, one founded on suppressing fire and the other on starting fire. As in Mexico, the U.S. fire controversies pitted formal knowledge against traditional practices. In the early twentieth century the antagonist had been Indigenous peoples and the newcomers who emulated their fire habits. The "light burning" controversy, as it became known, formally ended in the early 1920s. When a revival began in the 1960s, it drew on traditional burning by rural residents, mostly in the southeastern United States. The quarrel had less to do with decolonizing European expansion than with decolonizing the industrial conversion of a countryside occupied by European descendants. Still, the two visions tended to compete rather than complement each other. And prescribed fire could be hard to promote to urban publics no longer familiar with routine burning. Like reintroduced wolves and protected grizzlies, fires were welcome in the wild, but not in backyards.[54]

In Mexico the controversy revived both through contact with a reforming U.S. fire community and through encounters Mexicans had with their own lands. Exchanges, education at U.S. universities, and joint-venture field science in Mexico introduced a young generation of Mexicans to contemporary thinking about the need to live with fire. Equally, a handful of Mexicans had occasion, often by accident, to reconnect with their own heritage of landscape burning, which motivated them to missionize for modern fire management.

❋

Among the sources that contributed to a new perspective was a wave of joint ventures, Mexican and U.S., centered on reconstructing fire histories from fire-scarred trees. In a quirky way they were the science equivalent of maquiladoras, using Mexico to supply data that Americans wanted, while building up capacity in Mexico. What the Americans craved was what Aldo Leopold had sought 60 years earlier: evidence from landscapes that more closely approximated natural scenes, or at least more historic ones, than what existed north of the border. They wanted fire regimes characteristic of preindustrial times.

The world pioneer of pyrodendrochronology, the Laboratory of Tree-Ring Research, was housed at the University of Arizona in Tucson, and once it had surveyed the sky islands in the U.S. Southwest and assembled a master chronology, it was tempted to look south. Throughout the Southwest, a millennial rhythm of routine fires, accented by the cadences of ENSO, had abruptly broken around 1880. Widespread regional burning all but ceased. Were the reasons to be found in climate change or in anthropogenic activity? Since the U.S. firescapes were, by general consensus, a shambles, the fire community wanted to better understand just what standard they should try to restore landscapes to. Mexico's sky islands, subject to the same climate but a different human history, offered a seductive invitation.[55]

Beginning with a 1981 foray to the Sierra de los Ajos, but really quickening during the 1990s, graduates from the tree-ring lab, with Mexican collaborators, carried the technique to the Sierra Madre Occidental, to Baja California, to the Sierra Madre Oriental, to the Sierra Madre del Sur, and to select sites around the Valley of Mexico—some 3,000 trees from 100 sites, a panoramic cross section of Mexico's forested landscapes. By the 2010s Mexicans had assumed the task and were rewriting fire history in parks and preserves. A major presence in the early years was the Ecological Restoration Institute (ERI), established by Wallace Covington, at Northern Arizona University. Covington had developed a prototype protocol for restoring ponderosa pine that relied on mechanical thinning followed by burning, but restoration required a sense of what conditions the forest should be restored to. A good record existed for Flagstaff, Arizona, but there was pushback from some environmental groups that wanted better justification before intervening at a regional scale with chainsaws and torches. Mexico might have that data. Equally, Mexico needed baseline information to better manage fire in its nature preserves and to guide communal forestry. Pete Fulé oversaw the joint venture for the ERI south of the border.[56]

Especially in a Mexican context, restoration had to involve rural peoples, not just rural biotas, which meant cultural factors were as important as data excavated by natural science. Anthropologists and NGOs interested in communal forestry, Indigenous issues, and nature protection helped broker mutual projects. With its unusually long and widespread history of agrarian reform, Mexico attracted observers concerned with exporting comparable models to the developing world. Historically, outside observers had overlooked fire, or dismissed it as a quirk or as vandalism, but attempts to decolonize Indigenous experience and to celebrate traditional ecological knowledge, along with the gradual realization that the U.S. model of fire exclusion was at best an emergency intervention that over longer terms not only failed but aggravated environmental deterioration, created a new vantage point. Fire found standing as a theme that it had not had before. However much its shape-shifting might confuse reformers, fire was an enduring feature of the scene. Restoring Native management meant, in some way, admitting a perspective on fire different from that held by European elites. The fire-scar histories demonstrated unequivocally that agrarian reform through land redistribution had helped break the old order of fire.

The process took time. Americans considered the foundational state of nature to be a wilderness: U.S. ecologists sought to remove people in order to study the pristine processes involved before humans intervened and tainted the scene (the perspective also conveniently simplified the science). Mexicans regarded the foundation to be humanized landscapes, whether by the light hand of the Rarámuri or the onerous one that had transformed the Valley of Mexico from a lake into a polluted conurbation. As each group grew to better understand the other, Americans slowly accepted an ecological role for Indigenous people, and Mexicans gained an appreciation for the ecological role of fire even apart from people. As with their political systems and economies, each country retained sovereignty over their perspective on nature, but seepage between them moved their proposed solutions closer together.

FIRE'S CONTINUING EDUCATION

By the mid-1990s Mexico had both the need and the wherewithal to create a textbook specific to its circumstance. The first modern text on Mexican forestry, with a small section on protection (including fire), was Alfonso Gutiérrez Palacio's 1970 *Texto guía forestal,* with a separate volume by José Verduzco Gutiérrez on forest protection (*Protección forestal*) in 1976. Both reflect the impact of the early 1960s reforms. The next blossoming arrived in 1986 with a new forestry text by Javier Santillán Pérez, and a complementary volume on forest protection by Carlos Francisco Romahn de la Vega, Apolo R. Garcidueñas, and Ernesto Alvarado. At the time INIFAP employed 240 researchers, of which only 5 were dedicated even part-time to fire. In such a context, amid the environmental reforms of the Zedillo sexenio, the publication in 1996 of a 630-page survey, *Incendios forestales* (*Forest Fires*), arrived like a thunderbolt. It was the first Mexican (and Spanish-language) textbook dedicated to fire.[57]

Its author, Dante Arturo Rodríguez Trejo, was the son of a forester, a graduate of UACh, a veteran of the COCODER program, a participant in the second Marana course, and a recent master's recipient. In 1991, then a part-time professor at UACh, he had been approached by Alejandro Sánchez Vélez, director of UACh's Division of Forest Sciences, to write a "basic guide on preventing and fighting forest fires." Still at the beginning of his remarkable career, Rodríguez Trejo interpreted the assignment broadly.[58]

When it was published, the book's cover showed a surface fire burning under *Pinus patula,* a fire-adapted pine of commercial importance; the frontispiece featured Huehueteotl, the Old God of pre-Hispanic Mexico. The art was intended to show the "eternal relationship among fire, forest, and humans." The text opens with a typical survey of fire's extent and damages on Earth, but it segues into a more nuanced understanding by declaring that fire is a fundamental ecological fact, that "to claim to abolish fire is equivalent to claiming to abolish rain." The "ideal solution" is what some developed countries, notably the United States and Canada, were adopting. But in demonstrating the inextricable human role, *Incendios forestales* moved beyond the frontier of the other North Americans and toward a conception in which people could not be excluded.[59]

Yet there was also reason to pause. Northcote Parkinson once observed, semi-satirically, that great buildings tend to come after the primary era of achievement: they are, in a sense, monuments to a passing age, not the means to a future. Textbooks can have a similar role because they summarize the acquired state of knowledge. But if texts plow the known field of knowledge, even without breaking new ground, they also prepare for future planting. *Incendios forestales* did both. It orchestrated the existing knowledge of fire science, but by expanding its writ to include people, it also served as an introduction to Mexican fire history. It incorporated fire culture as well as fire science and technology.

Still, the jolt provided by the 1988 and post-Gilbert 1989 fires had passed. The cycle of fire institutions can resemble that of fires. Changes follow immediately after a fire, within a year, no more than two, then are lost as the preburn biota (and bureaucracy) reclaims the site. What begins in the ashes of one colossal burn is forgotten as institutional weeds regrow and more newsy topics crowd out competing claims for attention. Mexico's governmental makeover with each national election both abetted and blocked advancement. The año crítico was important, but without more conflagrations, it was easy to transfer political capital elsewhere. One point did not make a regression line, nor one event a narrative thread.

A decade later the fires returned.

VIDAS EN FUEGO

Sparks need tinder to kindle: there was little context for pursuing a career focused on fire. Mexico had no core program for fire research, and the United States had no desire to fund exchanges and joint projects on a permanent basis. What happened were scattered points of ignition, which, project by project, person by person, began to coalesce. The movement was, unsurprisingly, more individual than institutional. The story of Mexico's contemporary fire community hinges on the biographies of critical people.

THE *EJIDATARIO* WHO BECAME A *CIENTÍFICO*: ERNESTO ALVARADO CELESTINO

Ernesto Alvarado considered his career unremarkable, which was not true, and said it was not even his story but that of those who had accumulated knowledge over generations and made it possible for him to pass it along to the generations to come, which was at least partly true. His biography was a tangle of paradoxes: he was born in Boquilla de las Perlas, an ejido established in 1940 in Coahuila that offered formal schooling only through the sixth grade, yet he became a tenured professor at the University of Washington. It is, in some respects, a classic American immigrant story of rags to riches, a narrative framing that Ernesto discounts because he rejects the raw individualism behind that narrative in favor of a communal saga more appropriate to an ejidatario than a professor at a Research 1 university. He learned from Indigenous practitioners as well as from lab-based professors.[60]

His grandparents had migrated from Zacatecas and Jalisco to Torreón during the Mexican Revolution in search of security. His grandfather seasonally worked in the coal mines and the cotton fields at La Laguna, and as a bracero in Texas. His grandmother had trekked to La Laguna for two weeks by foot and burro. Wanting to control their destiny, desiring better conditions, they joined with others to create ranches, searching for the right locale. After a long struggle, which his grandfather led, the community acquired Boquilla de las Perlas as a legal ejido in 1940. The saga of their struggle—“una belleza extraordinaria”—became a source of inspiration and pride from Ernesto’s earliest years.[61]

His was a large family with eight brothers and two sisters. Ernesto spent much time with his grandmother, Tomasa, whom he regarded as a “second mother” and who taught at the village school. His father traveled north to work in cotton fields and construction. From him Ernesto learned to appreciate nature. Somehow the family came into possession of a collection of Jules Verne novels, which Ernesto devoured. The work in the fields was communal, as was charcoaling, both of which engaged Ernesto with fire in traditional forms.

He completed primary school at Boquilla and secondary school in Torreón, where he stayed with an aunt and uncle, then successfully took the admission exam to enter the Escuela Nacional de Agricultura at Chapingo in 1973, from which he graduated in 1980. The "ideology" of the Green Revolution informed the curriculum. The modernity promoted at Chapingo conflicted with the traditional knowledge he had absorbed as a child, but he met Efraím Hernández Xolocotzi, who defended some traditional practices and motivated Ernesto to find his own reconciliation. He graduated with a specialty in forest diseases and pests; an encounter with a U.S. scientist pushed him to pursue further study at the Colegio de Postgraduados, after which he went to the Instituto Tecnológico Forestal in El Salto, Durango. There he met prototype programs in communal forestry that returned him to the conflict between the formal knowledge he acquired at Chapingo and what he had learned at Boquilla de las Perlas. His mind was at the Instituto, his heart at the ejido.[62]

In 1983 he accepted a job at the Unidad de Manejo Forestal No. 12, a parastatal forest operation. Among his charges was fire protection. He found it easy to integrate the company brigades with the volunteers from the community, but not to integrate the company and the community in their approach to forests or fire. The company wanted maximum production; the community, maximum mixed use. The company and its technicians wanted fire removed; the community wanted it properly used. Ernesto brokered a compromise by appealing to prescribed fires, with which only a few experiments had been done in Mexico. The company had no experience in the practice, while the community had centuries. Guided by local knowledge, the experiment succeeded (Roberto Martínez visited and wrote a report). The experience caused Ernesto to "refocus" his studies. He completed a master's in forestry at the Colegio de Postgraduados in Montecillo on the subject of fire and determined to pursue a doctorate. In 1986 he published the first effort to apply the Rothermel fire-spread model to Mexico.[63]

Then, like his father and grandfather, he turned north, where the bulk of fire research and policy innovations lay. He accompanied the survey of Mexican fire conducted by González Cabán and Sandberg, who invited him to continue his studies at the University of Washington. It wasn't easy: he had an unhappy falling-out with the Colegio de Postgraduados; he struggled with modern equivalents to the trials his grandparents had endured. But he persevered and prevailed, and as Mexico spiraled into further economic crises, he secured a position at UW as a postdoc working on USAID projects in Brazil, Bolivia, and Mexico, and later projects funded by the USFS international forestry program. Eventually he became a tenured professor. Along the way he collaborated with some of his U.S. colleagues who were doing fire history, and with Mexican colleagues at the University of Guadalajara, UNAM, INIFAP, and the Universidad Juárez del Estado de Durango. But his particular enthusiasms, ironically grounded in his childhood experiences, had made him a fire cosmopolitan.

Though fire and smoke were his formal themes, his real strength was his familiarity with communal life and a sensitivity to traditional knowledge that made him a valued broker wherever the modern met the customary, because he saw the merit of each and the value of practical experience accrued by communities. He noted repeatedly that much of what modern science discovered had long been known, though in a different format. Even more unusual, he appreciated how Mexico's experience applied to the United States as well, and he lectured on Indigenous fire lore at the University of Washington. He became a founding member of the Research Subcommittee of the Intertribal Timber Council.

Over the years his connection with Mexico thinned, but he retained solid ties with those who were resolved to reintroduce fire, and he appeared at critical workshops and symposia. Then again, he didn't need to return for renewal, because what Mexico had to teach him was embedded securely within. He saw his life as continuous, with one grand narrative arc from an ejido in the Chihuahuan desert to global inquiries, just as knowledge was something passed through generations, not invented by Promethean individuals. Yet he was different. He promoted the worth of traditional practices at a time, certainly in Mexico, when the premise of higher education as handmaiden to the Green Revolution was to replace the archaic with the contemporary.

The border he crossed was not just political. The most important frontier was the one that segregated empirically based traditional knowledge from contemporary science. The story of many of his colleagues was to sojourn to the United States and carry back models of modernity and its perspectives to Mexico. Ernesto Alvarado's journey was to carry Mexico's lessons to the United States: that modernity had its limits and discontents and that the task was not to dismiss inherited lore but to translate it, even if that meant a mathematical model for fire behavior. "I have never done consulting work," he explained, "for the same reason and with the same taste I share with those who ask me without expecting a reward. Anyone can do what you do and what is the knowledge of many who have passed it to me unselfishly and so I pass it later, it is a community knowledge of several generations and many people who have surrounded me."[64]

In Ernesto's case those generations traced back centuries.

FIRE'S *INGENIERO*: JOSÉ GERMÁN FLORES GARNICA

José Germán Flores Garnica found his way into fire through the research needs of suppression. As Mexico's fire community thickened and spread, and as it learned more from the U.S. example, it appealed to research to lubricate technology transfer, to adapt modern techniques to Mexican circumstances. Germán spent nearly his whole career in INIFAP, the research arm of SARH (later, SEMARNAP). In this sense, Mexico recapitulated at an accelerated pace the evolution of U.S. fire research.

Born in the village of San Bartolo, Cuautlalpan, in the State of Mexico, from early childhood Germán wanted to be an agricultural engineer—the heroes of the Green Revolution. In 1977 he began preparatory work to enroll at UACh, where a serendipitous course on forestry redirected his imagination toward the woods. Seven years later he graduated in agricultural engineering, specializing in forests. He spent two months in Baja California on a project before joining INIF (which soon became INIFAP). He had an aptitude for math, and later computers; mostly he did inventories and mapping, none of which had anything to do with fire. Then a friend invited him to join a fire project at Tapalpa in Jalisco. It intrigued him, and when his friend resigned to join the private sector, the project fell to Germán. He began to read everything he could get his hands on; most of it was from the United States, but there were also studies from Canada and Australia, and among the handful from Mexico he found the work on fuels and prescribed fires done in the 1980s in Chihuahua by Jesús Sánchez, Héctor Alanís, and Guillermo Zerecero, before moving on to Celedonio Aguirre and Ernesto Alvarado, both graduates of Chapingo, and both of whom had gone to the United States for postgraduate work. But Sánchez had retired, and Aguirre and Alvarado had both stayed in the United States. It seemed Mexico might lose more fire scientists than it could produce.

Germán's turn came in 1990 when—as personnel exchanges were beginning with the United States—he was selected to spend a summer season at the Gila National Forest in New Mexico. The Gila had an active fire program, and by that time was an agency leader among western national forests in prescribed burning and the use of managed wildfires. It was a busy three months: Germán learned the value of integrating field with lab, appreciated that actual fires did not behave as mathematical models predicted, saw firsthand the latest technologies, and surprised his hosts with his facility with computers. It was revelatory: an in-the-field complement to a Marana course, all of it so different from what existed in Mexico. Germán was fascinated, and prospective research themes crowded into his mind, especially the need for better maps of forest fuels.

When he returned to Mexico, fire and fuel were a twinned informing theme—a closed loop in the sense that fires needed fuel, so mapping fuels was a way to improve fire control, yet controlled burns were a way to reduce the fuels that powered wildfires. Still, the fire protection establishment was not interested, nor were commercial companies that had to field brigades as part of their concession. It was hard to convince professional foresters, much less the general public. Equally, the people who really understood fire—those who had used it as an immemorial custom ("la gente de campo")—did not trust a scientist who came with no personal tradition of burning. The real problem was, Germán realized, that the campesinos knew more about fire than the researchers and that those in charge did not listen to those who knew the subject by living it.[65]

Meanwhile, he completed his fire survey of Jalisco in 1995 ("una labor titánica"). Here was revealed another indissoluble link: between agricultural burning and what were considered by the authorities to be wildfires. He realized that *quemas prescritas* could be considered a scientific upgrade of customary quemas. Moreover, Germán came to appreciate that burning involved ecological multitasking: fire did more than combust fuels. Fire behavior had to be complemented with fire ecology. All in all, the era turned his thinking about the forest fire problem in Mexico 180 degrees. He would have to convince the skeptics by overwhelming them with the science. But there was no place in Mexico to pursue his further education in the subject.[66]

So he created his own curriculum. He wrote to every possible university outside Mexico that touched on his theme of interest and could offer some support. He ended up studying ecology at the University of Wales, where he wrote his master's thesis on the behavior of prescribed fires. When he returned he lectured sporadically at the University of Guadalajara on fire, watersheds, and geographic information systems, and contributed fire expertise to studies at Bosque La Primavera, which served as the university's forest lab. He was invited to join the NAFC Fire Management Working Group, where he continued through 2018. The experience bonded him to the major leaders of fire management in Mexico, and also introduced him to the wider world in Canada and the United States. He learned about the fire program at Colorado State University (CSU), one of the few available anywhere. With support from the National Council on Science and Technology, he enrolled in the doctoral program.

With characteristic fervor Germán took every course he could—83 credits, double the 42 required by the program. It was overkill, but his fervor was characteristic of this first generation, all hungry for learning, all committed to matching agriculture's Green Revolution with a Red Revolution for forestry. His advisor, Philip Omi, specialized in fuels, fire behavior, and computer modeling, which was exactly what Germán wanted. He got further encouragement from Celedonio Aguirre, a Mexican who had emigrated north and worked for the USFS, but who had participated in the early prescribed-fire projects in Chihuahua. Aguirre suggested that Germán use that data for his thesis and connected him with the INIFAP people still there. Germán left CSU with his degree in 2001 and returned to Guadalajara.

The contours of his career were set: secure membership in the growing cadre of Mexican fire specialists and the NAFC Fire Management Working Group, participation in international fire conferences and travel throughout Latin America, and INIFAP-supported research, some of it long-term. By 2011 he completed what he had begun years before: a system for calculating forest fuels, maps of fire risk, and a system for predicting fire behavior. The National Council on Science and Technology sponsored further studies on forest regeneration; NASA, a complex project on fire emissions; CONAFOR, an inquiry into fire effects on soils; the National Fund for Protected Natural Areas and later the National Commission on Protected Natural Areas, research into fuels and fire regimes in select sites, with special attention to the Sierra de Manantlán. His models were propagated across Mexico and taken to

Honduras. He consulted on fire plans. In 2006 he and Dante Rodríguez Trejo edited the results of a fire symposium held in Guadalajara, which brought together nearly all of Mexico's fire establishment, into *Incendios forestales*, an updated survey of Mexican fire—what it was, how it was known, how it was managed. In 2011 he updated the project with *Impacto ambiental de incendios forestales* (*The Environmental Impacts of Forest Fires*).[67]

Like the others of his revolutionary cohort, Germán Flores remade something old into something contemporary. He belonged to the first generation that could pursue fire research as a dedicated career, not as something done incidentally or as a sidebar to primary projects. He knew everyone, participated in nearly every significant development, and contributed to the major themes in Mexican fire. Unlike Mexico's earliest fire researchers, he studied abroad but remained in Mexico. Like the others, he took an inherited tradition, forestry, and with indefatigable labor updated it with the data, maps, models, and other technical information its task demanded.

The Mexican forester was considered an *ingeniero* (engineer) who belonged to a tradition that traced to Europe, where foresters were a technical corps like civil engineers and mining engineers, playing the role of, as one British forester insisted, "a soldier of the state, and something more." Through his remarkable career Germán had achieved his ambition to serve as an ingeniero for Mexico's ecological infrastructure. But in transferring his goal from agriculture to forestry, and then from the woods to the fires that both threatened and enhanced them, he joined an avant-garde that created a Red Revolution to match Mexico's Green one.[68]

THE *ECOLOGISTA* WHO REDISCOVERED FIRE: ENRIQUE JARDEL PELÁEZ

Ernesto Alvarado and Germán Flores were a new generation of forestry graduates who committed to careers in fire. Enrique Jardel Peláez belonged to a generation of ecologists who finally, in spite of their education, saw the fires that landscapes had presented for aeons. Forestry had long engaged with fire, though as a threat. Biology had almost completely ignored it. In the 1980s and 1990s, especially as nature protection became a national objective, life scientists discovered that they, too, could neither understand nor manage ecosystems without incorporating fire. All the founders realized that Mexico's formal institutions had to accommodate its informal economy of fire, that they had to reduce the dissonance between what received science and administrative policies told them was necessary and true, on one hand, and, on the other, what they actually found on the ground. Enrique Jardel tracks that story.

He seemed destined for a career as the ecological equivalent of a forest engineer. Born in Cuernavaca in 1958, he studied natural resource management, with an emphasis on ecology, in the latter 1970s at the Universidad Autónoma Metropolitana in Xochimilco (UAM-X). The ecology that was taught was a doppelgänger of the forestry taught at that time: inherited texts that began with a pristine nature in balance, a biota able to be developed for human ends through applied science. Fire barely qualified as a footnote. This vision was wildly at odds with the Mexico outside the classroom, a landscape long and deeply shaped by humans in which fire and smoke were as common and regular as the rains. When Enrique graduated, he worked for a state agency identifying possible sites for nature conservation and protected watersheds. But when he criticized a proposed development (favored by the governor) that would harm a prospective reserve, he was fired. He found "refuge" as an assistant professor at UAM-X.[69]

He joined a research project in Campeche that examined primary production and decomposition, during which he realized that the work that organisms did in wet climates, fire might do in dry. He relocated to the Instituto Nacional de Investigaciones sobre Recursos Bióticos at Xalapa, where he pursued a master's in ecology designed by Arturo Gómez Pompa and Alfredo Barrera that integrated social and natural sciences. In Mexico it was possible to speak of landscapes without people, but only as physicists might appeal to an ideal frictionless surface. Real Mexican landscapes had people, and sooner or later, he realized, that meant they had fire.[70]

In 1984 he participated in a field course in a Zapotec community in the Sierra Norte of Oaxaca. Enrique's assignment was to document the history of forest management; others recorded the regeneration that followed a large surface burn. Both projects testified to the long history of fire in pine-oak forests. But Enrique was most drawn to the struggle between communal management and the intensive harvests by commercial forestry. For his thesis he translated the same skills to a study of the Cofre de Perote, a critical source of water for Xalapa and a national park since 1937. Specifically, he inquired into the effects of a 1950 ban on logging. He concluded that the ban only made the legitimate illegal and provoked land clearing for agriculture. Paradoxically, leaving the land in the hands of the locals did more for conservation than managing it by applied science and governmental decrees.[71]

In 1985 he had a chance to compare Mexico's experience with that of Costa Rica, where he interacted with Gerardo Budowski, who, like Gómez Pompa, insisted on the role of people, agriculture, and fire in the dynamics of the tropical ecosystem. What happened in Mexico was true for Costa Rica, and for all of Latin America, and it was true for the Cofre de Perote. For genuine conservation, the old ways—which, after all, had created the landscapes—were superior to the modern ones, which were destroying it.

That insight needed an institution to match it. The best option for Mexico was the biosphere reserve, whose design allowed space for old, new, and hybrids. In 1979 a previously unknown ancestral species of maize had been discovered in the Sierra de Manantlán, which led to investigations into the extraordinary diversity of the region, a pluralism both natural and cultural, much of whose ecological dynamics were inseparable from traditional agriculture. It remained for Enrique to cross the country and view the Sierra de Manantlán through the prism of the Cofre de Perote.

As part of a planned biosphere reserve, the University of Guadalajara, with help from the World Wildlife Fund, had established a research station in the Sierra de Manantlán, the Laboratorio Natural Las Joyas (later the Laboratorio Natural would become the Estación Científica, and then it would become the Instituto Manantlán de Ecología y Conservación de la Biodiversidad), in 1984. The station needed a director, and with his background, Enrique was urged to apply and was accepted. He assumed that with the data he would gather from the post he could pursue a doctorate; instead, practice replaced research. The place, the project, the planning—together they made a crossroads of nature, of history, of human land-sculpting, of knowledge, of conservation philosophy and management. What they did not do was allow time for a doctoral dissertation.

At its beginning the Las Joyas project prohibited almost all human activities save research and education. Fires were to be suppressed—there was no distinction made between traditional patch burning, fires caused by industrial logging, and fires caused by narcotraficantes. The consequences, almost all baleful, soon became apparent as fuels piled up, woods overwhelmed the inherited mosaic of shrubs, grasses, and trees, and the presence of *Zea diploperenni*, one of the justifications for the reserve, shriveled. When he entered the reserve, Enrique saw fire aplenty, but he also saw an exuberance of regrowth. He saw nature renewing itself without the salvage logging, replanting, and fire protection habitually employed by foresters.

Fire would not be denied. The 1988 fire season was, for Enrique, an epiphany that crystallized his years of impressions, like a precipitant dropped in a solution of dissolved chemicals. As director he was expected to lead the campaign against the flames. He quickly learned that the station was ill-equipped to cope, that locals knew far more about fire than he, that fire's management was not simply a technical skill like creating a fire line but a project of social engagement. The flames persisted for 45 days. It was, for global media, a "year the Earth burned" (the first of such pronouncements; other years would also claim the title), as Amazonia smoldered with land-clearing fires and Yellowstone National Park saw uncontrollable crown fires flash across its length. For Las Joyas other serious fire seasons followed in 1993 and 1996.[72]

Gradually, the Las Joyas staff recognized that there were many causes of fire, many kinds of fire and fire regimes, and many responses possible. Las Joyas had traditional agricultural fires, fires set to transform woods into fields (driven by misplaced government incentives), pastoral burns, fires for "cultivos ilegales" of marijuana and poppy, fires over land-use conflicts, fires as protest, and fires set to permit salvage logging of what burned. Such a diversity of causes could not be lumped together simply as "incendios forestales" and treated with a single policy of suppression any more than all the varied terrains of Las Joyas could be treated identically. A system of fuelbreaks and later fire blocks was devised, partly to assist fire control, partly to promote experiments in prescribed burning.[73]

For years the Reserva de la Biosfera Sierra de Manantlán (RBSM) had no director, and Enrique, as head of Las Joyas, had de facto responsibility for all the reserve from 1987 to 1993. The Zedillo sexenio rejuvenated and redirected the project—an "avance significativo," in Enrique's opinion, one that was broadcast throughout Mexico with the help of the Fondo Mexicano para la Conservación de la Naturaleza (FMCN). At Las Joyas a more nuanced basis for fire management commenced in 1995. The breakthrough, as everywhere in Mexico, came like a thunderclap with the 1998 fires.

The breadth of the burns shocked the country, and a serious investment by USAID helped spark a reformation. But it was one that emphasized firefighting, that sought to prevent another such catastrophe. The RBSM, however, needed a mix of fire practices; it needed to reinstate and redirect fires, not just extinguish them. Over the next decade it crafted management strategies to re-create the kinds of practices that had birthed the biodiversity for which the Sierra de Manantlán was famous. It had to counter the public perception, long inculcated, that fire was bad, to resist the pressure by the servicio forestal to continue a fire suppression program (now boosted by the performance enhancers injected after the 1998 blowup), to protest against "command and control" paradigms for nature protection.

As it matured, the RBSM became a model for other reserves, and an object lesson in the difficulty of sustaining a complex program of fire management. The U.S. example was troubled because it did not allow for a resident population that might enhance, not just trash, nature. When fire policy accepted the necessity of fire, it did so on the argument that fire was natural and that prescribed burning could serve as a surrogate for nature, helpfully reducing the fuels that hampered suppression. Anthropogenic fire was not considered a necessary part of how inhabitants might live on the land. U.S. preferences were to remove people and let nature have its way; eventually, the prescribed fire was challenged by the managed wildfire. It was a model that made sense in the context of the United States' vacant public lands and system of wilderness preserves. It made no sense in Mexico, where public land meant some variety of inhabited communal land. Mexico had to devise its own rationale for accommodating fire. The Sierra de Manantlán provided the critical proof of concept.

Enrique Jardel was the perfect outsider as insider. Alone among the founding fathers, he had trained in ecology, not forestry. He had access to sites for which fire restoration was critical, not just a nice idea or a handmaiden to fire control. He became the reliable, ever-present voice for fire pluralism. An active writer of model plans, an insistent proponent of moving fire research beyond the traditional topics related to suppression, the co-author of a CONANP manual for prescribed fire, he was a constant presence and reminder.

He never completed his planned doctorate. Instead, discoveries became experiments, trials became practices, and texts became landscapes. He wrote his dissertation in the earth—in soils, waters, pine and oak, grasses and shrubs, and flame. By the mid-2000s no survey of Mexican fire could ignore the alternative paradigm RBSM proposed. Las Joyas became as fundamental to Mexico's fire establishment as the central dispatch center at Zapopan.

THE UNSTILL VOICE AFTER THE FIRES

The catastrófica temporada of 1998 was the middle of three fire seasons that triangulate the fire history of contemporary Mexico. It overwhelmed efforts to contain it. There were too many fires, in too many places, with too much smoke to even map where exactly those fires were. No country could have coped. U.S. and Canadian advisors, and later those involved in the postmortem, freely confessed that the fires were beyond the capacity of the rest of North America. Environmental tumultos come and go, and Mexico had experienced what experts considered a chronic fire crisis for a century. But this time was different.

What changed was the response: in both amount and character, it leaped beyond what had been done in the past. The flush of postburn growth included all the institutions with some potential interest in fire. Universities and research institutes launched inquiries. The relevant authorities, including SEMARNAP, commissioned postfire reviews. All levels of government and NGOs planned rehabilitations and postfire restorations. Not least, the fires prompted a reconsideration of U.S. assistance. If the North American economies were to be bound by NAFTA, perhaps something comparable could hold for fires on their landscapes.[74]

The aftershocks rippled far beyond the servicio forestal. SEMARNAP outlined a massive forest restoration program aimed primarily at timber salvage, soil stabilization, and reforestation—all traditional responses, though they had never been mustered at this scale. With the FAO it cosponsored a Latin American forum, with addresses by Julia Carabias and Víctor Manuel Villalobos, dedicated to the "memory of the heroic combatants of the 1998 wildfires." With UNAM and the Instituto Politécnico Nacional, it sponsored a symposium on forest fires and agriculture that reviewed the underlying causes and prospective treatments. CONABIO sponsored reviews and symposia—rehabilitation requirements involved a change in kind as well as scale, not just replanting trees. The FMCN served as disburser of the USAID funds and oversaw a project for fire prevention and rehabilitation, which was the principal programmatic response to the crisis. Environmental integrity, not just traditional forestry, informed the response.[75]

New legislation sought to further regulate the use of fire in forests and on agricultural land, but skeptics thought the usual mechanisms for regulating costumbres, like passing new laws, would fail as they so often had in the past. Traditional farming was no longer economic (NAFTA made U.S. maize cheaper); farming techniques like quemas were not equal to those that underwrote the Green Revolution; nor were laws sufficient to overcome ancestral practices. "Those in the countryside," chided Eréndira Salgado, didn't read the Diario Oficial (and some couldn't read at all), nor did they rely on laws, but trusted in the implements "of their true heritage: some matches, a shovel, and a hoe, in addition to God's blessing."[76]

Modernity had come unequally to Mexico, in fire as in other matters: the developed lands burned fossil fuel, the lagging parts burned woods and fields. Mexican wildfire segregated the rich from the poor and the modern from the traditional. Interestingly, the United States showed a similar divide: its cities and commercial farmlands burned fossil fuel, while its public wildlands, from which fire had been excluded for decades, were awash with wildfires. Yet the impact of smoke (and more broadly, of greenhouse gases in the atmosphere) affected everyone. Metropolitan areas suffered through unhealthy air, and Earth endured a slow-cooking change in climate.

The U.S. response to the 1998 fires was significant but complicated. As smoke poured north, then swung east to Florida, which was enduring a historic fire year of its own, airports like those at Phoenix and Tucson were closed, public health concerns were voiced across the region, and politicians demanded the United States intervene. But the 1997 peso crisis and the U.S. response it provoked made the Mexican government reluctant to accept aid with conditions; Mexico would not ask for help. Instead, the United States "offered" emergency assistance without a formal request, and for the first time in living memory Mexico accepted. Nearly two decades of successful fire training and exchanges clearly greased the rails.[77]

Secretary of agriculture Daniel Glickman and USAID director Brian Atwood toured the fire region. "Certain national disasters are transnational now," Glickman declared at a press conference. "This is one of them." Atwood added that his agency was prepared to see the problem as an "opportunity." Under USAID's Office of Disaster Assistance, $7.4 million in equipment and personnel was channeled through the Forest Service, which also oversaw some Bureau of Land Management personnel and gear that joined the effort. At the June 1998 Binational Commission, Atwood signed a memorandum of understanding with Julia Carabias of SEMARNAP to implement a "Mexico National Fires Prevention and Restoration Program" to evaluate Mexico's fire preparedness and restore and conserve areas burned. The residual funds, which were significant, were routed through the FMCN for further training and other projects.[78]

The program's core was a North American Fire Management Team consisting of two representatives from the servicio forestal (Oscar Cedeño and Roberto Martínez, director of protection and subdirector for forest fires, respectively) and two from the United States, both from the San Bernardino National Forest (Paul Weeden, deputy forest supervisor, and Mike Conrad, deputy fire management officer). Serial reports were issued. Their essential conclusion was that Mexico had significant strengths and serious weaknesses, but "no country in the world has sufficient suppression resources to combat all the fires Mexico faced" during the May blowup. Never had Mexico, or perhaps any developed country, had to act on "so many large fires, burning simultaneously, over such a widespread geographical area, in ecosystems that had not been burned for decades or hundreds of years, in such remote and inaccessible terrain." Yet, in the aftermath of the disaster, the servicio forestal's funding for 1999 had actually decreased. Progress would depend on USAID funds.[79]

A comprehensive evaluation—a full-body MRI of Mexico's fire establish ment—followed, noting the immediate assistance granted during the emergency, the mid-term program aimed at building the capacity of government institutions, and the longer-term project to enhance all levels, not least local communities; the training program alone identified 35 needs that would "significantly upgrade Mexico's fire preparedness capabilities." That included offering Marana-style courses at national and state levels, and for agencies from SEMARNAP to SEDENA (the Defense Department) it meant implementing the incident command system, establishing a fire case system, improving helicopter (and helirappel) programs in southern Mexico, installing radio systems, and, after the development of a fire behavior course, offering training in fuels management and prescribed burning. Such courses were "the single most effective mechanism" in improving the overall program. Through the NAFC Fire Management Working Group, Canada contributed $100,000 in surplus funds from the 1997 International Wildland Fire Conference in Vancouver to the servicio forestal to further bolster its fire mission. In return, the U.S. Bureau of Land Management found Mexican assistance invaluable in helping train mostly Hispanic farmworker crews from the Snake River Valley that it used on temporary assignment.[80]

"In two short years," the report concluded, "Mexico has made quantum leaps in strengthening its overall organizational wildland fire organization." It had jumped to, or would soon have, rough parity with countries like Spain, Chile, Australia, Canada, and the United States. It was now a "vanguard" country, ready to disseminate modern fire management throughout Latin America and the Caribbean. (Interestingly, the United States, then in the midst of what the secretary of the interior Bruce Babbitt labeled "a national fire crisis" of its own, would authorize a National Fire Plan in 2000.) Mexico had a long history of well-intentioned failures in its drive to modernize, but its rapid buildup of fire suppression capability was a notable success. So, too, the United States and Mexico had a troubled relationship, too often compromised by mutual mistrust and uneven power, but their fire collaboration was an exception. In 1999, despite growing alarm over illicit border traffic, the two countries signed the Common Border Wildfire Protection Agreement, which allowed for mutual fire suppression within 10 miles of the border on either side.[81]

The fire plague of 1998 had come almost like a sudden pandemic, or a new sun rising across the land. It announced a pattern for the future, as fresh flames arose on ancient rhythms. In 1999 fires again followed storms in Quintana Roo (in the slipstream of other outbreaks in 1995 and 1996) that left not only woody slash but a highly combustible fireweed. Aerial fire control, in particular, reflected early benefits of the accelerated training, but funding continued to shrink relative to inflation and need. The new order of fire was distributing its flames across the countryside more rapidly than authorities outside the fire community appreciated or could respond to.[82]

Mexico's fire got less international attention than it might have expected because its record year got lost in an apparent pandemonium of planetary burning. Even U.S. interest would have been limited to the Forest Service had not the immense gyre of smoke forced airports to close and compelled public health officials to advise south Texans to stay indoors.[83]

The United States had its own record outbreak in Florida, in which every county had major fires and over 100,000 people were evacuated. The anomalous ENSO that powered Mexico's drought also propelled major fires in Indonesia, where land-clearing projects belched enough smoke to blanket Southeast Asia in what was euphemistically termed "haze," though it was dense enough to close Singapore's airport. Amazonia was again on fire, to world alarm. Cassandras of global warming used the fires to animate their warnings about climate change. Climate scientists first published the "hockey-stick" graph of global temperature increase. It was as though the planet was enduring a plague of combustion. The World Wildlife Fund dubbed the panorama "the Year the World Caught Fire."[84]

Mexico was but one star in a terrifying constellation: that was both the good news and the bad. It was, in a sense, joining the global league list of fire's developed countries, but as with its economy, that achievement came with serious downsides. Modernity was supposed to abolish relict primitivisms like open flame. It wouldn't in Mexico, but then, it was not expelling fires in any place that experienced natural conditions for fire. Even the United States, with a firefighting force and fleet of aircraft larger than the military of small nations, was unable to overcome its metastasizing wildfires. The public and politicians would call for more fire suppression, but the U.S. fire community confessed it could not do more with more of the same.

Humanity had called forth a new fire, and how people distributed it would decide the kind of future they faced. That humanity was transforming Earth at an unprecedented scale was obvious; in 2000 Paul Crutzen proposed the concept of the Anthropocene, a novel geologic epoch, to encompass it. It was clear that the burning of fossil fuels was giving afterburners to the epoch and unhinging climate, but no one was sufficiently clairvoyant to note that this fact was making climate history into a subnarrative of fire history. Humans had conjured not just a new fire but a new fire epoch. The Anthropocene might equally be called the Pyrocene. Mexico existed where those worlds crossed.

MEXICO'S RED REVOLUTION

The 1998 fires were a tipping point. These were not just burns in the hinterland that affected impoverished campesinos but flames that broke out everywhere across the country, spawned smoke palls on a subcontinental scale, and scorched even central Mexico, as shocking as an outbreak of cholera. The House of Deputies called upon Julia Carabias to explain why SEMARNAP had not extinguished the flames. President Zedillo decreed restoration of burned lands to be a national priority. Fire crews long accustomed to laboring in obscurity suddenly found themselves valorized as heroes in a battle to save the nation.[85]

It probably didn't hurt that Mexican politics experienced its own shock as the 2000 election, won by Vicente Fox Quesada and the Partido Acción Nacional, overturned 70 years of rule by the PRI and its predecessor parties. Centrifugal forces, always strong in Mexico, were pulling apart the bonds that had contained the Revolution. International finance and multinational corporations, narcotraficantes and local caciques, agrarian unrest and new political parties—all challenged the old order and could not be contained by simple suppression. That was as true for nature as for people.

Modern Mexico remained a fossil-fuel society. Oil provided the energy and petrochemicals behind nearly every dimension of modernity. Even fire protection relied on aircraft, engines, pumps, chainsaws, and electricity-powered radio and detection systems. The solution to more fire was, authorities argued, more firefighting capacity. With more roads, more crews, and more lookouts it might be possible to extinguish those fires that bolted or slinked from their patches of primitive farming. Had not the preponderance of the 1998 fires resulted from flames escaped from milpas and pastos? Had not the greatest number of deaths come from ill-equipped campesinos trying to control modern wildfires with rakes and hoes? Somehow Mexico needed to meld machete and engine, pump and *antorcha*. Over the next decade it attempted just that.

AFTER-ACTION REVIEW

A starburst of reviews and reforms followed the 1998 blowup. The critical agencies sponsored surveys, evaluations, and manuals—all drawn from Mexican sources, not imported experts from the United States or FAO. Like an emerging economy that shifts from exports to domestic consumption, Mexico's inchoate fire establishment had crossed a threshold, and while an international presence remained vital, Mexico had acquired sufficient mass in its domestic institutions to sustain its own momentum, conduct its own assessments, and write its own guidebooks. While its evolution to integrated fire management was full of pauses and hesitations, the reality was that no developed country had fully negotiated that transition.[86]

Quite apart from being overwhelmed by the 1998 blowups, Mexico had abundant reasons for reform. The old solution, to condemn burning, had failed because officials lacked the resources to eliminate fire (and always would), and because a suppression-only policy worsened conditions by leaving unburned fuels for the conflagrations yet to come. More radically, the suppression of traditional knowledge had left fire managers with plenty of tools but little counsel about how to use them to best advantage, much less how to understand fire as a relationship. No amount of scientifically informed research could substitute for empirical experience, long marinated in rooted societies, of living with fire on the land. Besides, fire abolitionism ran counter to the premises of the Mexican Revolution: it made little sense to grant rural populations the right to occupy land while denying them the means to live on it. They needed fire as much as they needed water, fertilizer, and seeds. Over and over, an official ban had only led authorities to attempt to suppress what couldn't be controlled, or to ignore as nuisances the burning that went on and the ecological dry rot that mocked the environmental justification for fire exclusion.

A compromise began with the shake-up that followed the 1998 season. In 1997 a comprehensive law had been passed (Norma Oficial Mexicana-015-SEMARNAP/SAGAR-1997, or NOM-015), published in 1999, to regulate all aspects of landscape fire. For the first time, controlled burning had legal standing. The process resembled the one demanded for forest use generally: it prescribed plans, forms, and official approvals; it retained the determination, however much decentralization was accepted, to maintain state control over the outcome. Still, the law established a foundation for official reform, and it sought to regulate burning that previously had been banned, criminalized, or simply ignored.[87]

NOM-015 was both legitimately "a historic law" and inevitably "a partial solution." It frustrated those who had no tolerance for fire or traditional burning, while encumbering the process in ways that discouraged ruralites from using official channels—this in addition to their customary wariness of elites and institutions and their long experience in evading mandates. But some condominium between private, communal, and public interests was inevitable. NOM-015 was both a symbol and a fulcrum—a remarkable achievement by those committed to nature protection. Between its passage and its publication, the 1998 fire season exploded to confirm its value. Subsequent efforts by a tiny but influential Nature Conservancy, a more confident servicio forestal, and a rising CONANP led to amendments in 2007 (published in 2009) that extended coverage to prescribed fire in forests.[88]

FONDO FIRE

In the transition that followed, the U.S. connection remained vital. During the emergency, USAID had contributed money, equipment, and personnel through its Office of Foreign Disaster Assistance. Some of the money remained ($1.2 million), and new monies became available for an initiative under the rubric of the Wildfire Prevention and Restoration Program, which was handled through USAID's regular Mexico mission. The agency sought direction regarding how to leverage that residual fund into an institutional future.

All parties agreed that "forest fires in many instances are merely symptoms of the larger issues of both fire and land-use in the Mexican landscape." Money, like hand tools and helicopters, had to work in ways adapted to Mexican society. USAID turned to a quick summary from its on-site team in May and June, which was absorbed by a larger overview provided by the North American Fire Management Team in 2000 and followed by a massive compilation of Mexican fire capabilities and accomplishments (dubbed the Big Black Book of Fire), personally presented to the government of Mexico by the chief of the U.S. Forest Service.[89]

What complicated the process was that the receiving institutions were themselves being reconstituted or invented. The 1992 reforms had redefined (again) the politics of land ownership and forestry. In 2001 the servicio forestal was relocated and reorganized into a National Commission on Forests (CONAFOR). The National Commission for Natural Protected Areas (CONANP) also reformed—the 1998 fires had shredded not only many of its plans but the assumptions on which those plans had been predicated. Not least, the U.S. funds needed a mechanism through which to disperse. In this case, SEMARNAP and USAID completed several agreements that specified a range of engagements and stipulated that the bulk of USAID funds would be routed through the Fondo Mexicano para la Conservación de la Naturaleza. And all of these efforts had to operate within yet another attempt to reconcile national and local control over forests.[90]

A second theme of the post-1998 upgrades involved community forestry and protected natural areas, which constituted another version of the rural countryside. On June 10, 1998, even as the blowups raged, SEMARNAP and USAID signed a memorandum of understanding for "cooperation on biodiversity [and] conservation, particularly in forest resources." The funded proposal that resulted was the Program for Forest Fire Protection and Rehabilitation of Burned Areas (PPIRA). Overseen by FMCN, the PPIRA sponsored 5 workshops and assisted 46 field projects. If SEMARNAP sought to upgrade firefighting capacity, FMCN sought to strengthen "institutional support and development," especially among NGOs. Some 29 organizations received funding—25 NGOs, 2 universities, and 2 research institutes (it seemed Mexico might achieve institutional megadiversity). Where possible, fire management could be folded into programs in community forestry. Given the havoc wrought by the fires in Chiapas, all parties targeted protected natural areas there for special attention.[91]

With its new funds, FMCN supported a constellation of projects. It helped expand CONABIO's interest in satellite surveillance as a way of estimating the damages to biodiversity caused by the fires into a general reconnaissance of active fires in real time. It served as a mechanism for channeling Canadian funds (monies left over from the 1997 Vancouver international fire conference) into adapting the Canadian fire-danger rating system for Mexican environments. It financed over 100 brigades in natural protected areas. It signed an agreement with the USFS and the University of Washington to create a photo series by which to evaluate fuel loading in core regions. It sponsored attendance by personnel from CONANP and nongovernmental agencies at courses offered in the SEMARNAP training matrix. It promoted regional-scale fire planning for such valued sites as the Sierra de Arteaga (Coahuila), El Ocote (Chiapas), and the Sierra de Manantlán (Jalisco)—this in cooperation with the Nature Conservancy, which extended its model of a "fire learning community" to Mexico. The USAID/Mexico review team regarded "this expansion of the fire management community" as something that filled an "institutional void" and would prove a "genuine and likely lasting achievement" of the FMCN's fire program.[92]

As with venture capital, not all investments would pay off, and the review team held firm to a suppression-themed version of "integrated fire management" that emphasized reducing catastrophic fire over encouraging ecological integrity. The assessment forcefully reminded everyone—as everyone had been repeatedly reminded since Quevedo's day—that Mexico's fire problem was symptomatic of larger issues of rural life that dated back centuries and had only been stalled by the Mexican and Green Revolutions. In his wide-sweeping chronicle of Mexican fire, Rodríguez Trejo noted that, while authorities had habitually blamed the costumbres of rooted ruralites for the backwardness of Mexico—in fire as in other matters—the long record of ill-treatment by elites gave those campesinos reason to suspect and ignore the new edicts, as they had others for centuries. Mexico's fire problem was not primarily a fire problem, and no amount of helicopters and imported Pulaskis could alter that reality. Responding to the fires that flamed episodically through the cracks where society and land failed to cohere was, as the USAID team reported, "an unpostponable element of the wider rural development agenda in Mexico."[93]

CONAFOR

The servicio forestal remained the keystone institution. It had long claimed fire protection as a mandate, it was the contact agency between the Mexican government and USAID, and NOM-015 extended in principle its capacity to manage fire on the countryside. It became the prime beneficiary of the post-1998 reforms.

Still, it was first among a throng of not-quite-equals, those fire-involved agencies that included the states and Federal District, the municipalities, the Mexican military, CONANP, civil defense, the weather service, timber concessionaires, the private sector, volunteer brigades, and institutions of civil society, along with nine federal bureaus that had some engagement with fire, from SEDENA to the secretariats for public education, biodiversity, rural development, social development, and prosecution for environmental infractions, each with its own perceived needs and capacity to contribute. If the aperture is enlarged to embrace research, then the fire community could also include universities, INIFAP, and the National Council for Science and Technology. Landscape fire was a sprawling empire, loosely held together mostly by crises. It didn't help that "forest fires" as a formal category did not include all the other varieties of landscape burning, not least agricultural fires. Ejidos and comunidades held 70 percent of Mexico's forests, while private owners held 26 percent and the public 4 percent. The servicio forestal had roughly a fifth of Mexico's crews and suppression resources. Its true power resided in its ability to orchestrate other agencies toward a common goal.[94]

In April 2001 the Fox administration created the National Forestry Commission (CONAFOR) and housed it within a refashioned SEMARNAP, now SEMARNAT (Secretaría de Medio Ambiente y Recursos Naturales). CONAFOR redistributed its responsibilities across 13 regions. An overlap in responsibilities between CONAFOR and SEMARNAT led to a full transfer of the National Forest Fires Program to CONAFOR in September 2002, along with legal responsibility for fire protection. The National Fire Control Center, along with the rest of CONAFOR, was moved to Zapopan (outside Guadalajara), which required staff to relocate from the Federal District. Many chose not to, and SEMARNAT retained other personnel and hoarded equipment.[95]

What had happened when SEMARNAP was initially created now repeated with the divorce of CONAFOR from SEMARNAT. The reorganization resulted in a 28 percent reduction in the fire budget, the loss of 60 percent of middle management, and a collapse in firefighting capacity, as many of those trained over the past two decades now found themselves outside CONAFOR or elected not to relocate to Guadalajara. The fire cache system was dismantled. The National Fire Control Center went from 24 officers to 8, and from 24/7 staffing to 8-to-5 hours five days a week. Even the toll-free number to report fires was canceled. Whatever its enhanced bureaucratic standing, the double downsizing weakened CONAFOR's ability to act.[96]

In March 2004 a team made up of U.S. fire specialists and Oscar Cedeño from CONAFOR submitted its "assessment and diagnostic review of fire training and management activities" funded by USAID in Mexico from 1998 to 2003. In essence, the report expanded the format of the North American Fire Committee's assessment immediately after the 1998 season and focused on the same topics: operational capabilities to improve fire protection with the "overall goal" of reducing "catastrophic effects of fire in the region." An assessment of previous history was not considered, since CONAFOR and FMCN were preparing their own assessments. Nor did the report consider the prevention and restoration programs that emerged from the USAID Environment Program and led to community-based projects under FMCN. Nor did it engage with issues of fire ecology and prescribed burning, which were addressed in an appendix submitted by Enrique Jardel.[97]

The results grouped under two themes. One continued the traditional training program with links from one nation's forest service to the other, what became known in shorthand as the "training matrix." Its emphasis was to bolster and upgrade Mexico's capacity to respond to wildfire—the classic goals for both the servicio forestal and the U.S. Forest Service, and a program well established and well regarded since the mid-1980s. Reviewers did not bother with an "evaluation of the past program" because "there is general agreement among the partners that the program has been effective."[98]

The second advanced more ecological goals and thus aligned with U.S. agencies' pivot toward "ecosystem management," as well as with Mexican interest, codified during the Zedillo regime, in more progressive thinking about conservation and biodiversity. Interestingly, both countries relied on NGOs—a realm Secretary Carabias knew well. In particular, the Nature Conservancy became a broker on both sides of the border, helping bridge with other communities and agendas beyond fire suppression.

Observers recognized that any meaningful movement had to blend two agencies, each with its own traditions, methods, and agendas. The shared code phrase was "integrated fire management." But, unsurprisingly, each side interpreted that expression differently. For foresters, IFM meant expanding activities to strengthen suppression. For environmentalists, it meant recognizing that controlled fires did ecological work beyond reducing fuels; that local communities, not just state agencies, managed fire as they did land and forests; and that traditional fire lore, not just formal science, had standing. It meant integrating fire with social, economic, and cultural conditions. Inevitably, the two groups competed in vision; and, not unexpectedly, funders and government agencies wanted suppression to have priority, reasoning that the Mexican public and officials demanded security from wildfire above all other agendas. Not everyone agreed—nine Mexican bureaus had some engagement with fire—but all shared a common goal to ensure that the programs complemented each other as far as possible on the ground.[99]

Whatever legal confirmation CONAFOR received through the reorganization, its fire program faced competition. States like Michoacán and Morelos were building their own forestry and fire capabilities. The FMCN was promoting pilot fire-management programs, many organized around more liberal understandings of what a fire agency should do. CONANP accelerated fire planning suited to its particular mission. NGOs added another institutional presence. What had been the stellar success of Mexico's fire efforts, higher-level training extended more broadly, was shaky, as courses appeared with instructors whose knowledge of fire was only what they had received in a previous course. The recommendations in the Big Black Book of Fire lay largely fallow. Historically, while Mexico's formal fire program had been a small agency that got handed around among larger bureaus and almost never had funding adequate to its task, it owned the subject. Now, others challenged what it did, how it did it, even how it imagined the problem.[100]

Yet there were good reasons, both political and environmental, to decentralize fire management. CONAFOR regrouped and refocused on what it did best—training, coordination, and response. The training matrix laid down a blueprint for furthering the legacy of past success, and there were enough critical fire folk who understood what needed to happen and how to do it that the National Forest Fires Program rebounded. The turmoil was worrisome enough, however, for outside observers to fret that, with the big fires blown out, Mexico was again downplaying its fire issues and letting the United States fill the gap. Still, the USFS considered Mexico among "the top six countries in the world in its capacity to respond to and suppress fire," and with the right strategic decisions, it could retain that standing.[101]

Beginning in 2004, mandated by the reforms that created it, CONAFOR arranged for a national evaluation of its fire activities. Program by program, capacity by capacity, state by state, a committee under the direction of a cadre from UACh tallied, analyzed, and graded CONAFOR's performance to determine how its efforts had translated into a reduction in burning, even comparing outcomes to those in North America, Latin America, Europe, and Australia. Only the United States had ever attempted anything so comprehensive, and its efforts had been diffused among agencies, themes, and years. The report identified topics and geographic areas to prioritize, debated the values of federalization, toted up strengths and weaknesses. It ranged from personal protective equipment, emergency response, and the role of women to ecological and economic impacts. The commentators were sympathetic but frank. As CONAFOR struggled to regain its footing, the running box scores furnished a quantitative measure of its improving abilities. The complex project continued through 2010.[102]

In the end, Mexico's needs were vast, it could ill afford to funnel all its energies into one program, and its U.S. cooperators were hardly the unassailable paragon they once had seemed to be. Earth was accelerating its spin into a mature Pyrocene—what had previously been adequate protection no longer was. The United States' fire agencies still struggled to implement at scale policies promulgated between 1968 and 1978, despite a major infusion of funds through the National Fire Plan passed in 2000. The same year that CONAFOR was assigned full responsibility for the National Forest Fires Program, historic fires—the largest on record—ripped through Arizona, New Mexico, and Colorado, and the term *megafire* was coined to express the country's metastasizing fire epoch.

CONANP

The problem was in some ways both easier and trickier for CONANP. The 1998 season had stunned the program as flames rambled over places thought immune to free-ranging fires, and damage assessment ranked high on postburn agendas. But the agency was less inclined to demonize fire itself. While nature reserves demanded protection against bad fire, they also required a healthy dose of good fire. Rather than denounce traditional burning, as national parks had done, or ignore it where nothing could be done to halt it, protected areas, notably biosphere reserves, sought to regularize and integrate the useful aspects of customary fires—at least in principle. The reserves could not banish rural populations, only arrange and guide them. The same proved true for rural fires.

Helpfully, appreciation for the role of Indigenous peoples in the megadiversity of Mexico was growing. Necessity became a virtue: biodiversity correlated with cultural diversity, so the political need to incorporate people could sync with ecological values. Community-based conservation could, in principle, complement community-based forestry. In 2008, Eckart Boege compiled a densely illustrated compendium that demonstrated how biodiversity, especially of agricultural species, aligned with cultural diversity. Yet while *The Biocultural Heritage of the Indigenous Peoples of Mexico* (*El patrimonio biocultural de los pueblos indígenas de México*) spoke eloquently of soils, rains, germplasm, and *paisajes manejados* in 342 pages of text, tables, and gorgeous maps, it failed to mention fire as a critical catalyst (although it did include photos of roza-tumba-quema in action).[103]

The omission is odd, not least because "cultural burning," as shorthand for a revival of Indigenous fire practices, was swelling into a global movement under the larger aegis of decolonization. In settler countries like Australia, Canada, and the United States, cultural burning aimed at restoring heritage, land, and treaty rights: it offered a fulcrum by which to acquire political standing. In Mexico comunidades already had the basics of political power. A restoration of traditional burning had to compete with other land uses like industrial forestry. The fact remained that, outside a small clan of academics, fire had little intellectual standing among anthropologists. Studies like Boege's, even when grounded in ethnobotany, might include photos of burned milpas, but they would not speak to fire's ecological role, or the vital place of humans in getting fire on the site. The emphasis was on people and endangered species, not on fire as an endangered process. Still, the potential for a coalition of interests was possible.[104]

CONANP had some advantages: a limited number of biosphere reserves, a physical and political geography that allowed some space for maneuvering, technocrats trained in ecology, a growing understanding of fire grounded not in binary categories (fire or no fire) but in the concept of fire *regimes*, and a resolve to ensure that the proper fires happened at the proper time—protected natural areas could be nimble and approach the fire "problem" sideways. Early on, CONANP staffers created exemplars like the Sierra de Manantlán biosphere reserve, and after the 1998 outbreak ravaged flagship sites like El Ocote and Los Chimalapas, they were animated by a sense of urgency. Not least, they had collegial critics who opposed traditional fire suppression as a sole policy.

Enrique Jardel was the best known. He submitted a lengthy appendix to the 2004 USAID/Mexico assessment that spelled out the legitimate place of fire, emphasized the fine-grained knowledge necessary to understand the role of fire in specific places, identified various kinds of fires, and addressed the sort of knowledge needed to manage those fires. He warned against taking fire expertise created in one setting and applying it to another. He accepted traditional fire know-how but also recognized that fire could be used poorly and contribute to ecological harm. Distinguishing among five kinds of fires, he skillfully parsed the good burns from the bad. And even as Mexico, with USAID assistance, was doubling down on fire suppression, and even as he admitted that Mexico was "no exception" to prevailing views that fire promoted environmental deterioration, a perspective Mexico had inherited from Europe and North America, he argued that agronomic, ecological, and ethnographic studies of rural and Indigenous fire practices offered an alternative understanding of fire as a "suitable and rationally fundamental" feature of land management.[105]

Still, better fire control was the foundation of the USAID/Mexico project, and the prime architects of the new order were unyielding about its primary importance. The United States had experienced several prescribed burns that escaped and sparked political firestorms. In 2000 one of these, kindled at Bandelier National Monument by the National Park Service, blew up and burned into the town of Los Alamos and the national laboratory where the country conducted most of its research on nuclear weapons (to the sardonic, it might seem that one blowup was assaulting another). Prescribed fire immediately became more expensive and bureaucratically cumbersome, and no one wanted anything comparable in Mexico.[106]

Rather, Mexico needed an upgraded suppression program "as an insurance policy and as a matter of national security" given the likelihood of a worsening fire future. Only with suppression secure could it proceed with "more science-based approaches." The official report accepted Jardel's vision but relegated it to an appendix, much as actual practice quarantined prescribed fire into natural protected sites.

Wildfires are an emergency that demanded a rapid response. Prescribed fires were maintenance that could be deferred. The recognition grew that landscape fire could not be restricted to the "forestry sector," that many of Mexico's preserves needed protection from fire's exclusion rather than from fire itself, and that "science-based" approaches could align with customary practice. The new concepts were becoming undeniable. They broke down the old fire dialectic between campesinos and elites, because now elites were looking to campesinos for inspiration and were arguing among themselves. Like St. Augustine's plea, Mexico seemed ready to move toward fire's management overall. But not just now.[107]

Between 2007 and 2012 CONANP wrote a national program for protected areas organized around six themes, one of which was fire. The pioneer was the Sierra de Manantlán Biosphere Reserve (RBSM). Its experiences, codified into principles and eventually a manual, were dispersed throughout the system. When a symposium or an anthology needed someone to speak about modern thinking regarding fire, most likely it was the Sierra de Manantlán Biosphere Reserve that provided that person. In 2010 those experiences were consolidated into a guidebook for planning fire management.[108]

The empirical reality of RBSM challenged the notion that all fire was bad, that all rural burning was harmful, that one perspective, however grounded in the natural sciences, was intrinsically correct. RBSM had many biomes, many social actors, and many kinds of fires, not all good, not all bad. The reserve was a *mezcla* of ownerships, of ecosystems, of purposes, and its fires reflected those circumstances. Some needed to be prevented, some suppressed, some kindled. To sustain what existed and restore what had degenerated required negotiations between campesinos, scientists, environmentalists, ejidos, timber companies, CONANP, municipalities, CONAFOR, the University of Guadalajara, the FMCN, pine-oak woods, alpine fir forests, and created pastures and milpas—fire manifested itself kaleidoscopically as these pieces tumbled into new patterns over time. Whatever scheme for fire management was adopted, it had to be adaptive. A successful plan required an integral strategy that considered all dimensions of fire and all of its human and institutional actors, which meant a plan was not just about policy but about politics as well.

The Sierra de Manantlán Biosphere Reserve demonstrated both the complexity of the problem and what a solution might look like. It was a model reserve for fire management, suitable for adaptation elsewhere. Thanks to the Las Joyas field station the reserve was hardwired to a major university, and in Enrique Jardel it had an eloquent and indefatigable advocate. It broadcast, as a mindset as much as a template, throughout much of Mexico's network of biosphere reserves. The ejido program addressed the problem of rural poverty and governance; the biosphere reserve program addressed the parallel problems of ecological impoverishment. Those reserves were to an age of sustainability what national parks and protected forests were to an era of conservation. The RBSM was its poster child.

THE GLOBAL FIRE INITIATIVE COMES TO MEXICO

Novel concepts and practices, however, do not come easily to legacy organizations. They are typically created elsewhere and field-tested by outside institutions before being absorbed. When CONAFOR's survey parceled Mexico's landed estate into fire-dependent, fire-influenced, and fire-sensitive biomes, and when it explored prescribed burns, it was appealing to a taxonomy and practice promoted by an NGO. That a small cohort was capable of countering the overweighted emphasis on fire suppression makes the saga of the Nature Conservancy's adventure in Mexico all the more astonishing.[109]

In the United States, a civil society had arisen to challenge the fire establishment in mostly healthy ways. The country's reformation in fire policy had been led by the Tall Timbers Research Station, a privately endowed facility outside Tallahassee, Florida. The fire ecology conferences it hosted beginning in 1962 were watershed events; that same year the Nature Conservancy (TNC) burned a prairie preserve in Minnesota. As TNC acquired lands, its fire program expanded—fire was too basic to too many biotas to be ignored—and went on steroids when the Conservancy moved into Florida, probably the most fire-friendly state in the nation. Before long TNC was burning as much per year as the National Park Service. It developed a role as a neutral broker, trained practitioners, organized demonstration sites, showed how to leverage cooperation into large outcomes, and generally served as a missionary arm for propagating the gospel of good fire. Amid the scramble that followed the 1998 fires, TNC found its way to Mexico.

The critical figure in both countries was Ron Myers, who transformed his personal conversion story into a continental one. A native Californian, he studied forestry, worked for the Forest Service and the National Park Service as a seasonal firefighter, and in 1969 served as a Missoula smoke jumper. In 1970 he enrolled in the Peace Corps and for two years taught at the National Forestry School at Siguatepeque, Honduras. But he reckoned he learned more from campesinos than he taught to students. He witnessed traditional burning in pine forests firsthand and was fascinated by what he saw. When his tour ended, he worked at Everglades National Park, then a cornerstone of fire restoration efforts. But he yearned to return to Latin America; he enrolled in the University of Florida, studied botany and ecology, did a dissertation on Tortuguero National Park in Costa Rica, and pursued a postdoc at UF, where he co-authored a book titled *Ecosystems of Florida* and wrote a fire management plan for the Archbold Biological Station. In 1986 he signed on as TNC's first dedicated fire staffer. Two years later he became its national director of fire management and research, responsible for fire plans and operations throughout TNC's far-flung archipelago of preserves. He worked out of Tall Timbers Research Station.[110]

TNC had accrued assorted interests in Mexico when a USFS colleague invited Myers to join the assessment that followed the 1998 fires. It was all about networking—using a small but dedicated staff to amplify effects through training and education. The Fire Learning Network that emerged in the United States following the 2000 National Fire Plan was adapted in still-leaner forms for Mexico; the workshops did for integrated fire management—the version of IFM based on social as well as ecological conditions—what the Marana courses did for suppression. In 2000 Myers hosted a workshop in Hermosillo, followed the next year by one in Chiapas.

In 2002 TNC joined with the World Wide Fund for Nature and the World Conservation Union to establish a Global Fire Partnership pledged to address "the causes and ecological and social consequences of altered fire regimes across the world." A workshop in 2004 (four of the founders of Mexico's modern era participated) performed a quick wellness exam of Earth's fire health, a project that segued into TNC's Global Fire Initiative. Granted TNC's tangible interests in preserves and Myers's enduring attachment to Mesoamerica, Mexico ranked high as a test case. (Myers also arranged for Mexican sessions at the 2003 and 2006 fire congresses hosted by the Association for Fire Ecology.) The program published two shockingly readable summaries of its findings and operating concepts.[111]

It made a difference. Myers proved charismatic. TNC's terms were adopted, its ability to fulcrum small numbers into big effects was emulated, its vision of how fire might be managed in ways that served both nature and people was propagated. The TNC program offered a practical counterweight to USAID's immersive baptism in suppression. It signed a cooperative agreement with CONANP to address fire-related issues. It figured by name in the USAID assessment, which even worried that its message, so compelling, might subvert fire prevention and control projects if misconstrued by the wrong actors. By rejecting industrial fire suppression as the sole solution, it did for Mexico's Red Revolution what critics of the Green Revolution did. It helped ensure that Mexico's national fire plan would include fire lighting as well as fire fighting.

TNC proved nimble and indefatigable. It sponsored prescribed fire workshops in Sonora. It midwifed four community-based fire-management projects at La Sepultura and El Ocote Biosphere Reserves. Its lessons influenced an eventual revision of NOM-015 that clarified the status of controlled burning. Many of the major figures in Mexican fire had contact with the program and incorporated its notions—TNC's grounded relationship to fire made sense in Mexico, given the country's ancient traditions of landscape burning. With Mexico as an anchor point, the project spread elsewhere in Mesoamerica and the Caribbean, and even carried the message back to the United States.[112]

Then the Great Recession struck. The private donors who had funded the program withdrew, while a changed TNC administration decided to target other themes. The Global Fire Initiative shut down. Personnel, including Ron Myers, were laid off. The program had lasted barely eight years, but it had as much impact on Mexico as any enterprise save the Marana courses, and more than any other lasting so briefly. No less surprisingly, its influence persisted. Dante Rodríguez gave it a full page in his compendium of Mexican fire, credited it with powerfully boosting the transition from fire control to fire management, and spoke admiringly of the "mystique" it had among its members.[113]

Yet it was a profoundly pragmatic program. For ideals to have meaning, they have to be transported into the field. Workshops led to plans, plans to burns. The ultimate expression of TNC's venture arrived when one of its Mexican staffers, Alfredo Nolasco Morales, who had been hired in 2004, succeeded Roberto Martínez as director of CONAFOR's National Fire Protection Program. There he oversaw a revision of NOM-015 to include forest (not just agricultural) burning, guided CONAFOR through the *grandes incendios* of 2011, and navigated the agency into a future of more pluralistic fire management.

RESEARCHING THE REVOLUTION

The Red Revolution swept over research as it did every other aspect of Mexican fire—it could hardly be otherwise. Old institutions like universities evolved, new institutions like the National Institute for Ecology appeared. Fire went from a subject as intellectually disreputable as the quemas set by campesinos to a field in which Mexico could claim international standing. Foreign scientists, primarily from the United States, found in Mexico a rich source of data and nurtured joint projects that could continue locally. Given the usual lead times for education, all this was an astonishing turnabout.[114]

Since the 1930s Mexico's research output around fire tracks three waves, each almost an order of magnitude greater than the previous. There is a tiny wave that accompanies the reforms of the 1960s. A second wave starts boldly in the early 1970s, crests in the early 1980s, then ebbs around 1990. In 1987 INIFAP restructured in ways that granted fire research a bit more autonomy (Jesús Sánchez marveled that only in the 1980s did forest fire, which had been around for thousands of years, get serious investigation). A third wave stutter-starts in the early 1990s, crests around 2005, then trails off around 2017. Looked at by decades, publications grew arithmetically during the 1960s through the 1980s, then exponentially during the 1990s and 2000s. By 1996 the subject had progressed far enough for Dante Rodríguez Trejo, then a part-time professor at UACh, to author the first Mexican (and Spanish-language) text exclusively devoted to the subject, *Forest Fires* (*Incendios forestales*). As its title indicated, this was a book that grew out of forestry, though it ranged beyond it and avidly discussed fire's human history.[115]

The outpouring in 1998 stands by itself, a publishing echo of the fires. After 2008 the numbers trail off, with the lowest output still roughly doubling that of the 1980s. What had been an afterthought or sidebar for scientific research, tucked away into niches isolated from major scientific questions, became a staple. Before 1998 UACh had 2 thesis students and no master's or PhD candidates studying fire. Between 2000 and 2021, it had 22 theses, 22 master's degrees, and 2 doctoral dissertations. These were in-country students, not those who still traveled to the United States or Canada for advanced degrees.[116]

The output had come in a gush, not exactly a blowout equivalent to Dos Bocas, but one that needed some taming. In 2003 the Center for Atmospheric Sciences at UNAM hosted a symposium that explored risks from fire disasters, with special interest in climate. Of the 23 participants, the usual doyens of fire research were present, but most of the attendees represented disciplines not traditionally involved with fire, and, significantly, nearly a third were women. Ten years after Dante Rodríguez had published the sprawling *Incendios forestales*, he and Germán Flores edited an updated survey that encompassed fire ecology, fire in nature preserves, and the social aspects of fire, together with examples of places that synthesized those themes into relatively coherent programs (nothing dealing with fire is ever wholly coherent). While the book's title remained unchanged, its focus had shifted in ways that could seem unintelligible to those trained on Verduzco's 1977 *Forest Protection* text.[117]

In 2007 Germán Flores and D. A. González presented an assessment to Cuba's Forest Congress. Noting the obvious, that research had blossomed, they confirmed that most projects had been initiated by individuals, a small cadre who knew one another, but without any central focus or systematic coordination, which could lead to duplication, missed opportunities, and a misallocation of what was still a preciously tiny corps of fire scientists. The primary themes were the traditional ones—fire behavior, fuel maps, fire danger rating—but newer projects looked to fire ecology, remote sensing, integrated fire management, air pollution, and climate change. New and old were not, however, well coordinated: too many fire behavior projects ignored fire ecology, and fire ecology projects too often overlooked fire behavior. Mexico needed better synchronization, especially if it had to fund research out of its own pocket. Flores and González argued for a national center dedicated to forest fire research.[118]

That was beyond the pale, but a year later the people and agencies involved with fire research—those who did it and those who needed it done—met in a grand convocation at UNAM's Morelia campus to review the state of the field. The first such assessment, in 1986, had featured 2 visiting U.S. researchers, 1 from the University of Washington and 1 from the USFS, under the auspices of the NAFC Fire Management Working Group. This time there were 3 Americans—1 from UW, 1 from the USFS, and a TNC staffer—and 47 Mexicans. The early movers of the research underscoring the Red Revolution were all present—Ernesto Alvarado, Dante Rodríguez, Germán Flores, and Diego Pérez Salicrup, along with Enrique Jardel. Of the 24 different institutions represented, 20 were Mexican, 3 were from the United States, and 1 was from Spain. Among the Mexican universities were UACh, the Universidad de Guadalajara, the Universidad Autónoma de Nuevo León, and various campuses of UNAM. Among the institutions present were CONAFOR, INIFAP, CONANP, CONABIO, and the National Center for Disaster Prevention. The Fondo Mexicano para la Conservación de la Naturaleza furnished funding, with help from the Center for Geographical Research and the Manantlán Institute of Ecology and Biodiversity Conservation.[119]

The participants wanted a guide to doing research that was better suited to Mexico's megadiverse environment and distinctive political history, and they wanted to help see that it got applied smartly. They appreciated the troubled history of taking notions devised in one part of the world, believing them transcendent, whether as Scripture or Science, and imposing them somewhere else. The immediate result was a volume coordinated by Germán Flores, *Impacto ambiental de incendios forestales* (*The Environmental Impact of Forest Fires*), which concisely summarized the state of knowledge about fire's influence on plants, animals, soil, water, air, human health, and the well-being of the countryside. A few years later a group from the workshop wrote a CONAFOR-sponsored field guide to fire management in forest ecosystems. More followed. Almost 20 years after he wrote *Incendios forestales*, Rodríguez Trejo published an encyclopedic two-volume summary of fire in Mexico, *Incendios de vegetación*. Complementing the text, Enrique Jardel, Diego Pérez, Ernesto Alvarado, and Jorge Morfín authored a field guide for managing fire in forest environments. At the time only Australia, Canada, Russia, and the United States could boast comparable syntheses.[120]

FIRE'S *GERENTE*: ROBERTO MARTÍNEZ DOMÍNGUEZ

Roberto Martínez Domínguez knew fire through politics—the politics that shaped bureaus and the politics that governed within them. His fire career began in 1983 during the transition from the López Portillo sexenio to that of Miguel de la Madrid, and he departed in 2000 amid what he regarded as another political makeover. In between, he lived within bureaucracies; unlike researchers, he had a relationship to fire that was not just intellectual but institutional. He had to take ideas and make them happen on the ground. He was present during virtually every development in Mexico's transition to being a world-class firepower, from the initial Marana courses to the post-1998 response.

Graduating from UACh in 1979 as a forest engineer, he had no special interest or experience in fire, and he was hired as a project analyst within SARH's directorate of forestry development, where he worked to design programs for several Mexican states. Three years later he was offered a position as head of department of the General Directorate of Forest Protection—a good promotion, but one hampered by a bureau that had a reputation for "a lot of corruption" and was politically hostile to the parent directorate. He asked instead to move into administration and became head of programming and budgeting.[121]

He escaped the ritual resignations that accompanied the onset of the de la Madrid administration, but with the stipulation that he leave his current position and take one in a technical area. That would mean displacing an existing engineer, which Roberto did not wish to do, not least for "fear of professional reprisals" later. Always there were political considerations. Yet there was one program that seemed devoid of any serious technical expertise. In December 1982 he became head of the Department of Protected Areas, with a specialty in fire control. His first assignment was to prepare a formal "diagnosis" of the forest fire protection program. His report was the first overall survey for the country. Its timing couldn't have been more opportune. The coming year witnessed the first air tanker in Mexico, which flew at the Parque Chipinque, and the inaugural Marana course—"a true technical impulse for professionalization." Roberto attended, along with most of the generation that would lead Mexico's fire community in the coming decades.[122]

That year shaped the contours of his career. He oversaw the elaboration of the Marana-style training courses throughout Mexico, and through Mexico to Central America. He midwifed Mexico's serious entry into aerial fire protection, through rentals from North America, through the training of military pilots, and through agency aircraft, especially helicopters. He facilitated exchanges that sent Mexican firefighters to U.S. hotshot crews and Canadian rappelling teams; he introduced U.S. firefighting tools, remote automated weather stations, Nomex clothing, and the incident command system; he coordinated with the various Mexican agencies that had some responsibility for fire. He participated in the NAFC Fire Management Working Group, becoming familiar with U.S. and Canadian fire programs. He accompanied the U.S. observers during their 1986 assessment of fire research. As director of fire for the servicio forestal during the año crítico, he was active in the post-1989 review. And, critically, he was manager during the 1998 season and the massive assessments that followed. By the time CONAFOR stabilized, Mexicans were able to do the job themselves, with evaluations contracted out to a group headed by UACh. Throughout, Roberto Martínez continually helped recalibrate the strategy of Mexican fire, then align technology with those goals.

Along with Gilberto Galeote Rivera and later Arturo Raygoza Martínez, both Marana graduates, he pushed steadily to move fire from the realm of *peones del campo* to that of people who had something like the professional standing of forest engineers. Roberto Martínez sought, in brief, to match goals with capabilities, always keeping the program progressing through reorganizations, relocations, and reassignments. He steadied it through the upheavals of the SEMARNAP (1994–95) and CONAFOR (2001–2) reforms, and then through the fires of 1989 and the firestorms of 1998. From 1988 to 2003, with political astuteness, he served as deputy director of forest fire protection, a 15-year tenure that he regarded as a "Guinness-like" record for Mexico. He was high enough in the organization to be influential, but not high enough to be a political target.[123]

He recognized that the organization also needed to see itself differently. With better training, with modern equipment, with disastrous seasons that caught politicians' attention came confidence, and with competence came pride. Roberto cited, for example, the "famous" Brotherhood of Prometheus, which beyond its "romantic" title was a team that "generated a spirit of service" in field and office, through both daily drudgery and emergency firefighting, and "a camaraderie, solidarity, and fraternal support that lasted for at least the last 30 years among a great majority of them." The founding generation was passing the torch to a second generation and then a third. The Mexican fire community became something more than an administrative afterthought or a begrudged duty. It saw itself as a special corps that could stand shoulder to shoulder with peers in the United States, Canada, Spain, Australia, and Russia.[124]

Most of his career promoted better fire suppression—that was the historic charge of the servicio forestal. This met the goals of the agency, but with the reforms of the Zedillo regime, there was a need to expand the domain of fire tasks to include fire's management. The new fire legislation regulated fire use, which meant that officials had to understand and manage fire in more sophisticated ways. As firefighting had advanced from farm implements to specialty hand tools, so fire lighting meant more than backfires, and it, too, demanded technical training. So prescribed fire was added to what was taught, mostly through courses and workshops sponsored by the Nature Conservancy. An amended NOM-015 granted controlled burning greater legal standing, similar to what U.S. agencies had enjoyed for 30 to 40 years. Roberto assisted this new transition, as planner, coordinator, and facilitator, as he had the old.

It was a long career. In recalling its highlights, he picked out for special praise those administrators who displayed integrity and a commitment to service and radiated a vision of community improvement: Don Manuel Verduzco Tapia in the Agriculture Secretariat, León Jorge Castaños Martínez in SARH, Gilberto Galeote Rivera and Oscar Cedeño Sánchez in SEMARNAT and CONAFOR. So it was with sadness more than irony that his own career in federal fire agencies ended with an administrative sanction in February 2009. Roberto dismissed it as politics—another regime wanted to strip away holdovers from its predecessor, and so it invented bogus charges to hide its true intent. He filed a formal protest, which, after five years of legal wrangling, ended in his favor. The charges were to be removed, he was to be reinstated, and his back wages were to be restored. Immediately after being reinstated, however, he was fired again (an attempt to void the decision), so he challenged his new "unjustified dismissal," this time with a legal lawsuit, and again won. CONAFOR appealed, and 10 years after his dismissal, after he had spent 28 years in public forestry, the case remained unsettled. Throughout the ordeal he continued most of his old jobs, but now as a consultant.[125]

Fires can be unpredictable—can change with a gust of wind, a cache of slash, a spot ignition outside the perimeter. The same holds for fire bureaucracies. The fire bureaucrat, too, faces hazards, and has to recognize the perils as well as the opportunities ever latent in a political landscape subject to flare-ups. For decades Roberto Martínez Domínguez had deftly avoided the threats, survived two quasi-purges, and advanced an agenda that flourished during an extraordinary time. His career spanned the life history of Mexico's Red Revolution, from a handful of ill-trained and haphazardly equipped crews in the Valley of Mexico to a national corps, from fire science and policy imported from elsewhere to a fire economy prepared to export its pertinent experiences, from an unworkable doctrine of simple suppression to mixed programs of fire management adapted to the particularities of landscapes and peoples.

He had lived through it all, and much of it he had helped make happen.

THE RED REVOLUTION'S HEROIC AGE

Between 2004 and 2008 every entity concerned with fire had conducted fundamental assessments of its mission and capabilities and published guidelines for management. More to the point, each agency, each initiative, each participant had interacted with the others. Suddenly, Mexico had not only a critical mass of researchers, institutions, and ideas but the synergy of their interplay. The Mexican fire community found itself in a historic sweet spot: a moment of political change and urgency had arrived; it had enough trained people to do the necessary work, but not so many that the major players could not all know one another personally and assemble in a single workshop; and it had connections to U.S. money, universities, and training exchanges, as well as a presence at international conferences. A radical revision of understanding was emerging, even at a global level, that argued for integrated fire management and an accommodation of the customary fire practices that, like its fabled biodiversity, put Mexico at the frontier of modern fire thinking.

How all these pieces shuffled into a new order was the story of the decade that followed the catastrófica temporada. In 1980 the servicio forestal was a drag on reform. By 2000, it had rejuvenated under the shock treatments of big fires, USAID-catalyzed upgrades, formal fire research, and the missionary effort of the Nature Conservancy. By 2010, after administrative reforms and the growth of a Mexican fire-research community, what might aptly be called a Red Revolution, it was a leader.

A group from UACh summarized the excitement for the international community: "By 2000, a new generation, comprised of chairmen of a few governmental agencies and non-governmental organizations and researchers, started to launch the basis and initial steps for an evolution in fire management of the region." Integrated fire management was a watchword; prescribed fire, a legitimate tool; traditional knowledge and governance, a wary companion to science and the state. In 2008 Rodríguez Trejo wrote a survey of Mexico's fire scene for the international readership of *Ambio*, emphasizing fire's relationship to the country's megabiodiversity. Some 50 percent of Mexico's ecosystems were fire-maintained, 40 percent were fire-influenced, and 10 percent were fire-sensitive, and all were affected by burning both interminable and occasionally explosive. The 1998 season demonstrated dramatically how biomes believed previously to be fire-impervious were now capable of burning. In 2011, along with several co-authors, Rodríguez Trejo returned with a state-of-the-field survey for *Fire Ecology*, the journal of the Association for Fire Ecology, reflecting on the challenge of reconciling traditional fire suppression with ecological fire use. Together the articles celebrated Mexico's coming-out party to the global fire community. Mexico had achieved standing with its aspirational peers.[126]

Which is part of the reason why the 2011 fires came as such a shock.

LOS GRANDES INCENDIOS

This time they struck Coahuila, burning more area than the 1998 season and completing a pyric trifecta. The 1989 fires had romped in the tropical Yucatán Peninsula. The 1998 fires had concentrated in the Sierra Madre del Sur and across central Mexico. Now the semiarid north, remote and largely ranching country, mostly matorral, grass, and mountain conifers, where fires were kindled by lightning rather than agriculture, blew up. Two great fires—La Sabina and El Bonito—blackened more area (316,954 ha) than the historic average for all of Mexico. The year's total (950,000 ha) exceeded that for 1998.[127]

The formula was familiar: hot, dry, and windy weather; remote and sparsely roaded or inhabited lands; abundant fuels, neither cropped nor routinely burned. Like historic burns, these came in the spring, from mid-March to the end of May. The immediate provocations were successive cold fronts, with some dry lightning, that kindled the La Sabina and El Bonito fires on March 16 and 17, respectively, and then sent wave after wave, week after week, of frontal winds across mountains and meseta. The intermediate contributors were ENSO rhythms and Hurricane Alex, which together made for a wet winter in 2010, fluffing fuels, followed by a cold snap in February that killed early growth, and ending with intense drying. Behind them, the deep drivers, were global changes in economics and climate. Grazing patterns altered, rural populations lost youth to migration and with them a dampening hand on the land, and sequential droughts recurred—or rather, a long, multidecadal drought deepened. In brief, fuels were abundant, weather favorable, and access onerous. Fires could start easily and spread unchecked.[128]

All observers acknowledged that fire was common and most of the biota of the Serranías del Burro had adapted to historic regimes. Satellite imagery existed from 2000 onward, which permitted a reconstruction of fire scars. The region of the El Bonito burn still showed the contours of major fires in 1998, 2002, 2005, 2006, and 2010; that around the La Sabina burn added 2008. Historic records noted significant burns in 1975 and 1989. Under normal conditions the fires could reduce fuels sufficiently to halt spreading fires for one to six years, after which the recovering vegetation could again carry flames. Only the most recent burns checked the newest ones. Historically, the mountains of Coahuila had lots of fire; in 2011, they had an order of magnitude more. And this time there were institutions to challenge the fires.[129]

The 1998 fires had been ruinous because they burned in places like cloud forests not accustomed to fire, because they afflicted rural areas, and because campesinos died when they gathered to fight them. The 2011 fires rambled through great ranches, around industrial sites, and across mountain ranges. Big landowners were influential and accustomed to doing things their way, ready to call President Calderón directly and initially dismissive of CONAFOR's expertise. Still, their capabilities and those of local municipalities dissolved before the implacable flames. The resulting mobilization outstripped that for any previous fire complex.

Fourteen federal institutions rallied under CONAFOR's direction. CONAFOR contributed 13 percent of the suppression force, SEDENA 33 percent, and SEMARNAT, CONANP, and other entities 27 percent. Local municipalities and volunteers filled out the muster. Machinery massed in unprecedented numbers. There were bulldozers, engines, backhoes, and front loaders. Assorted helicopters and air tankers flew 3,788 missions and dropped 11.3 million liters; even a Boeing 747, reconfigured for firefighting, dropped retardant. Satellite imagery recorded hot spots and mapped burns. President Calderón made an appearance at headquarters and declared the fires a disaster area. USAID's Office of Foreign Disaster Assistance dispatched three Forest Service experts to advise, along with U.S. air tankers. Researchers like Dante Rodríguez Trejo self-deployed to the fire to experience the spectacle firsthand. This time there were no fatalities. All those years of Marana courses, the struggle to create an emergency response center and a national disaster fund, and the classroom training in ICS had paid off. The contrast with 1998 and 1989 is striking.[130]

So is the comparison to the United States. These were Mexico's big burns and Mexico's response. USAID's investment was supplemental; Forest Service advisors were supportive. If Mexicans were unable to stop those fires, neither were their American colleagues across the Rio Grande. The same conditions that powered burns in Mexico affected the U.S. Southwest as well. Mexico had 930,000 hectares burn; Texas alone had 1.6 million (the Big Blowup of 1910 had racked up 1.3 million). This time Texas politicians were not demanding that Mexico stop its smoke; Texas was unable to stop its own. The air tankers dispatched south by USAID were soon recalled to assist at home. Mexico's mobilization was an exercise few countries could have matched. That it was not enough spoke less to Mexico's capabilities than to changing conditions. More material aid would not have helped Mexico any more than it could have helped the United States. Planetary pyrogeography was morphing in ways that made burns bigger, meaner, and more uncontrollable.

The inevitable aftershocks were absorbed within existing institutions. The 1989 fires had shown the need for reform. The 1998 fires had galvanized that reform and spot-welded programs into a system. The 2011 fires confirmed and toughened that system, then added to it.

The federal government created an interagency mechanism to coordinate among 20 federal bureaus. CONAFOR established a prescribed-fire training facility, modeled roughly on what the Americans had in Tallahassee, Florida. It discussed plans to supplement fire brigades sponsored by industry with volunteer brigades to assist rural communities generally. Even as the ashes of the Coahuila fires were cooling, CONANP completed two guidebooks for integrated fire management. One, a second edition of its manual for fire management, responded to a CONABIO study of Mexico's "natural capital" and introduced the general thinking about the reforms in fire policy. The updated strategy explicitly included climate change as one among the unruly gang of challenges that management confronted. The second response arrived in September 2011, in the form of a companion guide, which featured a general template for creating such plans. The publication was jointly sponsored by CONAFOR, FMCN, the U.S. Forest Service, USAID, and the German Agency for International Cooperation (Roberto Martínez participated as a consultant).[131]

Observers of the Mexican fire scene had described the 1998 fires as a *parteaguas*, a parting of the waters. Now the waters were parting again, and again. The firescapes of the past had been largely invisible. Now there was a legal matrix to define them, institutions to respond to them, and scientists to measure their effects. Yet even amid an era of big burns the 2011 outbreak seemed exceptional, almost apocalyptic. The Mexican fire community divided its history into an era before and an era after the fires—AC (*antes de Coahuila*) and DC (*después de Coahuila*). Mexico had crossed a divide. The waters had again parted, and this time they were not likely to close back up.

DC

More broadly, Mexico, like Earth, was dividing into two realms of combustion. One burned within living landscapes, as it had since the Silurian, subject to long-evolved ecological checks and balances—nature's costumbres. The other, more recent, burned lithic landscapes. This was the great disruptor, and as it passed over landscape after landscape, habitat after habitat, it reforged Mexico's pyrogeography and everything associated with it. It disdained natural bounds. It grew or withered as humans decided. By the twenty-first century it had so swelled that it saturated the global atmosphere and unmoored the climate that had prevailed through most of the Holocene.

As Mexico, relying on its oil, industrialized further, the knock-on effects became more pronounced. The effluents of lithic landscapes tainted land, polluted waters, blotted out the sky, and often sickened people. The new combustion reconfigured land use (fuels), weather (climate), and ignition. In modernizing, Mexico acquired the pathologies of modernity, of which megafires were an especially vivid expression. Even the state stumbled; instead of parastatal corporations, Mexico had what appeared to be a parallel state evolving out of criminal cartels and disaffected regions.

It seemed that Mexico would always have its rural poor and the simmering burning they practiced. Even when a mechanism existed to legally burn, "centuries of oppression," as Dante Rodríguez Trejo put it, left a deep sediment of distrust, and rural practitioners continued as before. Yet, paradoxically, there were fewer of them as more campesinos left for cities or migrated north. Much as was happening in Mediterranean Europe, the land was being abandoned, and, rather than recovering to an earlier, healthier state, the feral countryside was subject to more feral fires. Where lands were logged, industrial felling replaced milpa chopping and slash replaced fallow. Paradoxically, Mexico seemed to be trading a fire problem due to traditional use aggravated by overpopulation for the fire problems of modernity, worsened by depopulation.[132]

Even the United States was convulsed by waves of uncontrollable fires. California had the greatest concentration of firepower on the planet, yet in the years after the Coahuila fires it experienced serial conflagrations that fed on a punishing drought, insect-killed forests, and long-unburned landscapes. Megafires morphed into gigafires. Like the Washington Consensus that proclaimed itself a standard for macroeconomics, the U.S. model of fire protection was in danger of imploding. The more universally it was applied, the more it destabilized the conditions that it sought to ameliorate.

Mexico needed to find its own solution, and this being Mexico, it would almost certainly involve some degree of mestizaje. Once again Mexico's pyrogeography reassembled, entangling the new while retaining old roots. However it reconfigured, the fire world *después de Coahuila* could not simply return to what had prevailed *antes de Coahuila*.

THE *POETA* OF PRACTICE: ALFREDO NOLASCO MORALES

Alfredo Nolasco Morales was born in 1962 in a hut four meters long and four wide, in La Defensa, municipality of Misantla, in Veracruz, the second of eight children. The family never knew plenty, only scarcity, or what Alfredo later jokingly called "*sustentabilidad*." The floor was dirt; smoke from wood-fed cooking fires filled the room. His father was taciturn, a man of few words. His mother he adored. From his father he learned hard work; with his mother he knew a bond as great "as the cables of the Golden Gate bridge." In the woods and lush patches that crowded outside the shack's door, he found an enticing natural world of animals and plants, of interest and beauty. Alone among the six fathers of modern fire management in Mexico, he had a flair for poetry.[133]

It was an inauspicious birth, but Alfredo had grit and natural gifts, and he came of age during an extraordinary time for Mexican fire. He was 21 when the first regional Marana-type course was held (in Cancún). He was in Yucatán during the 1989 fires and in Quintana Roo during the 1998 crisis, and he was chief of fire protection for CONAFOR during the 2011 outbreak in Coahuila. When asked to reflect on his life in fire, he opened the memoir with an ode, "Añoranza," containing sharp, fond memories of the natural sounds and sights of his childhood.[134]

The family moved: his father was "*un nómada*." Eventually the family settled in Martínez de la Torre. There Alfredo's father worked as a watchman for the neighborhood. His mother became a midwife. One night his father was attacked and stabbed in the abdomen (another watchman was killed); he spent months in a hospital, and it was two years before he was able to return to work. Alfredo's mother washed clothes to survive, and when she took ill she was unable to seek help. Alfredo's early childhood was memorable, something to savor. But those years of ailing parents were best recalled sparingly.[135]

The moves disrupted schooling. His father had very little; his mother persisted until, at age 65, she completed secondary education. Even in primary school Alfredo's formal education had to compete with an informal economy. At the age of six he was a primary-school student, an errand boy, and a shoeshiner. "Early on," he recalls, he "joined the economically active child population in an informal way." Until he found his profession, that pattern continued.

> I worked as a fumigator of corn and beans with dangerous insecticides; cane cutter, mower with machete of irrigation canals, hoer, bricklayer's assistant, baker's assistant, butcher's assistant—personally I have only killed two pigs in my life, one at thirteen years old and another at fourteen, at which point I ended that adventure—seller of chili peppers, planter of crops with *espeque*, fisherman in *panga*, orange and tangerine cutter, fish seller in urban market, seller in floating market, stone splitter with sledgehammer, lumberjack, declaimer, orator, singer on a bus, gardener, foreman's assistant to establish cattle fences, street sweeper, card player, church singer, *rondallero*, member of a musical trio, drunk and bohemian and rural composer.[136]

Eventually the family moved to Liberación, to the north of Martínez de la Torre, where his father became an ejidatario. Here, Alfredo notes, were the years (1971–75) of his greatest happiness, and here he forged his "*carácter campesino*" and strengthened his passion for "the woods, the arroyos, the river, forest life, the trees, the land, and the work of the field." But bulldozers destroyed the holm oak forest, swiddeners slashed and burned a refuge of high woods, conflicts developed among the ejidos, and Leodegario Nolasco moved the family again. Alfredo's childhood paradise was gone. The scars of the destroyed woods branded his memory. "Things would never be the same again."[137]

Teachers early recognized his talents—and he recalled each teacher by name. When the family moved to the ranch, he lost a year of school, but again his abilities were appreciated, and he was subsequently advanced a year. He studied by oil lamp. Poverty promised a future of cane cutting, his father told him. Then two teachers visited his parents, argued into the night, and offered to bring him to Tlaxcala to take the admission exam for a scholarship to Escuela Tecnológica Agropecuaria No. 161. Leodegario asked Alfredo if he wanted to go. Alfredo said yes, he did. He passed the exam. The worlds of the field and the classroom, and the passage between them, "forged his character" and "defined the themes" that would guide him through life. Not least was the hope that his education might continue and help him someday remove his family from poverty.[138]

His choice for a career was to be a doctor, but the logic of his circumstances led him to apply to the Universidad Autónoma Chapingo. When he took the scholarship exam in 1979, there were 23,000 applicants, of which 1,300 eventually enrolled with or without assistance, and 650 graduated in 1986, Alfredo among them. For a while alcohol became a problem, but he overcame it, as he did so many other obstacles, by "an enormous force of will." He graduated as a forest engineer, with a specialty in silviculture.[139]

By then Mexico's economy had entered its cascading crises: a disconnect between economic theory and reality. But equally Alfredo observed a profound dissonance between silvicultural theory and practice. The august programs of management that restricted logging met a more unrestricted felling. Alfredo had many arguments with his bosses. In 1987 he decided to decamp to Quintana Roo, where he supervised 16 people in a rural development district under the Secretariat of Agriculture and

Water Resources. He loved the tropical woods. And he encountered for the first time in a sustained way the presence of fire.

Within a year Hurricane Gilbert had ripped across the Yucatán Peninsula, and with the following dry season came the fires of 1989. He was dispatched to a fire south of Tulum under the direction of Manuel García Gil, a veteran of the early Marana courses. At the time Alfredo had no training in fire suppression, no personal protective gear, no experience in fires of either woods or fields. With regard to fire he was, he confessed, an apprentice in everything and a specialist in nothing. It quickly became obvious that the fires were many and large, and firefighting resources few and scattered. The conflagrations were, for both Alfredo and Mexico, a baptism by fire.

The next year the Marana project came to Mexico, taught by graduates of the original courses (Roberto Martínez Domínguez, Gilberto Galeote Rivera, and Manuel García Gil), assisted by instructors from the U.S. Forest Service. Alfredo attended the second course, held at Cancún. To his surprise he ranked first in the final testing, was given a Smokey Bear pin, and discovered his life and profession had again changed. In fire he found a vocation; in Roberto Martínez, a mentor. Aftershocks of Hurricane Gilbert continued—fireweed flourished, burning was widespread for farming and pastoralism—so fires broke out again in 1991.

Throughout, Alfredo experienced what he regarded as another chasm between theory and practice, because fires free-burning in campo and monte were vastly more complex than they seemed in the classroom. On the K'atalox fire he struggled to align principles with practice. Day after day he walked, questioned, observed, and learned. He wrote up his thoughts in a lengthy study on fire protection in Quintana Roo that served as a thesis toward official recognition as a forest engineer. That reconciliation of formal and informal (call it "experience-based") learning set the pattern for his career.[140]

Alfredo participated in an FAO fire protection program for north Quintana Roo. He joined the study tour to British Columbia that introduced geographic information systems and aerial fire control—his first serious contact with foreigners and use of English. He attended Spain's echo of the Marana courses, hosted in Valsaín but attended by people from throughout Latin America. He took courses in emergency management featuring ICS, jointly offered by SEMARNAT, USAID, and the U.S.

Forest Service. He attended an intermediate fire-behavior course held in Ciudad Guzmán by CONAFOR and the USFS. He learned to serve as a coordinator of aerial fire operations. Meanwhile, 1995 was another serious year for fires in Quintana Roo, and then came 1998. From 1995 to 2004 Alfredo served as chief of protection for SEMARNAT for the region. By now Alfredo was the go-to guy for fire in tropical Mexico. Still, he struggled to find permanent employment. With political changes he was asked to resign. He had seen it coming and had already received an offer from the Nature Conservancy to become manager for fire in Mexico. TNC was then advancing its Global Fire Initiative, with Mesoamerica as a particular target. Soon, Alfredo learned about fire's integrated management, fire's ecology, and fire's social basis. He had to reimagine fire from being a threat to forests to being a necessary feature of many protected natural areas. He learned of early management experiments in reserves like the Sierra de Manantlán. This "new adventure" appeared to him "fascinating."[141]

His training restarted, this time focused on good fire. He met Dante Rodríguez Trejo, a professor at Chapingo, who had been initiated into prescribed fire while doing a doctorate at the University of Florida, and Ron Myers, inspirational in his commitment to adapting traditional burning to new purposes—a "great companion and mentor." TNC took him to a prescribed burning workshop in Arizona, an introduction to fire effects workshop in Minnesota, and the National Prescribed Fire Training Center in Tallahassee, Florida; it also boosted his formal qualification in suppression with courses on crew boss and initial attack leadership, the incident command system, ignition practices, and fire leadership. The training turned practical as he joined TNC for trial burns in Sonora and Coahuila. Its Global Fire Initiative took him to the Dominican Republic, Peru, Chile, Colombia, Brazil, and places throughout Mesoamerica for workshops. He learned that Earth was truly a "fire planet." By the time his new training cycled through, probably no other Mexican had his background in both traditional suppression and the new demands of integrated management together with his breadth of travel.[142]

Though the Great Recession killed TNC's fire initiative, the endeavor had, if improbably, helped shift the fundamentals of Mexico's legal and educational approaches to fire (integrated fire management was now taught at UACh). Most astonishingly, the new thinking got recognition in the amended NOM-015. In science, in fire strategies, in law, integrated fire management had become the mantra of fire progressives. What Mexico needed was what everyone else committed to the same ideal needed: a means to leverage its institutions into that transition. What it required was someone adept in both stopping and setting fires, someone who saw the disparity between formal learning and lived learning and who was undeterred in his desire to reconcile them.

In February 2009, the director general of CONAFOR offered Alfredo Nolasco Morales the job as national director of fire protection. In his resignation letter to TNC's Mexico program, Alfredo spoke of how the past few years had been full of challenge and learning, and said that during that time "everything has been good and every where I found a helping hand to help me grow as a worker and as a person." Now he hoped to continue those lessons and that relationship as he sought to redirect CONAFOR's fire mission fully toward a more pluralistic program.[143]

His strengths, he believed, were his technical skills and his capacity to make decisions. But a successful administrator also needed to understand finance and law, and had to establish good relations with institutional collaborators, and in none of these did he feel adept. But he learned. He was in charge when the two great fires, powered by an accumulation of 50-year-old fuels and sparked by dry lightning, blew across Coahuila. From that experience he learned the difference between fire as a technical problem, fire as a media event, and fire as a political provocation, and how to concentrate on the technical fire amid the distractions of the other two. He learned that perfection is impossible. "Something is always on fire."[144]

When the Calderón administration ended, the Peña Nieto sexenio began with the usual turnover in the upper levels of CONAFOR, but Alfredo survived; he explained the reforms under way in fire management to the new administration and continued to advance those goals, some symbolic, many pragmatic. The to-do list was long and ambitious. Make integrated fire management an institutional theme. Hold more fire management training courses. Upgrade prescribed-fire instruction. Promote a national day to honor firefighting. Establish national and regional dispatch centers. Renovate the motor pool. Upgrade radios. Improve decision-making tools for forecasting fire danger, priorities for protection, assessment of fire regimes throughout Mexico, fuel models, and a fuel management manual. Export fire crews to Canada. Fashion a strategy for improving gender equality. Confirm in policy the recognition that fire can both enhance and damage.[145]

The grand achievement was to incorporate modern fire management into the 2018 General Law of Sustainable Forest Development. While NOM-015 had recognized and laid down regulations for agricultural burning, which its amendment extended into forests, that achievement also came with a wary understanding that ideals are only as good as their expressions on the land and their rootedness in daily life. This was a man, after all, who had watched how the massive wildfires of 1989 had struggled to compete for media and political attention with the Miss Universe pageant in Cancún. Yet in unforeseen ways prescribed fire was where academic science and campesino experience met. It was where a Mexico keen to modernize might tap into its ancient roots.

We all want to leave a legacy, he observed, but it was important to relish the life you had, which meant doing what mattered to you. Alfred Nolasco Morales found his passion in fire. The child of nature savoring color and smell, the ardent student and young forester devouring training, the shrewd administrator reconciling the desirable with the possible—this was fire integrated and managed in a life. When he concluded his memoir, he rewrote, tongue in cheek, a love song by Nicola di Bari into an allegory of passion and consummation called "La quema prescrita."

POSTREVOLUTIONARY FIRESCAPES

Different eras of Mexican history had characteristic fires that seemed to speak for how Mexicans lived on their land. The Aztec New Fire Ceremony. The fire hunt. The fire of milpas. The fire of pastos. The cooking fire in a metropolis. The blowout well fire. The pyric transition was partitioning the national estate into two generic firescapes, one for developed Mexico and one for the rest of the country. The Red Revolution reorganized and recalibrated Mexico's megadiversity of landscape fires. It established a new suite of showcase fires, sites, and manuals. These were the exemplary fires, the problem fires, the fires that defined Mexico DC. They were the fires that made and unmade landscapes.

RESERVA DE LA BIOSFERA MARIPOSA MONARCA

Probably the best known internationally of Mexico's protected areas is the monarch butterfly reserve along the border between Michoacán and the State of Mexico. The monarchs quietly execute one of Earth's most astonishing faunal migrations, traveling over a two-year period from as far north as Canada to the high fir forests of central Mexico.

The monarchs were suffering from changes along their entire corridors of transit, but when the exact location of their wintering was identified in the early 1970s, a special burden fell on Mexico. A fixed abode, with millions of monarchs nesting in masses, was not only a special point of vulnerability, but a place where failure would be visible globally. In 1980 the critical sites were decreed a protected natural area. The sanctuary was expanded in 1986, and again in 2000, when the project was proclaimed the Reserva de la Biosfera Mariposa Monarca (RBMM).[146]

At all levels, management was complicated, with high stakes and little margin for error. Fire magnified the risks. "Key actors" included 7 institutions at the federal level, 3 at the state level, 18 municipalities, 59 ejidos, 13 Indigenous communities, 7 groups of local players, and 6 interagency groups, from fire brigades to local forestry councils. But intellectual challenges were no less complex. It was obvious that fire was an ancient disturbance, yet lightning ignitions were rare. That meant people had set the fires as a complement to other activities. The ironies compounded when the creation of the reserve itself upset the rhythms of the historic regime, and as tourism joined illegal logging as a provocation. Neither people, nor logging, nor fire could be banned. All the traditional tensions that accompanied Mexican efforts to promote certain kinds of fire practices and discourage others were present and magnified because the butterflies were charismatic, and the world was watching keenly whatever Mexico did or failed to do.[147]

A management plan was released in 2001 that identified forest fires as "one of the threats that most affected" the reserve. Although the reserve was on the border of the two Mexican states that historically recorded the most fires, little knowledge existed about the regimes at the reserve itself. About two-thirds of ignitions resulted from agriculture; another 28 percent were attributed to grudges between landowners, whose lands were now included in the reserve. Yet almost nothing solid was known about the particulars of how fire behaved on the protected land. In 2011 CONANP published its strategies for integrated fire management plans. But the RBMM had almost nothing with which to craft out of such model plans a practical one for the reserve.[148] Instead, researchers scrounged the literature for fire data based on similar forests that might serve as proxies, searched the Database on Fire Regimes and Fuels for Mexican Terrestrial Ecosystems, and during the 2012 season conducted "semistructured interviews" and visited 14 sites that burned that year. The conclusion: "information is so scarce" that researchers could not evaluate the merit of the established policy of suppression, yet "we know very little about the traditional use of fire in the [RBMM]." It was all informed guesswork, but the reviewers were sufficiently confident in the general principles to say that the reserve "should apply the new recommendations on fire management outlined by CONANP (2011)."[149]

CONANP revised its guidelines in 2013. In September 2014 the reserve convened representatives of the interested parties for a workshop to massage those principles into a working plan. Almost all the relevant information was known except the history and ecology of fire. That research soon followed, based on fire-scarred trees, and it inverted much of what had been working knowledge (if not received wisdom). There was little difference in fire frequency among the three types of forest: pine, fir, and mixed conifer all had fire-return intervals of around two years. There was no correlation between fires and tree establishment: other disturbances, notably small-scale logging, seemed to account for the forest regeneration pattern. Fire frequency had little connection with climate, or with ENSO rhythms of drought and deluge. The continuity of burning seemed to result from continuous ignition and a continuous if patchy array of surface fuels, both the outcome of traditional human practices. Pastoral burning and firewood gathering helped keep fires on the surface. Unlike other protected areas, the fire chronicle at RBMM did not rupture when the reserve was established. The Mariposa Monarca reserve was indeed special.[150]

There was no swift stroke that could guarantee the monarchs' survival. The best strategy was to have people continue to do what they had done historically. Yet too much was unknown, and too much was changing—socially, economically, environmentally, and not least climatically. Fire had a necessary place: excluding it would itself be a major disturbance.

Yet what exactly that ecological role was and how to maintain it was something locals, scientists, and administrators would have to learn by cautiously doing. They knew that they could destroy the habitat. They did not know, with guaranteed assurance, how they might sustain it.

The Codex Magliabechiano presents an image of butterflies emerging from flames in a kind of pyric chrysalis. Indirectly, that had been true, for fire had helped catalyze the habitat that supported them. How to fashion a modern version of this process, however, tested the skills and imagination of practitioners, and its reality was proving far less poetic.[151]

CHIAPAS CRUCIBLE

In Chiapas, both state and federal programs had converged on a consensus that fire on the landscape had to be controlled by law, technological replacement, or outright force. Yet the laws could contradict one another, and reforms depended on federal funding, due to which there was "the possibility this might not occur." Overall, fire legislation was "quite removed from social reality in the State of Chiapas." As one resident put it, "Without fire we cannot live. Fire is our life, it cannot fail." Still, the usual narrative would unfold in the usual way, as illustrated by a tale of two towns in southern Chiapas.[152]

Both towns had inherited a tradition of agricultural burning. But different political and legal circumstances meant that they reacted differently to the pulls of tourism and out-migration and to the push from fire's criminalization and a generational divide in which the young were less inclined to view fire favorably. Tziscao lay within Lagunas de Montebello National Park, fell under the authority of the federal government, and knew the attractions of modernity. Antelá lay outside the park and felt less pressured politically, economically, and culturally.

Unsurprisingly, Tziscao went further in distancing itself from fire. Its youth looked to the Green Revolution and appealed to fertilizers and biocides to do the work of flame. Being close to Lagunas de Montebello, their economic vision included tourism, and tourists didn't want to see burned land, didn't want smoke, didn't want to confront what they had been educated to believe was an environmental harm inflicted on a protected park. The criminalization of fire's use made roza-tumba-quema risky. Knowledgeable elders (*corredores de fuego*) were less able to transmit oral tradition; the ancestral knowledge of fire faded, as the inherited fire culture sloughed off in favor of a more modern upgrade. Meanwhile, beyond direct control by park and other officials, Antelá held longer to the old ways.

In brief, the two towns stood as a microcosm of the two Mexicos, one adhering to traditional practices, the world of second-fire, and the other to the principles of modernity, reforging a world based on third-fire. It was Mexico's essential tension, centuries old, here expressed in conflicts over competing economies, social orders, and visions of nature, for which fire was an index and emblem. The 1998 fires sharpened the differences between the two Mexicos by empowering the agenda of modernity. To authorities the way to prevent another wave of monster fires was to prevent all fires. But like early missionaries who understood that the best way to reform particular rites and beliefs was to reform society overall, authorities knew that eliminating fires meant eliminating a fire culture. Education, laws, reorganization or relocation—these were the time-honored methods that were now employed to proselytize against fire.[153]

Commentators on Latin America's agrarian poor had coined the phrase People Without Land (Sin Tierra), later expanded to include People without Water (Sin Agua), since water was needed to work the land. Laura Ponce Calderón characterized rural Chiapans as a People Without Fire (Sin Fuego). "Some alternatives offered by the government and transnational companies," she notes, are "to replace burning with agrochemicals (it generates a lot of wealth for many)." What's more, "they sell the complete package: the improved seed and the *quemacuat* (it burns the weeds with a liquid)." Such packages were too expensive for campesinos, who, "without many options left," were forced "to leave the countryside to dedicate themselves to being tourist guides or drug traffickers, [and] they migrate to large cities in Mexico," though they preferred to migrate to the United States. Meanwhile, the nominally protected areas became unstable relative to fire, unhinged by the disappearance of traditional burning, which led to a buildup of combustibles and to ecological deterioration caused by the loss of fire's catalytic presence. Equally, the People Without Fire suffered from social delamination and, with the loss of traditional knowledge, cultural deracination. "How much complexity is unleashed by the simple act of prohibiting the use of fire."[154]

✹

Old story, familiar narrative—it would seem that once again Mexico's modernizing projects might stumble, as national norms, codified in decrees and laws, clashed with local *usos y costumbres*. Yet once again, Mexico's peculiar liabilities could also become assets. What reformers needed was a site that could promote good fire and prevent bad. They found it in the Reserva de la Biosfera La Sepultura in the Sierra Madre de Chiapas. Ten years after the 1998 fires savaged prime sites, La Sepultura was a showcase for synthesizing community politics with integrated fire management.[155]

The reserve's fire program had passed through three stages. In 1997 it first considered fire as a serious issue for management. Then the 1998 fires blasted through, and the immediate response was to ramp up suppression capabilities. After the 2003 fires struck, the reserve recognized the need for an integrated fire program: fire as such could not be banned, but it might be redirected. Conservationists worried about pastoral burning that occurred too frequently to allow regeneration by the signature pine of the region, the otherwise fire-adapted ocote, and about escaped agricultural fires that, amid droughts, could rip through fire-sensitive forests. Locals had their own definitions. Bad fires (like ill-timed or ill-intentioned burns by neighbors) worsened habitability. Good fires freshened forage, readied milpas, eased travel, tamed pests, and promoted hunting.[156]

CONANP, FMCN, Espacios Naturales y Desarrollo Sustentable (an NGO), and the Nature Conservancy converged on La Sepultura to craft an approach that made sense in Mexico's context. The prime movers wanted not only a model fire plan for the reserve, but a methodology for bringing all the various communities together, a template for controlled burning, and a training program that would allow Mexicans to manage the reserve and export its experiences. TNC's ambition was to change the way CONANP, CONAFOR, and rural communities managed fire, even to reform those agencies and to amend laws like NOM-015 that regulated fire's use.[157]

In reality, the process went both ways. TNC sponsored training courses throughout the region, sent key La Sepultura staffers to the United States for advanced training, and arranged a fire study tour to observe the pine-oak forests of Arkansas, a showcase for prescribed fire within a modern fire system. Equally, Americans and CONANP officials learned to understand traditional burning. FMCN and TNC even arranged to fund a doctoral student at Colorado State University to document how all the participants understood fire and how the program's outcomes measured against its goals. Even as she translated customary burning into the technical lexicon of fire science—identifying 40 factors considered by local practitioners—Mary Huffman was struck by "the disconnect between the careful burning practices" she witnessed and the official reports that blamed slovenly agricultural burning for endless wildfires and various environmental ills.[158]

Neither knowledge system seemed adequate by itself: each had something to learn from the other. Traditional know-how needed tweaks to meet modern purposes and was in danger of disappearing, yet science was far from capturing (much less finding surrogates for) the full gamut of fire's biological presence. Most formal prescriptions, derived by defining fire as a physical-chemical reaction, reduced prescribed fire to flame and fuel. But local inhabitants wanted the endless ecological effects that fire stimulated. Park officials worried about the poor condition of ocote; scientists noted that regeneration issues with ocote were the result not only of annual burning in some locales but of the loss of flare-ups—moments when steep slopes, gusty winds, and a higher density of woods allowed flames to carve out canopy gaps above bare soil, a combination that made ideal seedbeds. Successful management only required tweaking (not abolishing) old techniques to encourage the necessary patchiness; some simple hand-raking and creating cover with green leaves could be enough. Similarly, the old practice of burning along ridgetops and allowing the fire to creep downslope introduced fuelbreaks that could be used to keep one ejido's fires from entering the lands of another, or to keep burns in pine-oak forests from breaching the border with tropical broadleaf woods. So, too, technology needed translation. Drip torches were useful, but they required a regular, and expensive, source of gasoline and diesel fuel, while traditional antorchas were abundant—an interesting micro-clash of Mexico's two realms of fire. The point of the La Sepultura project was to provide occasions for translation and adaptation between those two realms.[159]

Sponsors and organizers considered the project a success. By 2007 and 2008 the two target ejidos were implementing their plans, CONANP was preparing to disseminate the methods and example elsewhere in southern Mexico, and a local NGO, Ambio, was extending the program more broadly throughout agricultural landscapes. The principals were exploring proposals to furnish financial incentives (perhaps from carbon sequestration). The apex came when, as a "direct result" of the La Sepultura project, NOM-015 was amended to include prescribed fire in forest ecosystems. In 2018, under the inspired leadership of Pedro Martínez Muñoz, a retired fire officer from CONAFOR, and with help from other reformers in the avant-garde, a model manual for prescribed burning was developed for the municipality of Villaflores, Chiapas—the first in Mexico, and one that, it was hoped, would be an exemplar throughout Mesoamerica. TNC's Ron Myers regarded the outcome as a "paradigm shift" for successfully synthesizing community forestry with integrated fire management and as a "catalyst" for Mesoamerica.[160]

Unfortunately, that momentum could not sustain itself; the planets shifted out of their politically favorable alignments. The recession hit TNC hard and killed off the Global Fire Initiative, and of course it trashed Mexico's economy. New political crises diverted money and attention to other issues. CONAFOR had its budget cut. The amended NOM-015 was neutered. A change in the Chiapas government discouraged the hybrid regime. But the model endured.[161]

SIERRA JUÁREZ, OAXACA

There have to be exceptions, and the Sierra Juárez in Oaxaca state furnishes one. What begins as a typical fire history for a typical Mexican forest turns on itself. As local communities assume control over industrial forestry, the inhabitants themselves become the critics of traditional fire costumbres and even choose to forget that such practices once informed their relationship to the land.[162]

The Sierra Juárez has a typical pine-oak forest that exhibited the usual patterns of local practices based on patch burns for farming and herding, along with fuelwood gathering and charcoaling. The pine served mostly as timber, the oak as firewood. Episodic mining outbursts stimulated markets for food and mule-based transport, which expanded the ecological quilt of patches and the abundance of pasture relative to woods. During booms, forests were felled; during busts, they regrew. During the Mexican

Revolution, battles disrupted and violently burned the woods on a scale not known for centuries. The landscape's patches were rearranged with each new quilter, yet fire, ever a shapeshifter adapting to its setting, persisted. A chronology of fire scars preserved on trees from 1794 to 1933 gives a median fire-return interval of four years. The upshot is a fire history aligned with that generally common for Mexico's mountain forests.[163]

In 1913 Cassiano Conzatti, a botanist with an agricultural experiment station at Oaxaca who had traveled the region for 25 years, was dismayed that fires had burned the countryside during that entire era. It was a "pernicious" practice that he assumed must eventually disappear, but which continued to spill across the hills and fill the Valley of Oaxaca with smoke, a threat to forests, public health, and the national estate through eroded soils and the loss of trees. The fires lit the peaks even as he wrote; the only sites that didn't burn were those that had burned earlier. Everyone contributed, both the deeply illiterate *indio* and the self-seeking *hombre de razón*. The fires were the axioms from which any understanding of the degraded valley, state, and republic must derive.[164]

In 1930 the servicio forestal arrived, full of condemnation for rural burning. ("The fire control regulations were detailed," Andrew Mathews has observed, "and quite impossible to enforce.") More significantly, logging arrived, complete with roads, trucks, and a concern to protect timber from feckless fires. In 1942 a fire along the boundary between Ixtlán and Atepec led to an appeal to the governor for aid; he ordered all men over the age of 15 to report for fire duty. A major transformation occurred a dozen years later when several factors converged. Milpa agriculture melted in the face of imports from elsewhere in Mexico; internal combustion engines replaced mule trains for heavy hauling, and hence reduced the scope of pastures; out-migration kept local populations within boundaries.[165]

The servicio forestal had denounced traditional burning as backward and an impediment to proper forest management. But fire control was gnarly, with access often difficult, the state feeble, and fire protection duties delegated to community brigades, which, despite meticulous guidelines, still left capacity and commitment scarce. The servicio forestal lacked even the staff needed to process burning permits, much less to enforce measures to regulate burning and firewood gathering. In 1956 the underresourced servicio forestal yielded effective oversight of the sierra forests to Fábricas Papeleras de Tuxtepec, a timber company that obtained a concession allowing it to behave as a parastatal entity, one of whose duties was to organize a firefighting brigade.

A memo tallied the usual complaints: the forests "have suffered more or less intense exploitation, and all have suffered the impact of indigenous demographic pressure. Large areas have been deforested to dedicate them to a rudimentary and migrant agriculture; forest fires are intense and frequent, destroying the young trees and therefore the regeneration capacity of the forest. Uncontrolled grazing by goats and sheep completes this picture of the enemies of the forest." Restoration was possible and would follow "technically directed exploitation." A technocratically informed company could both do good to the land and do well for itself. So far, so familiar—the Sierra Juárez was reading a script common throughout forested Mexico. Traditional fire practices began to fade as the shift from pastos to clearcuts, and from milpas in the mountains to farms in the valley, got under way.[166]

Then something new appeared under Mexico's fifth sun. A younger generation saw few of the promised benefits from the timber concession: the threat they faced came less from a customary agriculture that fewer of them practiced than from the havoc wrought by the timber company. It was trashing the forest and failing to deliver the promised social benefits. Thirty years after it arrived, amid fervent protests, the company yielded its concession and local communities assumed control over the woods. The Sierra Juárez became a model both of community forestry and of industrial logging.

The new overseers imagined fire differently. An older generation had seen fire as a part of land and life, essential for field and hearth, more friend than foe. A younger cohort came to see it the way state and industry defined it: a danger to life and property, and a threat to the economy that now informed their lives. The generational conversion spared the Sierra Juárez the usual "institutional misfit," as Claudia Monzón termed it in Campeche, "between a state regulation, the burn calendar, and the physical conditions and cultural needs at the local level." Younger *serranos* were more interested in stopping fires than starting them.[167]

The goal of reform had been to replace the parastatal corporation with a community-based regime. The resident community wanted a change in managers, not in management policy. In assuming the role of the servicio forestal and Fábricas Papeleras, serranos accepted modernity's condemnation of traditional burning, thus removing official objections to their autonomous oversight of the woods. They were doing what both government and company had claimed needed to be done but had failed to do. They had co-opted the arguments. They could do forestry more effectively and more fairly. They could eliminate folk burning and impose fire suppression more thoroughly than the state or its proxies; in fact, they insisted they had a right to manage the forests because they could control fire better than "ignorant" or "incompetent" neighbors. The reformation was, as Andrew Mathews notes, "very unusual for Mexico as a whole."[168]

The episode gives a curious twist to Mexican fire history. The issue is not just that the recurring fires of millennia are gone, or that the serranos have likely traded the surface fires of milpa and pasto characteristic of a preindustrial economy for the episodic rewilded fires typical of an industrial order. The issue is that, in discarding the traditional torch for an industrial axe, serranos forgot their past as keepers of the flame; they suppressed their memory of fire. Returning land to traditional owners does not here mean returning that land to management by customary practices. A second twist further complicates the moral. The serranos were committing to fire control at roughly the time a new generation of fire-interested scientists was urging a revival of historic costumbres in the name of ecological integrity. At this point the narrative arc of fire history in the Sierra Juárez starts to resemble a Möbius strip.

NORTHERN *FRONTERA*

As it modernized, Mexico's northern region attracted maquiladoras in border towns, logging in the mountains, irrigation agriculture in the floodplain coasts, and ranching in the high plains. The climatic rhythms of wet and dry (once in spring and summer, again in fall and winter), grew and readied fuels. People and lightning supplied sparks. Altogether that peculiar configuration of land and life gave the borderlands their own regimens of fire. Yet, although the Madrean Archipelago, the meseta, and the seasonal monsoon spanned the international frontier, different human histories yielded contrasting landscapes and fire histories.[169]

Commercial ranching, as always, found itself in a veritable pinball machine of competing forces. As with Mexico's economy generally, success meant reducing population (in this case of livestock) or increasing productivity, which is to say forage. The small, semi-subsistence rancho; the low-value local breed of cattle; the low-grade productivity of the Sonoran and Chihuahuan Deserts—it seemed an opportunity to devise a livestock echo of the Green Revolution in which better-bred cattle, higher-value forage, and mechanical land clearing could fashion a commercial cattle economy to export beef throughout Mexico as well as outside the country. Range science would supply the requisite engineering.[170]

One ingredient was buffelgrass. In effect, an exotic forage was being introduced to feed exotic cattle. But one element couldn't be changed without causing ripples throughout the larger landscape. Buffelgrass required landscape clearing, usually by tractors and bulldozers, which removed native vegetation and disturbed soils; in 1973 buffelgrass pastures amounted to 7,700 hectares, and by 2000 some 140,000 hectares. The project required land and capital, which favored the big landowner over the small. The costs of reforging the biota meant the livestock had to be specially bred, which removed the desert-hardy *criollo*. Not least, buffelgrass was a pyrophyte, like cheatgrass on steroids, which furthered the assault on native vegetation by sparking a cycle in which buffelgrass could encourage more intense fires, which created conditions for more buffelgrass, eventually overturning landscapes.[171]

As with the Green Revolution, there were immediate strides in efficiency and output ("buffelgrass reaches its highest productivity worldwide on Mexico's arid rangelands"), but the collateral damages eventually asserted themselves ("it still remains inferior to native grasses in nutrition, protein content, and palatability"). A significant fraction of planted pastures failed, and those that flourished often had to be reseeded again, or several times. Native vegetation had provided reserve fodder in the form of forage, fruits, and seeds that had helped ranchers survive droughts; these were now stripped away. The new breed of cattle, like specially bred wheat, demanded additional inputs to be worth their price. Weeds took over razed and abandoned fields. The unhinged fire regime encouraged a flush of woody plants, which, in the right conditions, could fuel major burns outside the adaptive range of the ecosystem. Removing the planted invasives required more machines and petrochemicals. Oil, it seems, was now a part of beef as well as maize.[172]

✵

Forestry avoided some of the sky islands as being too secluded, too small, and too sparsely stocked with commercial timber. Throughout the twentieth century a few mountain ranges had experienced neither significant grazing nor logging, while receiving plenty of lightning. All this invited cross-border comparisons with regard to fire history. The U.S. side showed a massive disruption in the latter nineteenth century due to overgrazing and the removal of Indigenous fire, followed by formal fire control. The Mexican side held to presettlement patterns much longer, and in a few cases had seemingly escaped axe and hoof.

In the early twenty-first century, the Madrean mountains of southern Arizona became the site of experiments in fire restoration. The FireScape project considered each mountain range as an integrated unit, which encouraged fire officers to appeal to managed wildfire, rather than distinct patch burns, to do the ecological work. In Mexico mountain ranges became sites for extending the integrated fire management model being developed for CONANP. The relative lack of formal development meant that there were fewer competing interests to reconcile and fewer ecological disruptions to counter. The sky islands have been a frequent source of fire chronologies derived from scarred trees, and in Mexico they yield a minimum fire-return interval of 4.7 years.[173]

The border between Sonora and Arizona was especially attractive. The Nature Conservancy hosted one of its first workshops in Hermosillo; that led in 2005 to active prescribed-fire plans at the Rancho Los Fresnos in the municipality of Santa Cruz. Signing off as planners were Mark Haberstich of the U.S. Fish and Wildlife Service and Alfredo Nolasco, then with TNC; Haberstich served as burn boss, with Ron Myers of TNC as overall fire manager. It was a model plan, the basis for "training academies" and prescribed fires at Cuenca Los Ojos near Agua Prieta, the Sierra de los Ajos, and the Sierra San Luis (a national forest and wildlife reserve where New Mexico, Chihuahua, and Sonora converge).[174]

Isolation was both boon and bane. There were patches of land that had not much felt the strains and stresses of the twentieth century. Other sites could be left alone: simply ending efforts to stop fires was a boost. But that same remoteness—combined with proximity to the border—could make the region inviting to illegal or sublegal operatives. Farther south, throughout the Sierra Tarahumara, milpas for marijuana were displacing milpas for maize; narcotraficantes were crowding out Rarámuri. Several polygamous sects that had long split from the Mormon church had compounds around Bavispe, with uneasy relations with neighbors. Research, not to mention active reserve management, was becoming difficult, if not outright dangerous. Thanks particularly to TNC, northern Sonora was an early adopter of integrated fire management. But model plans and training still had to confront issues that were not normally part of checklists that burn bosses had to sign off on.

In November 2019 the Association for Fire Ecology hosted a conference in Tucson on the theme of "cultivating pyrodiversity." Participants came from the United States, Canada, Australia, and Europe, and there was a substantial contingent from Mexico. There were sessions on comparing data across the border, on pastoral burning and introduced grasses, on satellite surveillance, and on Mexico's national fire-danger rating system. Citlali Cortés Montaño delivered a keynote address on gender diversity (by then 5 percent of CONAFOR's fire personnel were women). Diego Pérez Salicrup spoke about the Mexican fire scene and was prepared to lead an overnight field trip to the Cuenca Los Ojos in the Sierra San Luis, where prescribed burning had been introduced. CONAFOR officials would meet the group.[175]

That was scheduled for November 23. Two weeks before the conference began, perhaps 50 miles south, three SUVs from the LeBarón compound at La Mora were ambushed and killed in what was characterized as a quarrel with drug cartels, though old and newly morphed conflicts were involved. Three women and six children, all holding dual citizenship, died. The cars were burned. The two sites—two expressions of anthropogenic fire at its best and worst—were too close. A large group of Americans poking around territory contested by cartels, for reasons that would seem incomprehensible to traffickers, could invite confusion at best. The field trip was canceled.

Enthusiasm for fire ecology had blossomed in the United States out of a passion for restoring a natural process. Reintroducing fire was as symbolic as reintroducing wolves or other charismatic, if often predatory, species. Yet more and more the field had recognized that the keystone species was humanity, for whom fire was an ecological signature.

Mexico had always known this. Americans forgot, then had to restore that memory as well as the flames. Even in a place as seemingly sequestered from the modern world as the Sierra San Luis, fire history could not be unknotted from the endless entanglements of human history.

BOSQUE LA PRIMAVERA

Mexico's national fire narrative rudely traced the transformation from traditional burning to industrial combustion. But that general arc splintered into a thousand geographies, some of which continued the storyline while many others arrested it, leaving the "two republics" of fire mingled without truly mixing. Large metropolitan areas had *colonias* of rural migrants; national parks and biosphere reserves held ejidos and comunidades. Charcoal and firewood persisted in cities. Agricultural burning and fire suppression coexisted uneasily in reserved enclaves and along borders. It all made for a metastable compound, perhaps nowhere better expressed than where Guadalajara and the Bosque La Primavera met and arguments over appropriate fire practices morphed from controversies into conundrums.

Bosque La Primavera is a rumpled volcanic dome, a scrubby oak-pine forest, and a protected natural area that abuts Guadalajara. It spans the climatic transition between hot and temperate zones, hosts geology unique in North America, and holds a biota emblematic of that in Jalisco overall. It furnishes increasingly valuable ecological goods and services, most preciously water and open-air recreation. Not least, it is an island refuge amid demographic pressures that are transforming natural scenes into agriculture, agricultural lands into industrial zones, and everything into the relentless conurbation that is Guadalajara. Interest in its protection formalized in 1973 with a study by SARH, which led to a formal decree in 1980 that set aside 36,000 hectares.

Its proximity to Mexico's third-largest city is both a bane and a blessing. Even as its assets are easily visible, and valued, given its status as a protected natural area and biosphere reserve, Bosque La Primavera (BLP) suffers endless insults from its awkward linkage to the Guadalajaran metropolis. If it provides a representative sample of the Jaliscan landscape, it also puts on display the politics of Mexican land tenure and use. Some 15 percent of BLP is public, 50 percent private, and 35 percent "social" (ejidos). In 1997, governance over the reserve passed from federal to state responsibility, which includes not only the landowners within the reserve (the government of Jalisco, private owners, and ejidos), but also CONANP, the Jalisco department for the environment and development, the municipalities of El Arenal, Guadalajara, Tala, Tlajomulco, and Zapopan, and all the entities invested in conservation, from federal agencies to civil society to institutions of research and education. The University of Guadalajara took a special interest, and in 1988 it published a proposal for proper management, later adopted into law in 2000. An accord among various agencies was signed in 1996. The forces encouraging degradation were remorseless. But nothing is guaranteed to rouse public, political, and media alarm like fire and smoke.[176]

La Primavera is prone to burn—that is its essential nature. Guadalajara and its affiliated landscape is designed, in principle, not to burn. As the space between them shrinks, natural landscape and cityscape find in fire a soft weld. People light fires, people fight fires. Wildfires kindled in neighboring settlements rush into BLP, wildfires within the BLP threaten adjacent (and invasive) colonias. That linkage creates a kind of neurotic environmental codependence that, while not unique to Mexico—Barcelona, Sydney, Cape Town, and of course Los Angeles all show similar pathologies—speaks with a strong Mexican accent. The BLP fire scene is another expression of modernity, if a macabre one: Mexico is showing the same kind of problematic wildfires as its aspirational peers.[177]

Wide-ranging fires have become a nearly annual event. Most have clustered around the perimeter, mapping closely with settlements. A few have bolted across the reserve. There were large burns in 1998, 2000–2001, and 2005 that overwhelmed the local capacity to fight them—the 2005 fires burned 8,487 hectares out of the 36,000 that made up the reserve. Between 1998 and 2019, 90 percent of the BLP's land had burned at least once (and most of it more than once). Such fires threatened the integrity of the reserve as much as they did the communities around it. Politicians demanded the fires be extinguished. The media broadcast the firefights with breathless drama.

The 2005 fires frightened the city's residents, obsessed the media, and prompted an all-out campaign. Fires had broken out on April 11, 23, and 24 where town and park met, but were contained. Then two fires began six kilometers apart at 2 p.m. on April 25—investigators subsequently classified them as "intentional," but no particular persons were identified or any motive given. They quickly bolted beyond control and flooded the city with smoke. Schools closed. Businesses reduced hours, as did public agencies. Some 1,200 firefighters rallied to the lines; one helicopter pilot died in a crash. For the next four days officials confronted an "environmental disaster without precedent."[178]

The park's managerial structure had crystallized in 1997; the 1998 fires soon overwhelmed it, but they also inaugurated the modern era (reliable statistics become available, among other reforms). A management plan for BLP was published in 2001. A fire plan based on prevention, control, and active suppression followed in 2004, to be overrun by fires the next year. The plan identified rural burning, campfires of travelers, cigarettes, and intentional ignitions—the traditional litany disapproved of by authorities—as the primary sources. The 2005 outbreak burned in the footprint of the 1998 and 2000 fires. Even the most skeptical observers realized that fire was not something that visited BLP once in a while, as outbreaks did cities, but was intrinsic to the character of the preserve. (That the smoke was visible from CONAFOR's national fire center in Zapopan was an unspoken irony.) Whether by choice or chance, BLP's fires of spring were here to stay.[179]

Still, Bosque La Primavera was as vital to Guadalajara as Desierto de los Leones or El Ajusco to Mexico City. CONAFOR commissioned a history to preserve the "great lessons" of the 2005 fire, though one told, in good journalistic fashion, with human-interest hooks from those affected by it. Manuel Reed Segovia, director general, wrote a brief prologue. The publication ended with a long interview with Enrique Jardel Peláez, who cautiously introduced a counternarrative. The burn had been widespread but of low severity; recovery was vigorous—all it took was the seasonal rains. The fire panicked the city but renewed the park, and, by peeling away combustibles, it would lessen the severity of future burns. This was a bold pronouncement, but it was issued humbly, not as a manifesto. Even for someone who had made a vocation of fire ecology, who understood how nature regenerated after the most alarming burns, Jardel Peláez admitted to a lurking fear when you "confess that you let go of a fire."[180]

The core problem was not Bosque La Primavera but its unsettling relationship to urban growth. The drivers behind the bad burns were fundamentally social and political, and they were not resolved by more engines and helicopters. But if the city brought problems, it could also contribute institutions to inquire into and resolve them. The preserve, after all, was in the backyard of the University of Guadalajara, CONAFOR's fire center, and INIFAP. After the 2005 fires the university organized a forum to assess the state of the reserve and its threats, and to bring together all the actors with an interest in BLP—its experience with the biosphere reserve at the Sierra de Manantlán provided a loose model. The forum invited all the varied agencies of government concerned with BLP, a range of academic institutions, and civil society. Of the 59 papers presented, 36 were from the University of Guadalajara, not only from scientific departments but from the arts, humanities, and social sciences as well; the fires engaged Mexican culture, not just its fire agencies. When serious fires returned in 2008, a second forum followed. More environmental insults, more fires, more alarms—this time, in 2017, Jardel Peláez was enlisted as a consultant, and in late 2019 a research contract was signed with the University of Guadalajara to prepare an analysis of Bosque La Primavera's pyrogeography and recommend courses of action. A third forum followed.[181]

The assignment was a query Enrique Jardel knew well from his years at Las Joyas. But the outskirts of Guadalajara introduced considerations that had not been pertinent in the remote Sierra de Manantlán. For one thing, public scrutiny was intense: for urban residents and media commentators, wildfire was not only a pressing crisis but a vivid emblem of the degradation of the reserve. When fires broke out in La Primavera, he lamented, the media lavished overwrought terms like "destruction and devastation"; "experts," including some university researchers, spoke of the fires' "threats"; environmental groups fretted over water and browse for wildlife; foresters rushed to "restore" the damage by replanting; and then the issue would be forgotten until flames again appeared on the horizon and smoke steeped the city in foul air. Fire was denounced as "the worst enemy of the woods," when the true threats were urban sprawl, air pollution from industrial combustion, and climate change. These were classic urban-based perspectives, but BLP was as much a park in a city as a city in a park. Increasingly, the quarrel was less between urban and rural understandings than between competing visions among urban elites, one group that recycled historic judgments and another that sought to fuse ecological science with traditional knowledge about fire.[182]

The final draft of the University of Guadalajara's report appeared in September 2020. Applying the lessons learned over the past 20 years, it distinguished between "fire" and fire's regimes, it discriminated between classes of severity in the places that burned, and it distributed causes beyond the dismissive appeal to rural ignorance. What the Bosque La Primavera needed was a program of integrated fire management, though one adapted to the presence of an urban environment along its border (and in some cases within it). Instead of simply beefing up suppression or seeking a delusionary ban on all fires, it needed a nuanced program of targeted prevention, selective suppression, prescribed burning, mindful rehabilitation, and pointed risk mitigation along the rural-urban border, together with research, training, capacity building, and interagency cooperation. And it required public education so that the citizens of Guadalajara could understand that what was proposed would actually improve the protection of both the city and the reserve.[183]

This was a complex program, a cameo of the Red Revolution, and it demanded a transition that few nations, even in principle, had attempted. The likelihood was that, in Bosque La Primavera, as in the comparable parklands around Southern California, Catalonia, Provence, New South Wales, and the Eastern Cape, political urgency would emphasize firefighting and immediate threats to public safety and leave the fundamental reforms for a putatively more favorable time. Eventually the realization would diffuse through the informed public that the most favorable time had already passed, that conditions would worsen, that suppression's costs could only escalate if the causes behind those threats were not reduced. Modernity had its pyric pathologies.

PROYECTO AJUSCO

Airshed. Somehow the center held. Since 1950 Mexico City's metropolitan zone had increased arithmetically from 3.5 million inhabitants to 22 million in 2020. Its land area had swollen from 262 square kilometers in 1950 to 1,495 square kilometers in 1990. It was the largest metropolitan area in North America, and the sixth largest in the world. By 2000 it held 20 percent of Mexico's entire population and claimed over 30 percent of its industrial output. Its intense urbanization concentrated and magnified every feature, like a diesel piston compressing fuel and oxygen into an explosion. As it had been for centuries, the valley was the center of Mexico, and the ills of its industrialization a microcosm for the country's halting modernization.[184]

Since the Conquest every element of the Valley of Mexico had been remade on a gargantuan scale. Water that had flooded the valley and created a mosaic of marshes, lakes, shorelines, and islands was gone, leaving the exposed soil to be paved over or blown away. Earth was converted from agriculture to city. Air was no longer seasonally filtered with the filmy smoke of milpas and cooking fires but suffused with the effluent of industrial combustion, which the enclosed basin made stagnant and foul, among the most polluted atmospheres anywhere.

And fire? Fire had supplied the energy for most of this transformation, and in keeping with Mexico's character, it was a mestizaje of old and new, with the relics of ancient fire practices still lingering amid the evolving order of industrial combustion. The persistence of agricultural burning complicated modern management, but the drivers behind the valley's wretched air were literally the ones steering cars and other engines of a fossil-fuel society. In 1998 internal combustion engines contributed 75 percent of air pollutants, with the other 25 percent coming from factories, services, and "natural sources." In the late 1990s, efforts to replace old vehicles with newer ones outfitted with catalytic converters, and to substitute for sulfur-rich gasoline, measurably improved air quality, until the 1998 wildfires temporarily overwhelmed that success. It would not be so easy, it seemed, to replace one fire with another.[185]

The optics, as publicists like to put it, were not good—literally. The particulates of woodsmoke fill the visible spectrum of light: they are eminently visible in ways that greenhouse gases are not. Depending on the time of year and the extent of burning, they can similarly saturate the sprawling basin that is the Valley of Mexico. A major study in 2006 found that "organic aerosols" contributed more than half the mass of fine particulate matter, that landscape burning was responsible for 66 percent of those particulates, and that nitrous oxide emissions were two to four times greater than those "typical for fires."[186]

The reasons were unclear. So was the implication. Landscape burning does not go on all year, as fossil-fuel combustion does, and while eliminating the sources would cut smoke during the dry season, "fire suppression efforts may have environmental costs as well as benefits," such that a ban on burning "may be inconsistent with proper forest management practices." In the end, for all the high-tech and multinational science brought to bear, the implications of the study were "not straight forward." While it is possible to outfit cars with catalytic converters and to replace industrial combustion with renewable energy sources, no technological surrogate exists to replace the ecological alchemy of fire. Some landscape fire not only would remain but had to remain.[187]

The Valley of Mexico's airshed further complicated the issue because the emissions from any and all fires mingled promiscuously, with many consequences, not all of which could be removed and some of whose removal might further destabilize the scene. It was not possible to centrifuge out the desirable from the undesirable components. The valley's plume of pollutants—aerosols, ozone, nitrous oxides—even perturbed regional weather downwind, and by adding to the greenhouse gases that were filling Earth's atmosphere, they were helping warm the planet. Modern Mexican fire would not be viewed through a glass, darkly, but through a sky, murkily.

Fireshed. There were plenty of efforts to elucidate the Valley of Mexico, which remained Mexico's center—its most populated, most environmentally disturbed, most politically potent landscape. Since 1917 the federal government had sought to protect the soils and the watershed of the volcanic peak El Ajusco, with special attention to fire. Ajusco's compromised landscape was widely regarded as a visible symbol of the region's environmental ills. So it is not surprising that here in 2000 a major research program commenced. It had a champion in Dante Rodríguez Trejo, who in the early 1990s conceived the project—what might accurately be called a vision quest—to describe the fire ecology of Mexico. At the time a Mexican literature on fire could hardly be said to exist.[188]

The project had a point of entry with the experimental forest that UACh had in Izta-Popo National Park, and it had a predecessor in the late 1970s: a group led by Antonio Sierra Pineda, a Chapingo graduate, had convinced the regent of Mexico City, Carlos Hank González, to create a commission to coordinate agricultural development (COCODA) and help reverse the adverse effects of deforestation by farmers and herders in the valley's southern rim of mountains. The primary focus was reforestation, but the program included a fire protection component (it was the setting for the 1985 experiments in prescribed fire). In 1988 the program was absorbed into a broader-themed coordinating commission for rural development (COCODER). The program became a hothouse for germinating national projects—fuel inventories, dispatch centers, a synthesis of international information about fire behavior and ecology, out of which came experiments with expert systems (EXTINGE) and even artificial intelligence. Not least, COCODER launched Dante Rodríguez Trejo in his stellar career, ultimately as a professor-researcher at UACh.[189]

The project's conception was bold, and befitting its prominence, Ajusco's dome was the highest point in Mexico City, adjacent to Parque Nacional Cumbres del Ajusco. Its announced goal was to investigate ways to integrate fire ecology with human communities and traditional fire practices, "to conciliate the ecological role of fire in this ecosystem with the communities' use of fire and fire prevention and control." What happened in La Sepultura or the Sierra San Pedro Mártir would most likely stay in Chiapas and Baja California. What happened amid the peaks of Ajusco would be visible to the 23 million people who inhabited the Valley of Mexico and to the politicians and bureaucracies that governed the rest.[190]

It had to begin with research: Mexico's authorities, both political and intellectual, didn't appreciate fire's inextricable role in Mexico. Proyecto Ajusco would provide that foundational data. Though the circumstances were hardly unique to Mexico, the situation was bizarre: it was as though no one had thought to scientifically study Mexico's waters or winds. What investigations had occurred had focused on fire behavior, the better to control fire. Proyecto Ajusco promised to reorient what was known and to create what was lacking so that a comprehensive understanding of fire would be possible.

The project was also a training program for the next generation of fire managers and scientists, a demonstration site for prescribed fire, a working example of many institutions coming together to perform a practical task, and a public demonstration of what integrated fire management could be. It tested a full gamut of prescribed fires in various settings and intensities. It recorded the astonishing adaptations to fire apparent in the mountains—and the florescence so many plants featured after the rains arrived. It confirmed the value of the right fire to stimulate pasturage and promote biodiversity. With tree-ring scars it established an average fire return interval of five years. It showed to officials, influential ejidatarios, students, and fire crews how to do controlled burns and what such fires could mean for all of them. It supported a flourishing, full-spectrum educational ecosystem of student theses. Even as fire interest propagated among research agencies and universities, notably UNAM and the Universidad de Guadalajara, Proyecto Ajusco reestablished UACh as a major center for fire studies.

Granted its standing as one of the largest megacities on Earth, roughly half of Mexico City was still rural, or at least not wholly urbanized. Since most fires started from traditional usage such as pasturage, flame and smoke were a perpetual reminder of Mexico's incomplete transition to a developed country, and of the political accommodations even public lands had to make. Increasingly, the lands served another transient population: tourists and recreationists. On the first issue, Proyecto Ajusco showed that routine surface fire did no irreparable damage, but could actually promote ecological health as well as reduce combustibles that could stoke major fires. On the second, it found through surveys that scenes recently burned and renewed were viewed more favorably by visitors than unburned sites. There was no justification for a total fire ban, only discussions over what form fire should take.[191]

The project inevitably devoted an immense amount of attention to the Mexican mountain pine (*Pinus hartwegii*). No other Mexican species has had so much research devoted to its multiple adaptations to fire. Situated on a volcanic mountain subject to routine disturbances, it should surprise no one that a tree that has evolved to withstand wind, drought, quakes, and eruptions should also accommodate fire. But the project went further. It made the Mexican mountain pine a poster child for integrated fire management, what Rodríguez Trejo considered Mexico's answer to the longleaf and ponderosa pines that advertised fire in the southeastern and southwestern United States.[192]

Proyecto Ajusco was too vibrant and valued to remain tethered to Mexico City. It provided a platform and model that quickly expanded throughout central Mexico. Then it went further, taking on assignments that spanned Mexico overall, supporting students from Chiapas to Chihuahua and from Guerrero to Quintana Roo. The more it did, the more it inspired. From the summit of Ajusco it was possible to see the pyrogeography of Mexico, and from Mexico, large swaths of the previously invisible fire world that was Earth.

ERUDITO ON FIRE: DANTE ARTURO RODRÍGUEZ TREJO

If a single career could stand as a synecdoche for Mexico's Red Revolution, it surely belongs to Dante Arturo Rodríguez Trejo. Not only did he live through the revolution as a student, researcher, consultant, professor, and participant, from the earliest Marana classes to the Coahuila megafires, but whatever he personally didn't experience, he knew those who had, many of whom he worked with and not a few of whom he had taught. He mentored students, worked with administrators, translated fire science to Mexican conditions, and hosted international visitors. But above all, he wrote. Unlike most fire people and scientists, he liked publishing. Through presentations, journal articles, and books, he became not only one of the founding fathers of the Red Revolution, but its encyclopedist, synthesizer, and chronicler.

His father, Maximino Rodríguez Aguilar, hailed from Oaxaca. Maximino had grown up without a father but was determined to study, and with the help of an uncle he attended a boarding school, then kept on struggling in order to continue to the National School of Agriculture, where he graduated in 1949 as a forest engineer. Dante's mother, María Dolores Trejo Vázquez, was born and raised in Texcoco, where she worked in a beauty salon, and it was in Texcoco that she and Maximino met and married. Dante recalled Maximino as "rigorous, disciplined, and fair"; María Dolores was "very loving, and pampering, but also disciplined." They had ten children.[193]

Dante considered his childhood a privileged one. His parents enrolled him in kindergarten when he was two, and in primary school at age five. They loved to read, kept books in the house, and named him for Dante Alighieri, all of which nudged him toward writing. When he was six, Mexico City hosted the Olympics, which inspired him to dream of a career as a sprinter. When he was seven, he began accompanying his father on forestry assignments. Maximino traveled throughout the country, and at one point the family spent three years at Huixtla, Chiapas, before returning to Texcoco, where Dante completed secondary schooling and college prep before following in his father's footsteps at UACh, graduating in 1983 in forestry. Meanwhile, the national economy had collapsed, and Dante's dream of becoming a forester seemed as delusional as his aspirations for the Olympics.

He got hired to work for COCODA (later, COCODER). Antonio Sierra Pineda was organized and visionary, and became for the young Dante "a second father." At COCODA Dante was introduced to fire as a practical subject. In October 1984 he joined the Mexican contingent to the second Marana course. So much of what he learned fascinated him: applied meteorology, the intricacy of forest combustibles, terrain effects on fire behavior, crew safety, financing, prescribed burns, fire suppression equipment, strategy and tactics, the Albini nomograms for predicting fire spread rates (which seemed "magical")—all of it "a world of new information." It was his first international trip, and with it he found his vocation and a cohort to share it with. "If my father motivated me to be a forester, Sierra Pineda and this course kindled a professional flame." At 22 he resolved to dedicate himself to the study of fire.[194]

The course aroused his interest in two fields of study in particular: forest ecology and physics, expressed as fire behavior. Attending the course allowed him to reposition himself within the Comisión de Recursos Naturales (CORENA, which had absorbed COCODA), to engage with fire as both a participant and an instructor, and to experience his first prescribed burn. He co-directed with Sierra Pineda the country's most extensive inventory of forest fuels (a "memorable experience"). The two men then worked to outfit the first fire dispatch center in Mexico. He oversaw the academia de ecoguardas, drawn from students at the centers of environmental education, as they worked on the lower flanks of Ajusco. In 1989 he was accepted into the National Academy of Forest Sciences as an academic collaborator. It was a second schooling, in which he learned about fire on the ground, not in a classroom—fire was hardly taught at all, anywhere. But his scholarly instincts were not dormant. "Always I have liked to publish." All this activity led to his first publications—a synthesis of the literature on Mexico's fire ecology and, harking to the enduring appeal of athletics, a study of the physical attributes suitable for firefighting.[195]

It all made a strong résumé, and because of it he was offered a position as subdirector. He declined, as he declined similar offers throughout his career. Why? His secret dream was to become a research professor. He relished the university; he was at heart a teacher and scientist, not an administrator. Moreover, there was so little known about fire in Mexico—it was "practically virgin" territory—and fire's ecology was the foundation for everything else. Sierra Pineda, now with Probosque, invited him instead to take responsibility for fire protection in Texcoco, which allowed him to work toward a master's degree at the Colegio de Postgraduados. For his thesis he analyzed the economic efficiency of fire management in Mexico City. That took him into the library as well as the field, and it confirmed that a university professorship was his most desirable, if not yet an attainable, dream.[196]

The next fork in the road came with an offer of a temporary post at UACh to teach while writing a guide to fire prevention and suppression. The university he regarded as "a noble institution," at which he could help others, as UACh had helped his father and himself. Regardless, he declined the invitation and from 1992 to 1995 worked on the guidebook as an academic collaborator in the National Academy of Forest Sciences. Taking the offered temporary post, he felt, would disrupt the career he had advanced over the past eight years, though the ecologist in him noted that "perturbation and regeneration are related."

Meanwhile, the text had lengthened over the course of three drafts and nearly four years. In 1994 *La lucha contra el fuego* (*The Fight Against Fire*) appeared, followed two years later by a 630-page book that summarized all aspects of Mexican fire. When *Incendios forestales* was published, the world literature could almost be tallied on one's fingers—one book dealt with the fundamentals of fire ecology, two books with the fire ecology of North America, one with the fynbos of South Africa, two with Australia's biota, and two with the circumpolar boreal forest. The United States had three general texts on fire management (one outdated but still in print), and Australia had one. Mexico joined a select club of countries that could boast of having not only a serious fire scene but a scholarly treatment of it.[197]

Despite the book's bulk, Dante knew its limitations. There simply wasn't enough known about fire in Mexico, not even a cursory survey; that information would have to be created. Dante moved on to a doctorate at the University of Florida, where he looked at regeneration in longleaf pine, with study plots at the Joseph W. Jones Ecological Research Center. Anything to do with longleaf involved fire, usually sooner rather than later, so in researching fire and reforestation Dante was studying, in a North American context, the twin obsessions of Mexican forestry. Not least, Florida was the epicenter of prescribed-fire enthusiasm in the United States, and it was the University of Florida, and later the Nature Conservancy in Mexico, that gave Dante the final push to convert to integrated fire management. Though fire ecology and prescribed fire were, so to speak, in Florida's air and water, it is revealing that even at the University of Florida there was no graduate course dedicated to fire ecology.

The doctoral program lasted from 1996 to 2000. The university was a "great stimulant," but Dante's years there were undeniably hard. He was married, with one child, and spent these four years away from his close family; he was in a foreign land where Spanish was not the prevailing language; even his mother died his first year away—the doctorate brought "unexpectedly great challenges" and left a "permanent mark." But those years were critical and fondly recalled, and they positioned him ideally to ride the wave of money and interest in fire that followed the 1998 outbreaks, which occurred halfway through his stay in Florida. (Even Florida had a record fire season that year; it was during a small symposium on the fires that Dante met Ron Myers of TNC.) Not least, those years launched him on his vision quest to find where possible, and to create where necessary, all the information required to understand fire's ecology in Mexico, and to write it up. Until it was written, it wouldn't be real to those who made policy.[198]

He was indefatigable. When he returned to UACh, he conceived the notion of Proyecto Ajusco, taught courses on fire ecology and management, accepted students for advanced degrees, and dedicated himself in particular to the fire-adapted Mexican mountain pine. He connected with the other pioneers of the new era—it was not a large fraternity. With two Mexican colleagues and one American, he wrote a short introductory text for students, *Educación e incendios forestales* (2002). With another U.S. colleague, Pete Fulé, he wrote about fire management and restoration in Mexico's pines. With Roberto Martínez he surveyed the state of fire in Mexico and Mesoamerica; for *Ambio,* he offered a tutorial for a global audience. He was a part of the UACh cadre that evaluated CONAFOR's fire capabilities from 2004 to 2008 and generated a supplemental assessment of the agency's corps of firefighters. With Germán Flores he coordinated publication of the 2006 symposium organized by CONAFOR. He knew Ron Myers, whom he regarded as a pillar of integrated fire management throughout the region; he frequented TNC's "legendary workshops"; and he participated in the 2006 Berkeley gathering at which TNC conducted its reconnaissance of fire on Earth. With Myers he wrote an article about using oaks as a guide to restoring fire in pine-oak forests, and a chapter on tropical pine ecosystems for a major text. He was active in the reforms that put Chiapas in the vanguard of Mexican fire management, with special focus on Villaflores. He wrote a guide to prescribed burning for FMCN, CONAFOR, and CONANP. He presented Mexico's fires and their management to international symposia. Out of Proyecto Ajusco came a future generation of Mexican fire specialists. He worked with Roberto Martínez and Alfredo Nolasco to amend the legal standing of controlled burning. He even served as coordinator for the fire brigade organized at UACh in 2008, which was active in the 2017 Tláloc fire. He self-deployed to the 2011 Coahuila conflagrations and offered a preliminary assessment—fire in arid zones was not well understood. That Mexico, "with a bit of luck," had withstood the fires he took as a measure of the lessons it had learned since 1998.[199]

Then, in 2014 and 2015, he published his *summa summarum*—a magisterial two-volume, 1,705-page panorama of Mexican fire; all fire's ecosystems, all its regimes and behaviors, all its history. As the title indicated, *Incendios de vegetación* (*Vegetation Fires*) ranged beyond forestry, and as its bulk testified, it sought to be as comprehensive as possible. Fifteen institutions contributed funding toward its publication—UACh and the Colegio de Postgraduados, of course, but also all the Mexican agencies with fire responsibilities, and even the U.S. Forest Service and USAID. Only someone who had grown up with and participated in the grand swell of reforms and research—had absorbed it bit by bit, wave by wave as it evolved—could have imagined a project at such a scale or dared to undertake it.

It's a strange book, unlike anything else in the global literature on fire, nearly all of which has emerged out of scientific research and reads like compilations of data. Dante's bursting tomes stand by themselves. Perhaps their closest analogues are the immense volumes penned by Spanish scholars who lived in Mexico in the sixteenth century—the *Cosas de Nueva España*, the *Historias de Indias*—that recorded everything from events and places to flora and fauna. Vital or tangential, everything pertaining to fire Dante included, and what didn't exist but was needed, he would create through the research embedded in Proyecto Ajusco. The works of Oviedo, Las Casas, Durán, and the others had disappeared into archives, only to be published centuries later as historic relics. Dante's were released to the world.

Like those predecessors, Dante had encountered a previously undiscovered world, a world of fire, and he was determined to describe it in all its strangeness, practicality, and power. He assembled information not only from the natural sciences, but from myths, paintings, fire behavior and fire ecology, and the history of fire protection and fire research, with photo galleries of every significant personage—all in all, every imaginable aspect of fire in Mexican land and life.

Yet he knew how fragile reforms could be in Mexico, where institutions could be spun about or blown away with each sexenio. Too many reforms were still impermanent, and too much funding could be arbitrary. Paradoxically, CONAFOR's successes led to reduced budgets; after 2011 its finances were cut by 30 percent. Dante knew that "practically all of the population, urban or rural, visualizes only the negative side of the fire." New officials viewed prescribed fire with suspicion. The state government in Chiapas, presented with a world-class model for integrated fire management, sought to restrict, not expand, controlled burning. The 2019 fire season was the fourth-worst on record, with eighteen fatalities.[200]

Dante had lived through it all, and always he had written (and not just about fire; there was a short popular book on Mexico's trees, and, completing the track loop, the former runner even published a book on athletics at UACh). Always, in writing as in teaching, he felt it was important to communicate "con pasión." Yet passion needs discipline and tenacity, and these traits Dante learned from his parents; those virtues were reinforced as he traversed one crossroads after another. That, as much as timing, explains his achievements, as the would-be sprinter became the marathoner of Mexican fire research; the eager student, an honored professor; the aspiring writer, the great encyclopedist of Mexico's fire ecology and the chronicler of its fire history.[201]

MEXICO JOINS THE NEW WORLD ORDER ON FIRE

In the handful of years that followed the 2011 fires, Mexico's Red Revolution climaxed. Integrated fire management was established as the consensus concept. Model research programs and fire demonstration sites existed. Prescribed fire in forests was legalized. Wildland fire was taught in universities. Fire suppression had capability. CONAFOR's fire program had a handsome center for dispatch and operations in Zapopan. Rodríguez Trejo's *Incendios de vegetación* brought the whole subject—its data, its methods, its history, its people—into formal scholarship. An enterprise that 30 years earlier struggled to cope with fires, that had almost no knowledge of fire's behavior and effects outside of personal experience, that shunted fire among agencies that didn't care about it, had leaped into the top tier internationally.

But Mexico achieved global standing just as the globe was in a fire-informed upheaval. The developing world glowed with burns from traditional practices, amplified by massive land-clearing fires in rain forest and peat. The developed world rediscovered wildfires it thought it had banished or stuffed into tidy boxes for filing in archives. Even Mexico's aspirational peers found themselves nearly helpless as megafires overran all their vaunted machines. Australia suffered an outbreak of epic bushfires in 2019–20 that blasted its eastern seaboard, submerged Sydney, Canberra, and Melbourne in deadly palls of smoke, and sent a smoke plume around the globe. Even before it invaded Ukraine, Russia was helpless before immense fires in Siberia. Fires and heat waves hit Britain, Sweden, Germany, and France like a returned plague. Subsquently, fires continued to hammer Mediterranean Europe with appalling casualties; Greece and Portugal burned with a savagery worse than anything Mexico had endured in its catastrófica temporada. Fires blasted over tourist meccas in Maui and Rhodes. Even Greenland burned. Dozens of disciplines that had ignored fire as a subject now raced to absorb it.

The notion was broached that humanity's fire practices, both from living and lithic landscapes, had become so powerful that they were informing a fire age. Earth was experiencing a matured Pyrocene.

As with other countries that had undergone the transition, the world Mexico entered was not the world it had expected. Originally, the competition between the two realms of combustion was welcomed: fossil biomass was the means to power and modernity that would lift the republic and its peoples into wealth and well-being. But the many fire Mexicos coalesced into two Mexicos. Then, as fossil biomass replaced second nature with a third nature, reshaping land use and fluffing the atmosphere with greenhouse gases, the two realms of combustion began to collude. Mexico displayed not only the powers of those countries it compared itself to, but their pyric pathologies as well.

Certainly, the U.S. example looked troubled. Megafires were becoming the norm; prescribed fires struggled to achieve a scale sufficient to influence wildfires; suppression expenses were hollowing out the Forest Service, as multibillion-dollar firefighting costs mounted; frontline fire crews were overworked and underpaid and ranks went unfilled. California suffered cascading conflagrations that burned into urban centers like Santa Rosa, and in 2018 burned over the town of Paradise with 85 fatalities—as many as the Big Blowup of 1910 had caused, and more than Mexico had suffered in 1998. Yet the United States seemed as incapable of dealing with mass burnings as with mass shootings. Its political system spiraled into dysfunction to such a degree that a majority of one of its two political parties believed the 2020 election fraudulent. A mob rallied by the political loser stormed the Capitol as election results were being confirmed. The American way of fire seemed as problematic as the Washington neoliberal consensus.

Among the pushbacks against modernity's experiment with fire was an interest in traditional fire practices, or what was termed "cultural burning." Fire protection had been a high-modern project to replace messy, inherited, largely oral knowledge and the palimpsest of local landscapes such knowledge produced with what elites regarded as a rational reconstruction. As the failures of the enterprise mounted, observers turned to legacy practices that had worked for hundreds, if not thousands, of years. The explanations locals gave might seem like nonsense to educated elites, but what they actually did worked, where science-based projects often failed. The old knowledge was not just about data and technique, but about how live on the land with fire as a companion.

Europe's second-wave expansion had been a dual project: it colonized, most dramatically by means of settlers, and it brought industrialization. In terms of fire history, it interbred Europe's second-fire practices with those of Indigenous peoples, and it began replacing second-fire with third-fire. Now fire felt the riptide pull of decolonization. Australia reexamined Indigenous fire practices and sought opportunities for joint management (national parks were fertile sites). Tribes in Northern California like the Yurok and Miwok became celebrities for offering a critique of modernity's fire suppression strategies and found themselves consulted as local prescribed-fire councils strove to put fire back into the hands of landowners and private citizens. Catalonia looked to traditional agriculture, with its cultivated fields and orchards, its choreographed flocks, and its heritage of controlled fire as a way to contain metastasizing fuels in the countryside around Barcelona. In the face of rural depopulation and rampaging fires, Portugal investigated how to re-create traditional landscapes in more contemporary ways, perhaps amenable to ecotourism. As Europe showed, the struggle was not a question of settler colonialism but of coping with the third nature that third-fire had helped birth.

The conversion to a fossil-fuel society had been incomplete, most notably in the countryside. The pyric transition ultimately occurs when the old practitioners leave, either through migration or death. Old habits and oral traditions depart with them, while the newcomers (if any appear) assume third-fire as the reality. Mexico's partial transformation had left many rural poor still in place, along with revolutionary principles that enshrined their right to land, while at the same time the commitment to fossil fuels disrupted historic fire regimes and imposed its own deleterious side effects. Yet just when Mexico had cause to welcome its enclaves of traditional fire knowledge, a slowing birth rate, the pull of urban centers, and emigration were drawing down the pool of rural practitioners. It was likely that many landscapes would be emptied of the fires that had, in one form or another, recycled and renewed them, or rather, if no surrogate were fashioned, those fires would be replaced by feral flames.

Living with a fire age would not be easy. Mexico's elites continued to view fire as a disaster, as something intermittent, at best seasonal, dramatic but not comparable to the chronic, seemingly endemic catastrophes that plagued Mexico, like its economic malaise, or its criminal cartels and the loss of state control, or its troubled politics. They viewed landscape fire as a relic from an archaic past: curious, irrelevant, an anachronism like burning wood and charcoal for cooking, quaint, sometimes burdensome, and occasionally threatening.

Elites, the media, and most of the urban public saw fire's spasms, not its quotidian integration with Mexico's natural estate and social order. They saw it as an ecological riot or uprising that force could quash or that economic development could erase, not as a necessary feature of ecological integrity and megadiversity, which was also a measure of cultural diversity, not as something to be celebrated only in ceremony and myth but integral to the character of Mexico. They saw it, when properly hidden in machines, as a sign of progress, and, when flaming in milpas and montes, as a badge of primitivism. They knew fire, when it combusted petroleum, as a source of raw energy, the prime mover of Mexico's economy. They knew it sublimated into electricity or petrol and the petrochemicals that bolstered the Green Revolution. They only vaguely viewed it as the great recycler and renewer of Mexico's natural economy. They might come to appreciate it in nature reserves, and they might admit that an unhinged climate would require them to replace oil-fed energy with renewables, but they wanted flame and its smoke nowhere around the cities they lived in. They did not comprehend that third-fire had not only befouled the atmosphere with greenhouse gases but bequeathed a deep fire debt with regard to second-fire. They did not appreciate that the more people succeeded in ratcheting down the burning of lithic landscapes, the more they would have to ratchet up their burning of living landscapes.

The one certainty about the future was that it would have fire, probably a lot more fire, and what those fires would look like would depend on what people did and did not do.

EPILOGUE

BETWEEN THREE FIRES

THE PLAZA de las Tres Culturas in Mexico City celebrates the three epochs, and three primary peoples, of Mexican history. One part is Tlatelolco, a pre-Hispanic city-state, here partly excavated. A second is the Colegio de la Santa Cruz de Tlatelolco, the oldest institution of higher learning in the New World, built by Spanish friars in 1534. The third is the former offices of the Secretariat of Foreign Affairs, later repurposed into a memorial museum dedicated to the victims of the Tlatelolco Massacre of 1968. Together the plaza distills into a single square the Indigenous, European, and Mexican eras that have, in succession, occupied the land.

No comparable memorial exists to honor the three fires that have illuminated and powered Mexican history. But perhaps no monument is needed since the fires permeate all aspects of life. Instead, consider the customary brasero used for cooking, which, like Mexican fire, stands on three legs. One is the fire of nature—fire that exists due to Mexico's natural estate. A second is the fire that humans tamed from nature, with which people could expand and reorganize nature's fire regimes. Those regimes changed kaleidoscopically as people came and went, as empires and trade routes arrived and receded, as new species appeared and old ones vanished, as inhabitants modified the vegetative cover and chose where and when to start their fires. Anthropogenic fire completes Mexico's enduring fire triangle: biodiversity, cultural diversity, pyrodiversity. Fire within living landscapes was elastic, but for both first- and second-fire there were bounds and borders. For third-fire there were none.

Third-fire involved the combustion of lithic landscapes. This had no existence apart from people, and no bounds save what people imposed. By 2020 it was no longer new. (Fuel theft even joined the other lucrative activities of Mexico's criminal cartels.) As Germán Vergara notes, "Mexico had lived under a fossil energy regime for roughly 70 years—about the same time it took to shift to fossil fuels." The new regime created a slate of novel wildfires caused by oil well blowouts, explosions at gas pipelines and plants, and even seagoing tankers. But its most powerful consequences were indirect: it empowered people with a new variety of combustion which they used to reconfigure Mexico's land, air, waters, and biota. As raw energy, third-fire powered the Mexican Miracle. As fossil fallow, it made the Green Revolution possible and shifted biodiversity from agricultural fallow to protected natural areas. As a prime mover for machines that could counter fire, it made possible Mexico's Red Revolution. As a source of essentially unbounded pollution, it tainted every aspect of Mexico it touched. It contributed to global pollution and, through greenhouse gases, to planetary climate change. It catalyzed a third nature that was not only powered by third-fire but stuffed with plastics, asphalt, and petrochemicals. It made Mexico a developed country, with all the assets and liabilities that designation conveys.[1]

It meant Mexico no longer passed between two fires but between three.

The Desierto de los Leones—the central nature preserve abutting Mexico City, and the site of Mexico's earliest experiments in reforesting, of its first dasonomic survey (by French foresters), and of its first national park—has long been an arena for controversies over landscape fire and for experiments in controlling it. It lies in the backyard of Mexico City—an enduring emblem of threats and hopes. Here, if anywhere, might be found a plaza of the three fires, or, to mix metaphors, the cosmic race of combustion that distills Mexico's fire history.

I had the opportunity to visit the Desierto shortly after the 1998 fires with Dante Rodríguez Trejo and Antonio Sierra Pineda as guides. The

Desierto offered a veritable outdoor museum of the astonishing adaptations of central Mexico's biota to disturbance generally and to fire specifically. Among the renewing tussocks, the budding flowers, and the resprouting branches and trunks of pine and oak lay a downed log. It had been split by lightning, scorched by flame, and killed, prior to the recent fires, probably by polluted air.[2]

Here, it seemed to me, was a cameo of, and a material testimony to, Mexico's fire saga as it moved from natural fire to anthropogenic burning to industrial combustion. The source of the flame in this particular case was unknown, and the tree's cause of death not certain, but the scene worked as allegory. No one knows how long the epoch of the Fifth Sun will last or how widely its new fire will propagate. But that thrice-struck log in the Desierto de los Leones inscribes a deep chronicle. It can stand as a synecdoche for Mexico, as Mexico can stand for Earth.

AUTHOR'S NOTE

FIVE SUNS tells the story of fire in Mexico, and the story of Mexico as refracted through fire. I became interested in the project after touring scenes of the 1998 fires at the invitation of Dante Rodríguez Trejo. That visit resulted in an essay. A decade later an invitation from Roberto Martínez to consider a book-length work (along with a visit to CONAFOR's fire center in Zapopan) resulted in a promise that I would write that book when I could. My retirement from classroom teaching made the project possible.

It's easy to say what the book is not. It's not an archive-driven history; the archives hold very little on fire, which, apart from the occasional disaster, was simply not a category of interest that mattered to chroniclers, academics, administrators, and even naturalists until well into the twentieth century. Documenting it would be like noting the morning sunrise or an afternoon breeze. Not until the mid-twentieth century was there even an agency that took some routine interest in landscape fire, much less in the totality of fire in its natural and anthropogenic manifestations. Fire was simply there. Until the last quarter of the twentieth century, documentation is sparse. Fashioning a narrative requires a lot of interpretive interpolation.

Researching had other difficulties. A planned trip to launch the project in late 2018 was aborted by the need for cancer treatment. Fortunately,

I was able to divert my funds toward a collection of autobiographies by six of the founders of modern fire management in Mexico, which has been published as *Seis vidas en fuego: Seis cuentos de México*. The stories proved so compelling that I included abbreviated profiles in *Five Suns*. Meanwhile, COVID-19 shut down travel and libraries. And to further the difficulties, I can read Spanish adequately but speak it abysmally. Instead, I have relied on my fluency in fire.

The book may sit oddly within Mexican and Latin American historiography. The fact is, I didn't write it to answer questions posed by current or past historians who have influenced the field. *Five Suns* takes fire as its narrative axis, around which—as around a *fogata* or *hogar*—peoples and topics gather. It uses fire to illuminate other themes. It contains all the power and frailty such an approach entails. It completes a promise I made many years ago to Mexican colleagues that I would write their national fire epic as best I could.

Special thanks go to Dante. Even sluggish readers will recognize the attention granted him as the concluding profile in the final chapter. I make no apologies. At every step, Dante has been a prod, a source of tangible assistance, and an inspiration. Twice he helped me hire a research assistant from among his students to explore the Archivo General de la Nación and then to track down documents that I could not find or was unable to request through interlibrary loan. Eduardo Cepeda and Ramón Guadarrama deserve my thanks, and Dante merits the attention I have extended.

So many others have contributed, especially during my preliminary reconnaissance of the topic—Christopher Boyer, Emily Wakild, Mary Huffman, Ron Myers, Alexandra Zamecnik, Pete Fulé, Citlali Cortés, Diego Pérez Salicrup, Laura Ponce Calderón, Baruk Maldonado, Paul Weeden. The Interlibrary Loan group at Arizona State University was inventive and indefatigable in tracking down documents; the project would have been impossible without their tenacity and courtesy. Matt Toro overcame his initial doubts and converted his wizardry with GIS into two maps infused with a touch of cartographic aesthetics. Matt Gleeson edited my nonexistent Nahuatl, my reading-competent Spanish, and my often vague English (I seem to have a flair for untethered pronouns) with both mastery and grace. And Allyson Carter (and the University of Arizona

Press) deserve special thanks for waiting patiently while I worked my way through a queue of projects to arrive finally at Mexico.

Five Suns completes my North American trilogy of fire histories, from the tropics to the tundra. Given my age I expect it will be the last of the country-scale (much less continent-scale) fire studies I attempt. It feels like a good angle of repose.

ABBREVIATIONS

BLP	Bosque La Primavera
COCODA	Comisión Coordinadora para el Desarrollo Agropecuario
COCODER	Comisión Coordinadora para el Desarrollo Rural
CONABIO	Comisión Nacional para el Conocimiento y Uso de la Biodiversidad
CONAFOR	Comisión Nacional Forestal
CONANP	Comisión Nacional de Áreas Naturales Protegidas
CSU	Colorado State University
ENSO	El Niño–Southern Oscillation
ERI	Ecological Restoration Institute
FAO	Food and Agriculture Organization (of the UN)
FMCN	Fondo Mexicano para la Conservación de la Naturaleza
ICS	incident command system
IFM	integrated fire management
INIFAP	Instituto Nacional de Investigaciones Forestales, Agrícolas y Pecuarias
NAFC	North American Forestry Commission
NAFTA	North American Free Trade Agreement
NGO	nongovernmental organization
RBMM	Reserva de la Biosfera Mariposa Monarca
RBSM	Reserva de la Biosfera Sierra de Manantlán

SARH	Secretaría de Agricultura y Recursos Hidráulicos
SEDENA	Secretaría de Defensa Nacional
SEMARNAP	Secretaría de Medio Ambiente, Recursos Naturales y Pesca
SEMARNAT	Secretaría de Medio Ambiente y Recursos Naturales
TNC	The Nature Conservancy
UACH	Universidad Autónoma Chapingo
UAM-X	Universidad Autónoma Metropolitana Xochimilco
UF	University of Florida
UNAM	Universidad Nacional Autónoma de México
USAID	U.S. Agency for International Development
USFS	U.S. (USDA) Forest Service
UW	University of Washington

NOTES

PROLOGUE

1. Fray Bernardino de Sahagún, *Florentine Codex: General History of the Things of New Spain*, trans. and ed. Arthur J. O. Anderson and Charles E. Dibble, *Book 7: The Sun, Moon and Stars, and the Binding of the Years*, Monographs of the School of American Research 14, part 8 (Santa Fe: School of American Research; Salt Lake City: University of Utah Press, 1953), 25–32.
2. For my understanding of the solar calendar, I rely on Eduardo Matos and Felipe Solís, *The Aztec Calendar and Other Solar Monuments* (Mexico City: Conaculta-Instituto Nacional de Antropología e Historia / Grupo Azabache, 2004); on the creation of the calendar, see p. 72.
3. Matos and Solís, *Aztec Calendar*, 70.
4. Matos and Solís, 72.
5. On the temple, see Eduardo Matos Moctezuma, *The Great Temple of the Aztecs: Treasures of Tenochtitlan*, trans. Doris Heyden (London: Thames and Hudson, 1988). On the range of the Old God, see Hasso von Winning, "The Old Fire God and his Symbolism at Teotihuacan," *Indiana* 4 (1977): 7–62.
6. An excellent summary, which ties the New Fire Ceremony to political developments, is Catherine R. DiCesare, "A New Sun Emerges: The Aztec New Fire Ceremony in Word and Image," *Word and Image* 38, no. 3 (2022): 190–206. For the archaeological basis behind current thinking, see Christina M. Elson and Michael E. Smith, "Archaeological Deposits from the Aztec New Fire Ceremony," *Ancient Mesoamerica* 12 (2001): 157–74. On the early Aztec myths of a First Fire and its fusion with the New Fire, see Christophe Helmke and Ismael Arturo Montero García, "Caves and New Fire Ceremonies in the Central Mexican Highlands: The Case of the Cerro de la Estrella, Iztapalapa, Mexico," *Contributions in New World Archaeology* 10 (2016): 55–100.

7. Fire was everywhere in ceremonies and rituals. The indispensable reference source is Fray Diego Durán, *Book on the Gods and Rites and the Ancient Calendar*, trans. and ed. Fernando Horcasitas and Doris Heyden (Norman: University of Oklahoma Press, 1971).
8. Silvia Limón Olvera, "El dios del fuego y la regeneración del mundo," *Estudios de Cultura Náhuatl* 32 (2001): 51–68. See also Christian Roy, *Traditional Festivals: A Multicultural Encyclopedia, Volume 1* (Santa Barbara: ABC-CLIO, 2005), 315–16. The variety of fire avatars can be bewildering. A helpful playcard (though perhaps too rationally organized?) can be found in "Fire Deities," in *The Oxford Encyclopedia of Mesoamerican Cultures*, ed. David Carrasco (Oxford: Oxford University Press, 2001).
9. Eduardo Matos Moctezuma and Felipe Solís Olguín, *Aztecs* (London: Royal Academy of Arts, 2002).
10. On the relation to Ometeotl, see Michael D. Coe and Rex Koontz, *Mexico: From the Olmecs to the Aztecs*, 6th ed. (London: Thames and Hudson, 2008), 206. I have restricted my account to central Mexico. Other peoples had their own fire gods—for example, the Purépecha had Curicaueri, who mixed with other gods as Michoacán history unfolded. See Yólotl González Torres, *Diccionario de mitología y religión de Mesoamérica* (Madrid: Larousse, 1995); and for a very elaborate explication, Claudia Espejel Carbajal, *La justicia y el fuego: Dos claves para leer la Relación de Michoacán*, vol. 1 (Zamora: El Colegío de Michoacán, n.d.).

FIRST SUN

1. The fullest, early summary of the evidence is Walter Alvarez, *T. Rex and the Crater of Doom* (Princeton, N.J.: Princeton University Press, 1997). The earliest serious linkage between the impact and widespread fires is Jay Melosh et al., "Ignition of Global Wildfires at the Cretaceous-Tertiary Boundary," *Nature* 343 (1990): 251–54. On more recent research, see Sean P. S. Gulick et al., "The First Day of the Cenozoic," *Proceedings of the National Academy of Sciences* 116, no. 39 (September 9, 2019): 19342–51, https://doi.org/10.1073/pnas.1909479116. Or for more popular versions, see Nicholas St. Fleur, "Drilling into Chicxulub Crater, Ground Zero for Dinosaur Extinction," *New York Times*, November 17, 2016, https://www.nytimes.com/2016/11/18/science/chicxulub-crater-dinosaur-extinction.html?searchResultPosition=1;%20Jonathan%20Amos,%20the%20day%20the%20dinosaurs%27%20world%20fell%20apart; and Jonathan Amos, "The Day the Dinosaurs' World Fell Apart," *BBC*, September 10, 2019, https://www.bbc.com/news/science-environment-49651406. For an update that climaxes a general history of dinosaurs, see Stephen Brusatte, *The Rise and Fall of the Dinosaurs: A New History of a Lost World* (New York: William Morrow, 2018).
2. I found a very handy summary of Mexico's physical estate in Richard Rhoda and Tony Burton, *Geo-Mexico: The Geography and Dynamics of Modern Mexico*

(Ladysmith, B.C.: Sombrero Books, 2010). On the central volcanic zone, see California Institute of Technology Tectonics Observatory, "The Unusual Case of the Mexican Subduction Zone," last modified May 14, 2009, http://tectonics.caltech.edu/outreach/highlights/mase/.

3. Carl O. Sauer, "The Personality of Mexico," in *Land and Life: A Selection from the Writings of Carl Ortwin Sauer*, ed. John Leighly (Berkeley: University of California Press, 1963), 104–17.

4. Again, I found Rhoda and Burton, *Geo-Mexico*, a helpful distillation of Mexican climate (ch. 4).

5. For a handy introduction, see Sarah E. Metcalfe et al., "Records of Late Pleistocene-Holocene Climatic Change in Mexico—A Review," *Quaternary Science Reviews* 19 (2000): 699–721.

6. On lightning densities among countries, see https://interactive-lightning-map.vaisala.com/.

7. A thorough examination of biodiversity in Mexico is available in Antony Challenger, *Utilización y conservación de los ecosistemas terrestres de México: Pasado, presente y futuro* (Mexico City: Comisión Nacional para el Conocimiento y Uso de la Biodiversidad, 1998). See pp. 34–35 for particulars, but the entire volume traces the evolving history of Mexican biodiversity from prehistory to 1990. For a distilled survey, see Dante Arturo Rodríguez Trejo, "Fire Regimes, Fire Ecology, and Fire Management in Mexico," *Ambio* 37, no. 7–8 (December 2008).

8. Annual fire statistics from CONAFOR: https://snif.cnf.gob.mx/incendios/; see figure 5–5 for graphic representation over the past 50 years.

9. Roberto Martínez Domínguez and Dante Arturo Rodríguez Trejo, "Forest Fires in Mexico and Central America," in *Proceedings of the Second International Symposium on Fire Economics, Planning, and Policy: A Global View*, General Technical Report PSW-GTR-208 (Albany, Calif.: U.S. Forest Service, 2008), 710. For the argument that lightning fires are undercounted, see Ronald L. Myers and Dante Arturo Rodríguez-Trejo, "Fire in Tropical Pine Ecosystems," in *Tropical Fire Ecology: Climate Change, Land Use, and Ecosystem Dynamics*, ed. Mark A. Cochrane (Chichester, UK: Springer, 2009), 584–85. Roberto Lindig-Cisneros, "Ecological Restoration in Mexico: The Challenges of a Multicultural Megadiverse Country," *Ecological Restoration* 28, no. 3 (2010): 232. On the exotic ignitions, see "Meteor Starts Forest Fire," *New York Times*, April 27, 1910, 1; and "The Mexican Earthquake," *New York Times*, May 22, 1887, 1.

10. The fullest survey of fire ecology in Mexico is Dante Arturo Rodríguez Trejo, *Incendios de vegetación: Su ecología, manejo e historia*, 2 vols. (Guadalajara: Colegio de Postgraduados, 2014–15). Burned area statistics come from a very recent overview in Laura E. Montoya, Rogelio O. Corona-Nuñez, and Julio E. Campo, "Fires and Their Key Drivers in Mexico," *International Journal of Wildland Fire* 32 (March 14, 2023): https://doi.org/10.1071/WF22154.

11. On Hurricane Gilbert, see Edward N. Rappaport and Colin J. McAdie, preparers, *Hurricane Gilbert (1988) in Review and Perspective*, NOAA Technical

Memorandum NWS NHC 45 (Coral Gables, Fl.: National Hurricane Center, 1991); Dennis F. Whigham et al., "The Impact of Hurricane Gilbert on Trees, Litterfall, and Woody Debris in a Dry Tropical Forest in the Northeastern Yucatán Peninsula," *Biotropica* 23, no. 4, part A (1991): 434–41.

In 2007 Hurricane Dean, another Category 5 storm, with maximum winds at 315 km/h, repeated the scenario. It was estimated that the winds affected 5,054,539 ha, 25% at high severity, 20% at medium severity, and 55% at low severity. Jackstrawed fuels ranged from 40 to 137 tons/ha. See Rodríguez Trejo, *Incendios de vegetación*, 1:52–54; and Dante Rodríguez Trejo et al., "Modelaje del peligro de incendio forestal en las zonas afectadas por el Huracán Dean," *Agrociencia* 45, no. 5 (2011): 593–608.

12. Myers and Rodríguez-Trejo, "Fire in Tropical Pine Ecosystems," 557–605.
13. On the northern fire histories, see Jed Meunier, William H. Romme, and Peter M. Brown, "Climate and Land-Use Effects on Wildfire in Northern Mexico, 1650–2010," *Forest Ecology and Management* 325 (2014): 49–59; John H. Dieterich, "Cronología de los incendios forestales de la Sierra de los Ajos, Sonora, México," *Revista Dasonomía Mexicana* 50, no. 3 (1985): 20–24. On Los Chimalapas, see Y. Wård, A. Malmer, and H. Asbjornsen, "Historical 14C Evidence of Fire in Tropical Montane Cloud Forests in the Chimalapas Region, of Oaxaca, Southern Mexico," in *Tropical Montane Cloud Forests: Science for Conservation and Management*, ed. L. A. Bruijnzeel, F. N. Scatena, and L. S. Hamilton (Cambridge: Cambridge University Press, 2010), 575–78; Heidi Asbjornsen and Bart Wickel, "Changing Fire Regimes in Tropical Montane Cloud Forests: A Global Synthesis," in Cochrane, *Tropical Fire Ecology*, 606–28.

SECOND SUN

1. Karl Taube, *Aztec and Maya Myths* (London: British Museum Press, 1993), 47. On fire and water as metaphor for war, see Michael D. Coe and Rex Koontz, *Mexico: From the Olmecs to the Aztecs*, 6th ed. (London: Thames and Hudson, 2008), 164 (figure).
2. Diego Durán, *The History of the Indies of New Spain*, trans. Doris Heyden (Norman: University of Oklahoma Press, 1994), 197, 267.
3. Durán, *History of the Indies*, 267, 269.
4. There are many popular accounts. The standard geologic report, begun within weeks after the cone originated, is Jenaro González Reyna and William F. Foshag, "The Birth of Parícutin," in *Smithsonian Institution Annual Report 1946* (Washington, D.C.: Government Printing Office, 1947), 223–34, with pp. 226–27 including a first-person account from Dionisio.
5. Quote regarding the hearth from Coe and Koontz, *Mexico*, 10. Coe and Koontz make an interesting observation on the theme of duality: "The bewildering multiplicity of Mexican gods was to these thinkers but an embodiment of one cosmic principle of duality: the unity of opposites." *Mexico*, 205. On Xiuhtecuhtli

as day glyph for water, see Gisele Díaz and Alan Redgers, *The Codex Borgia: A Full-Color Restoration of the Ancient Mexican Manuscript* (Garden City, N.Y.: Dover Publications, 1993), xix.

On fire and the Sacred Cenote, see Clemency Chase Coggins and Orrin C. Shane III, eds., *Cenote of Sacrifice: Maya Treasures from the Sacred Well at Chichén Itzá* (Austin: University of Texas Press, 1984), 29. Other sources: Taube, *Aztec and Maya Myths*, 45–47; and "Fire Deities," in *The Oxford Encyclopedia of Mesoamerican Cultures*, ed. David Carrasco (Oxford: Oxford University Press, 2001).

6. I follow Coe and Koontz, *Mexico*, 19–22; Susan Toby Evans, "Time and Space Boundaries: Chronologies and Regions in Mesoamerica," in *The Oxford Handbook of Mesoamerican Archaeology*, ed. Deborah L. Nichols and Christopher A. Pool (New York: Oxford University Press, 2012), 114–26; and Guillermo Acosta Ochoa, "Ice Age Hunter-Gatherers and the Colonization of Mesoamerica," in Nichols and Pool, *Oxford Handbook of Mesoamerican Archaeology*, 130–40. On the recent findings in Zacatecas, see Ciprian F. Ardelean et al., "Evidence of Human Occupation in Mexico Around the Last Glacial Maximum," *Nature* 584 (August 6, 2020): https://doi.org/10.1038/s41586-020-2509-0.
7. Acosta Ochoa, "Ice Age Hunter-Gatherers," 130–34.
8. Douglas J. Kennett, "Archaic-Period Foragers and Farmers in Mesoamerica," in Nichols and Pool, *Oxford Handbook of Mesoamerican Archaeology*, 141–50.
9. Andrés Moreno-Estrada et al., "The Genetics of Mexico Recapitulates Native American Substructure and Affects Biomedical Traits," *Science* 344, no. 6189 (June 13, 2014): 3–4, https://doi.org/10.1126/science.1251688.
10. Quote from Moreno-Estrada et al., "Genetics of Mexico," 5.
11. For a summary of aboriginal burning patterns, see Stephen J. Pyne, *Fire: A Brief History*, 2nd ed. (Seattle: University of Washington Press, 2019).

    A note: I use "aboriginal fire" as a generic term for hunting and foraging fire practices in the same sense I use "agricultural fire" for the miscellany of fire practices associated with farming and herding. If capitalized, the word refers to a specific people, although the trend these days is to replace *Aborigine* with *Indigene*.
12. Rhys Jones, "Fire-Stick Farming," *Australian Natural History* 16, no. 7 (September 1969): 224–28. See also H. T. Lewis, "The Role of Fire in the Domestication of Plants and Animals in Southwest Asia: A Hypothesis," *Man* 7 (1972): 195–222. Serra quote from Henry T. Lewis, *Patterns of Indian Burning in California: Ecology and Ethnohistory*, Ballena Press Anthropological Papers 1 (Ramona, Calif.: Ballena Press, 1973), xxi. Lewis's book has been elaborated and expanded in M. Kat Anderson, *Tending the Wild: Native American Knowledge and the Management of California's Natural Resources* (Berkeley: University of California Press, 2005).
13. Sherburne F. Cook, *The Historical Demography and Ecology of the Teotlalpan* (Berkeley: University of California Press, 1949), 33.

14. Adolf Bandelier, ed., *The Journey of Alvar Nuñez Cabeza de Vaca* [. . .] *1528–1536*, trans. Fanny Bandelier (1905; repr., New York: AMS Press, 1973), 92–93; Gonzalo Fernández de Oviedo y Valdés, *Historia General y Natural*, part 1 (Madrid: Imprenta de la Real Academia de la Historia, 1851), 600; Bartolomé de Las Casas, *Historia de las Indias*, vol. 5 (Madrid, 1542), 301. Andagoya quoted in Carl Ortwin Sauer, *The Early Spanish Main* (Berkeley: University of California Press, 1969), 244. M. Hartman [Carl Hartman], "The Indians of North-Western Mexico," in *Congrès International d'Americanistes: Compte rendu de la dixième session* (Stockholm: Impr. I. Haeggström, 1897), 117. On the La Brea tar pits, see F. Robin O'Keefe et al., "Pre–Younger Dryas Megafaunal Extirpation at Rancho La Brea Linked to Fire-Driven State Shift," *Science* 381, no. 746 (2023).
15. D. S. Matson and A. H. Schroeder, eds., "Cordero's Description of the Apache—1796," *New Mexico Historical Review* 32 (1957): 343–44, 348.
16. José de Acosta, *Natural and Moral History of the Indies*, ed. Jane E. Mangan, trans. Frances M. López-Morillas (Durham: Duke University Press, 2002), 273–74.
17. See Francisco Javier Clavijero, *Historia antigua de México*, 5th ed. (Mexico City: Editorial Porrúa, 1976), 234.
18. Carl O. Sauer, "The Agency of Man on the Earth," in *Man's Role in Changing the Face of the Earth*, ed. William L. Thomas Jr., vol. 1 (Chicago: University of Chicago Press, 1956). I've taken a little liberty with Sauer's observation, based on his many comments elsewhere. In this essay he says, "Wherever primitive man has had the opportunity to turn fire loose on a land, he seems to have done so, from time immemorial; it is only civilized societies that have undertaken to stop fires" (55).
19. Hartman, "Indians of North-Western Mexico," 117.
20. Hartman, 18.
21. I found the simplest summary in Coe and Koontz, *Mexico*, 26–38. Also helpful was Dolores R. Piperno and Bruce D. Smith, "The Origins of Food Production in Mesoamerica," in Nichols and Pool, *Oxford Handbook of Mesoamerican Archaeology*, 151–64.

    Despite its growing age, the classic account by Teresa Rojas Rabiela, "La agricultura en la época prehispánica," in *La agricultura en tierras mexicanas desde sus orígenes hasta nuestros días*, ed. Teresa Rojas (Mexico City: Editorial Grijalbo, 1991), 15–138, remains a marvel of concentrated information.
22. Carl O. Sauer, "The Personality of Mexico," in *Land and Life: A Selection from the Writings of Carl Ortwin Sauer*, ed. John Leighly (Berkeley: University of California Press, 1963), 107–8.
23. For an excellent and accessible summary, see Michael Blake, *Maize for the Gods: Unearthing the 9,000-Year History of Corn* (Berkeley: University of California Press, 2015). For thumbnail distillations, see Coe and Koontz, *Mexico*, 35–37; and Kennett, "Archaic-Period Foragers and Farmers," 155–57.
24. Quoted in Rojas Rabiela, "La agricultura en la época prehispánica," 53–54.

25. For detailed descriptions of the best-known variations, see Rojas Rabiela, "La agricultura en la época prehispánica," 54–77. For an earlier study of a later century, see Teresa Rojas Rabiela, *Las siembras de ayer: La agricultura indígena del siglo XVII* (Mexico City: Secretaría de Educación Pública, 1988).
26. On the date of the earliest swidden, see Piperno and Smith, "Origins of Food Production," 156.
27. I found the best survey of Mexican agriculture's variety to be in Rojas Rabiela, "La agricultura en la época prehispánica."
28. Odum quote from Howard T. Odum, *Environment, Power, and Society* (New York: Wiley Interscience, 1970), 116.
29. Piperno and Smith, "Origins of Food Production," 156.
30. My primary source has been Campbell W. Pennington, *The Tarahumar of Mexico: Their Environment and Material Culture* (Salt Lake City: University of Utah Press, 1963).
31. Pennington, *Tarahumar*, 28; Peter Z. Fulé et al., "Fire Regime in a Mexican Forest Under Indigenous Resource Management," *Ecological Applications* 21, no. 3 (April 2011): 766.
32. Pennington, *Tarahumar*, 50. Hartman quote from Hartman, "Indians of North-Western Mexico," 118. Fulé et al., "Fire Regime," 772.
33. Carl Lumholtz, *Unknown Mexico: A Record of Five Years' Exploration Among the Tribes of the Western Sierra Madre, in the Tierra Caliente of Tepic and Jalisco, and Among the Tarascos of Michoacan*, vol. 1 (New York: Charles Scribner's Sons, 1902), 213. On goats, see Christopher R. Boyer, *Political Landscapes: Forests, Conservation, and Community in Mexico* (Durham, N.C.: Duke University Press, 2015), 49.
34. An interesting description of primary fire uses in the forest is available in "Causas de frecuentes incendios originados en los bosques de la Alta Sierra Tarahumara," *Boletín Forestal: Órgano de la Unión de Madereros de Chihuahua* 1 (April–May 1949): 4–5; but look also at Antonio H. Sosa, "Exploración forestal en la Alta Sierra Tarahumara del estado de Chihuahua," *Boletín del Departamento Forestal y de Caza y Pesca* 3, no. 8 (September–November 1937): 187–235.

    But also see Pennington, *Tarahumar*, 90; Fulé et al., "Fire Regime," 770; Lumholtz, *Unknown Mexico*, 180. For very recent times, see Boyer, *Political Landscapes*; and Iain J. Davidson-Hunt, "Journeys, Plants and Dreams: Adaptive Learning and Social-Ecological Resilience" (PhD thesis, University of Manitoba, 2003), 137–50.
35. Lumholtz, *Unknown Mexico*, 184–85.
36. Fulé et al., "Fire Regime," 769, 764.
37. On timber statistics, see Fulé et al., "Fire Regime," 771. On marijuana, see Davidson-Hunt, "Journeys, Plants and Dreams," 126.
38. Frances M. Slaney, "Double Baptism: Personhood and Ethnicity in the Sierra Tarahumara of Mexico," *American Ethnologist* 24, no. 2 (1997): 279–310.

39. It's hard to know how to interpret population estimates. Apart from the lack of censuses, the wave train of epidemics affected the numbers by the time Spaniards willing to count arrived. I rely on Pennington, *Tarahumar*, 23–24; and Hartman, "Indians of North-Western Mexico," which estimates 20,000 to 30,000 in 1890; and for context, Carl Sauer's meditations in *Aboriginal Population of Northwestern Mexico* (Berkeley: University of California Press, 1935). For the Maya, again, estimates vary wildly; I relied on Michael D. Coe and Stephen Houston, *The Maya*, 9th ed. (New York: Thames and Hudson, 2015), in which the authors maintain a running tally of guesses as their chronicle of the Maya unfolds.
40. Drought dates from Coe and Houston, *Maya*, 32.
41. For general background, see Richard Rhoda and Tony Burton, *Geo-Mexico: The Geography and Dynamics of Modern Mexico* (Ladysmith, B.C.: Sombrero Books, 2010); and Jerzy Rzedowski, *Vegetación de México* (Mexico City: Editorial Limusa, 1998). Other background information came from the various essays in Peter D. Harrison and B. L. Turner II, eds., *Pre-Hispanic Maya Agriculture* (Albuquerque: University of New Mexico Press, 1978). A succinct summary is available in Coe and Houston, *Maya*, 16–22. Also useful is the classic account of tropical swidden in Mesoamerica, O. F. Cook, "Milpa Agriculture, a Primitive Tropical System," in *Annual Report, Smithsonian Institution, 1919* (Washington, D.C.: Government Printing Office, 1921), 307–26.
42. Quote from Rhena Hoffmann, "La importancia ecológica y económica de las tecnologías tradicionales en la agri- y silvicultura en áreas de bosque tropical húmedo en México," in *Changing Tropical Forests: Historical Perspectives on Today's Challenges in Central and South America; Proceedings of a Conference Sponsored by the Forest History Society and IUFRO Forest History Group*, ed. Harold K. Steen and Richard P. Tucker (Durham, N.C.: Forest History Society, 1992), 113.

    On the post-Gilbert land rush, see Kelvin Hirsch, "Wildland Fire and Its Management—A Look Towards the Future," in *Fire in the Forest*, ed. Peter A. Thomas and Robert S. McAlpine (New York: Cambridge University Press, 2010), 190. The resulting fires were a major event that affected Mexico's *servicio forestal* and which will be examined in detail in chapter 5.

    John L. Stephens, *Incidents of Travel in Yucatan*, vol. 1 (1843; repr., New York: Dover Books, 1963), 137. Benjamin Norman, *Rambles in Yucatan, or, Notes of Travel Through the Peninsula: Including a Visit to the Remarkable Ruins of Chi-Chen, Kabah, Zayi, and Uxmal*, 3rd ed. (New York: J. & H. G. Langley, 1843), 254.
43. A concise summary is found in, again, Coe and Houston, *Maya*, 19; and Hoffmann, "La importancia ecológica y económica," 110–22. Both, however, lack the sympathetic and rich understanding available in such sources as the works by Teresa Rojas Rabiela. An excellent, and sympathetic, recent summary is Ronald Nigh and Stewart A. W. Diemont, "The Maya Milpa: Fire and the Legacy of Living Soil," *Frontiers in Ecology and Environment* 11, no. s1 (2013): e45–e54, https://doi.org/10.1890/120344.

44. A nice account of burning, nested within the larger agricultural cycle of highland Chiapas, is given in George A. Collier, *Fields of the Tzotzil: The Ecological Bases of Tradition in Highland Chiapas* (Austin: University of Texas Press, 1975), 30–32.
45. J. Eric S. Thompson, *The Rise and Fall of Maya Civilization*, rev. ed. (Norman: University of Oklahoma Press, 1966), 301. Chilam Balam quote from p. 201.
46. I've relied on A. Barrera, A. Gómez-Pompa, and C. Vázquez-Yanes, "El manejo de las selvas por los mayas: Sus implicaciones silvícolas y agrícolas," *Biotica* 2, no. 2 (1977): 47–61; and for a shorter version in English, Arturo Gómez-Pompa, "On Maya Silviculture," *Mexican Studies / Estudios Mexicanos* 3, no. 1 (Winter 1987): 1–17; quotation is from the latter, p. 2. Also helpful was Arturo Gómez-Pompa, José Salvador Flores, and Victoria Sosa, "The 'Pet Kot': A Man-Made Tropical Forest of the Maya," *Interciencia* 12, no. 1 (January–February 1987): 10–15. Unfortunately, only after the manuscript was in production did I discover the marvelously concise summary available in Anabel Ford and Ronald Nigh, *The Maya Forest Garden: Eight Millennia of Sustainable Cultivation of the Tropical Woodlands* (London: Routledge, 2015), which seems destined to become a standard for the topic.
47. See Gerardo Budowski, "Fire in Tropical American Lowland Areas," in *Proceedings, Fifth Annual Tall Timbers Fire Ecology Conference* (Tallahassee: Tall Timbers Research Station, 1966), 5–22; Myers quote from p. 9. Budowski's geographic range ends at the Mexican border, but his observations apply. Other sources for tropical fire include Mark A. Cochrane, *Tropical Fire Ecology* (New York: Springer/Praxis, 2009); and the classic, densely empirical volumes of H. H. Bartlett.
48. An interesting effort to estimate the quantities of wood cut and burned for various purposes is Elliot M. Abrams and David J. Rue, "The Causes and Consequences of Deforestation Among the Prehistoric Maya," *Human Ecology* 16, no. 4 (1988): 377–95.
49. As an example, see Abrams and Rue, "Causes and Consequences," 377–95.
50. Arturo Gómez-Pompa has been especially prominent in promoting the positive ecological virtues of Mayan land use. See, for example, "Tropical Deforestation and Maya Silviculture: An Ecological Paradox," *Tulane Studies in Zoology and Botany* 26 (1987): 19–37; Gómez-Pompa, Flores, and Sosa, "Pet Kot"; and Barrera, Gómez-Pompa, and Vázquez-Yanes, "El manejo de las selvas por los mayas." The "magnificent way" quote is from p. 47 of the latter.

    New discoveries continue to be made, as often as people look beyond what textbooks say. See, for example, a study that adds the role of biochar to the litany of desirable milpa outcomes: Nigh and Diemont, "Maya Milpa."
51. Barrera, Gómez-Pompa, and Vázquez-Yanes, "El manejo de las selvas por los mayas," 47.
52. Bernal Díaz, *The Conquest of New Spain*, trans. J. M. Cohen (New York: Penguin Books, 1963), 225–27. Animals in parentheses from Lane Simonian, *Defending*

the Land of the Jaguar: A History of Conservation in Mexico (Austin: University of Texas Press, 1995), 26.

53. Figures from Simonian, *Defending*, 24–25.
54. The roster of pre-Columbian conservation activities paraphrases Simonian, *Defending*, 24–25. See also H. Torres C., "Netzhualcoyotl forestal," *México Forestal* 1, no. 2 (1923): 11.
55. I found very helpful the panoramic survey by Thomas M. Whitmore and B. L. Turner II, "Landscapes of Cultivation in Mesoamerica on the Eve of the Conquest," *Annals of the Association of American Geographers* 82, no. 3 (September 1992): 402–25, particularly because of their concern for waterworks, which served as a counterpoint to my instinctive attraction to fire.

    Also invaluable is, once again, Antony Challenger, *Utilización y conservación de los ecosistemas terrestres de México: Pasado, presente y futuro* (Mexico City: Comisión Nacional para el Conocimiento y Uso de la Biodiversidad, 1998), 75–118. Though dating, the bibliography is a treasury of references.
56. Cortés comment from George McCutchen McBride, *The Land Systems of Mexico*, American Geographical Society Research Series 12 (New York: American Geographical Society, 1923), 180.
57. On languages, see Coe and Koontz, *Mexico*, 14–15.
58. Coe and Koontz, 62.
59. On the book burning, see Camilla Townsend, *Fifth Sun: A New History of the Aztecs* (New York: Oxford University Press, 2019), 33.
60. Coe and Koontz, *Mexico*, 188; Townsend, *Fifth Sun*, 33.

THIRD SUN

1. Bartolomé de Las Casas, *The Devastation of the Indies: A Brief Account*, trans. Herma Briffault (Baltimore: Johns Hopkins University Press, 1974), 132.
2. Las Casas, *Devastation*, 28. Pedro de Córdoba quoted in Carl Ortwin Sauer, *The Early Spanish Main* (Berkeley: University of California Press, 1969), 293–94. J. M. Cohen, ed. and trans., *The Four Voyages of Christopher Columbus* (New York: Penguin Books, 1969), 117.
3. Miguel León-Portilla, ed., *The Broken Spears: The Aztec Account of the Conquest of Mexico*, trans. Lysander Kemp (Boston: Beacon Press, 1962), 4–9. I have created a composite of two of the sources quoted. Given Spanish control over the narrative of the Conquest, it is unclear whether the portents really occurred or were interpreted as omens at the time, or whether they are a construction by later chroniclers.
4. Alfred Crosby Jr., *The Columbian Exchange: Biological and Cultural Consequences of 1492* (Westport, Conn.: Greenwood Press, 1972); and Alfred Crosby Jr., *Ecological Imperialism: The Biological Expansion of Europe, 900–1900* (Cambridge: Cambridge University Press, 1986). In the two paragraphs that follow, I hew closely to Crosby's insightful and wry account.

That said, I have found extremely useful the many studies published by Karl W. Butzer, and his summary of the Columbian encounter is more systematic than Crosby's. See Karl W. Butzer, "Biological Transfer, Agricultural Change, and Environmental Implications of 1492," in *International Germplasm Transfer: Past and Present*, Crop Science Society of America Special Publication 23 (Madison, Wis.: Crop Science Society of America, 1995), 3–29. A constant reference, too, has been Antony Challenger, *Utilización y conservación de los ecosistemas terrestres de México: Pasado, presente y futuro* (Mexico City: CONABIO, 1998); for the colonial era, see pp. 119–56.

Helpful for placing Spain's exertions in a broader context is John F. Richards, *The Unending Frontier: An Environmental History of the Early Modern World* (Berkeley: University of California Press, 2003).

5. Carl Sauer, "The Personality of Mexico," in *Land and Life: A Selection from the Writings of Carl Ortwin Sauer*, ed. John Leighly (Berkeley: University of California Press, 1963), 105.
6. Thomas M. Whitmore and B. L. Turner II, "Landscapes of Cultivation in Mesoamerica on the Eve of the Conquest," *Annals of the Association of American Geographers* 82, no. 3 (September 1992): 402.
7. The most comprehensive historical survey is the collection of essays compiled by Lucina Hernández, *Historia ambiental de la ganadería en México* (Xalapa: Instituto de Ecología, 2001); quote by Guevara from "Presentación," 1.
8. An excellent summary of the transfer process is available in Karl W. Butzer, "Cattle and Sheep from Old to New Spain: Historical Antecedents," *Annals of the Association of American Geographers* 78, no. 1 (1988): 29–56.
9. Butzer notes the use of fire in dehesas: "Cattle and Sheep," 44. For a review of their modern status, see Leopoldo Olea-Márquez de Prado and Alfonso San Miguel-Ayanz, "The Spanish Dehesa: A Traditional Mediterranean Silvopastoral System Linking Production and Nature Conservation" (opening paper delivered at the 21st General Meeting of the European Grassland Federation, Badajoz, Spain, April 2006), http://www2.montes.upm.es/Dptos/dsrn/SanMiguel/APUNTES_PRESENTACIONES/PASCICULTURA%20Y%20SAF/SPANISH_DEHESA_XXI_EGF.pdf.
10. The plague reference derives from its use as a metaphor in Elinor G. K. Melville, *A Plague of Sheep: Environmental Consequences of the Conquest of Mexico* (Cambridge: Cambridge University Press, 1997), specifically the 10 plagues listed in the epigraph.
11. Butzer, "Biological Transfer," 9.
12. Manuel Gamio, "Foreword," in George McCutchen McBride, *The Land Systems of Mexico*, American Geographical Society Research Series 12 (New York: American Geographical Society, 1923), ix.
13. Quoted in Lesley Byrd Simpson, *The Encomienda in New Spain: The Beginning of Spanish Mexico*, rev. ed. (Berkeley: University of California Press, 1950), xvi.

14. The actual diseases and demographic losses are a constant source of inquiry. Good numbers on demographic change are hard to come by and harder to interpret. Probably the most detailed inquiry, which also reviews the estimates of others, is Robert McCaa, "The Peopling of Mexico from Origins to Revolution," in *A Population History of North America*, ed. Michael R. Haines and Richard H. Steckel (Cambridge: Cambridge University Press, 2000), 241–304. I've avoided particular estimates in favor of ratios, which seem to have a better consensus.

The classic study is S. F. Cook and L. B. Simpson, *The Population of Central Mexico in the Sixteenth Century*, Ibero-Americana 31 (Berkeley: University of California Press, 1948). Mostly I have followed Crosby in the causes of collapse, but for an inquiry into possible indigenous diseases and vectors, see Rudolfo Acuna-Soto et al., "Megadrought and Megadeath in 16th Century Mexico," *Emerging Infectious Diseases* 8, no. 4 (April 2002): 360–62.

On the global consequences, see Alexander Koch et al., "Earth System Impacts of the European Arrival and Great Dying in the Americas After 1492," *Quaternary Science Reviews* 207 (2019): 13–36.

15. Karl W. Butzer and Elizabeth K. Butzer, "The Sixteenth-Century Environment of the Central Mexican Bajío: Archival Reconstruction from Colonial Land Grants and the Question of Spanish Ecological Impact," in *Culture, Form, and Place: Essays in Cultural and Historical Geography*, ed. Kent Mathewson (Baton Rouge: Department of Geography and Anthropology, Louisiana State University, 1993), 175–77.
16. Simpson, *Encomienda*, xvi. A divided populace was not uncommon in colonial settings. Canada had its "two solitudes" between Anglophone and Francophone spheres. The United States had two cultures that proved so irreconcilable that the country was ripped apart by a civil war, whose scars still endure. Those, however, were distinctions among the settlers. Like Australia or New Zealand, Mexico segregated Indigenes from Europeans.
17. Karl W. Butzer and Elizabeth K. Butzer, "Transfer of the Mediterranean Livestock Economy to New Spain: Adaptation and Ecological Consequences," in *Global Land Use Change: Perspective from the Columbian Encounter*, ed. B. L. Turner (Madrid: Consejo Superior de Investigaciones Científicas, 1995), 153–91.
18. Quoted in McBride, *Land Systems*, 39, 122.
19. Again, I rely on McBride, *Land Systems*, 100.
20. The classic study remains Robert Richard, *The Spiritual Conquest of Mexico*, trans. Lesley Byrd Simpson (Berkeley: University of California Press, 1966).
21. Figures from California Missions, San Juan Capistrano, Key Facts, https://missionscalifornia.com/san-juan-capistrano-mission/, accessed May 3, 2020.
22. Sauer's "Personality of Mexico" is still a useful digest of impacts; see especially pp. 114–16.
23. On the silver age, see François Chevalier, *Land and Society in Colonial Mexico: The Great Hacienda*, trans. Alvin Eustis, ed. Lesley Byrd Simpson (Berkeley: University of California Press, 1966), 4.

24. Alexander von Humboldt, *Political Essay on the Kingdom of New Spain*, trans. John Black, unabridged, vol. 3 (London: Longman, Hurst, Rees, Orme, and Brown; H. Holburn; and W. Blackwood, and Brown and Crombie, Edinburgh, 1811), 212–13.
25. Fire disaster did of course occur. As late as 1953, the Angangueo silver mine had a fire that killed 25 miners and led Mexico to authorize the entry of two American experts to help extinguish it. See "Mexico Calls U.S. Firefighters," *New York Times*, May 2, 1953, 8.
26. All statistics and quotes come from Daviken Studnicki-Gizbert and David Schecter, "The Environmental Dynamics of a Colonial Fuel-Rush: Silver Mining and Deforestation in New Spain, 1552 to 1810," *Environmental History* 15 (January 2010): 94–119, esp. 96, 99–100, 110. This article is the only study of mining I know that interprets its environmental impact on (and economic viability based on) fuel availability.
27. I found very helpful the account of colonial developments in the Sierra Madre Occidental in Michael M. Swann, *Tierra Adentro: Settlement and Society in Colonial Durango*, Dellplain Latin American Studies 10 (Boulder, Colo.: Westview Press, 1982).
28. Frances Calderón de la Barca, *Life in Mexico* (Mexico City: Ediciones LARA, n.d.), 471.
29. See the remarkable article by Studnicki-Gizbert and Schecter, "Environmental Dynamics," 105–8.
30. Maturin M. Ballou, *Aztec Land* (Boston: Houghton, Mifflin, 1890), 128.
31. To my knowledge, the only study that considers built environments in terms of fire regimes is Stephen J. Pyne, *Fire: A Brief History*, 2nd ed. (Seattle: University of Washington Press, 2019), ch. 6.
32. Gabriel Alonso de Herrera, *Agricultura general de Gabriel Alonso de Herrera: Corregida segun el testo original de la primera edición publicada en 1513 por el mismo autor, y adicionada por la Real Sociedad Economica Matritense*, vol. 1 (Madrid: Imprenta Real, 1818). On the persistence of quemas in Spain, see, for example, Fritz Krüger, "Cosas y palabras del noroeste ibérico, I. Un sistema de cultivo arcaico: La quema del monte," *Nueva Revista de Filología Hispánica* 4, no. 3 (1950): 231–53.
33. Birgitta Leander, "Mestizaje ecológico en Mexico (etnobotánica de algunas plantas alimenticias)," *Revista de Indias* 95–96 (January–June 1964): 87–135. For an explicit association of plants and people, see p. 88: "La ecología del México actual presenta el mismo cuadro de mestizaje que otros aspectos de su cultura."
34. Francisco Javier Clavijero, *Historia antigua de México*, 5th ed. (Mexico City: Editorial Porrúa, 1976), 233–34.
35. Friar Diego de Landa, *Yucatan Before and After the Conquest*, trans. William Gates (1937; repr., New York: Dover Publications, 1978), 93, 38–39. Antonio de Ciudad Real, *Tratado curioso y docto de las grandezas de la Nueva España, Tomo II* (Mexico City: Universidad Nacional Autónoma de México, Instituto de Investigaciones Históricas, 1976), 318.

36. Juan de Torquemada, quoted in Teresa Rojas, coord., *La agricultura en tierras mexicanas desde sus orígenes hasta nuestros días* (Mexico City: Editorial Grijalbo, 1991), 53–54.

    The Mayan milpa was the pivot for reassessing traditional practices, particularly in the work of Efraím Hernández Xolocotzi. For a collective summary and (implied) tribute, see Efraím Hernández Xolocotzi, Eduardo Bello Baltazar, and Samuel Levy Tacher, eds., *La milpa en Yucatán: Un sistema de producción agrícola tradicional*, 2 vols. ([Chapingo?]: Colegio de Postgraduados, 1995).
37. Quotes from Harley H. Bartlett, *Fire in Relation to Primitive Agriculture and Grazing in the Tropics: Annotated Bibliography* (Ann Arbor: University of Michigan Botanical Gardens, 1955), 477; Bartlett is quoting or paraphrasing Oscar Schmieder, *The Settlements of the Tzapotec and Mije Indians, State of Oaxaca, Mexico*, Publications in Geography 4 (Berkeley: University of California Press, 1930). On the Chiapas Highlands, see George A. Collier, *Fields of the Tzotzil: The Ecological Bases of Tradition in Highland Chiapas* (Austin: University of Texas Press, 1975), especially pp. 30–32 for the annual almanac of burning.
38. Frans Blom and Oliver La Farge, *Tribes and Temples: A Record of the Expedition to Middle America Conducted by the Tulane University of Louisiana in 1925*, 2 vols. (New Orleans: Tulane University, 1927), 60, 239.
39. John Lloyd Stephens, *Incidents of Travel in Central America, Chiapas and Yucatan*, vol. 2 (1842; repr., New York: Dover Books), 237–39.
40. Calderón de la Barca, *Life in Mexico*, 323.
41. The classic compendium, drawn from historical sources, remains Bartlett, *Fire in Relation to Primitive Agriculture*. For a focus on Mexico and Mesoamerica, see Ronald L. Myers and Dante Arturo Rodríguez-Trejo, "Fire in Tropical Pine Ecosystems," in *Tropical Fire Ecology: Climate Change, Land Use, and Ecosystem Dynamics*, ed. Mark A. Cochrane (Chichester, UK: Springer-Praxis, 2009), 557–605.
42. Charles H. Harris III, *A Mexican Family Empire: The* Latifundio *of the Sánchez Navarros, 1765–1867* (Austin: University of Texas Press, 1975); see ch. 2, pp. 28–57.
43. For a survey of fire in Mediterranean Europe, see Stephen Pyne, *Vestal Fire: An Environmental History, Told Through Fire, of Europe and Europe's Encounter with the World* (Seattle: University of Washington Press, 1996), 81–147, with Spain-specific observations on pp. 99–103.
44. See Olea-Márquez de Prado and San Miguel-Ayanz, "Spanish Dehesa."
45. Alfred H. Siemens, "Land-Use Succession in the Gulf Lowlands," in Turner, *Global Land Use Change*, 212.
46. See William Howard Dusenberry, *The Mexican Mesta: The Administration of Ranching in Colonial Mexico* (Urbana: University of Illinois Press, 1963).
47. Quoted in C. Raymond Clar, *California Government and Forestry*, vol. 1 (Sacramento: Division of Forestry, 1959), 7.

48. Clar, *California Government and Forestry*, 8–9.
49. I follow Andrew Sluyter, "The Ecological Origins and Consequences of Cattle Ranching in Sixteenth-Century New Spain," *Geographical Review* 86, no. 2 (April 1996): 161–77. A more archival—perhaps methodological would be a better term—survey is available in Andrew Sluyter, "Ganadería española y cambio ambiental en las tierras bajas tropicales de Veracruz, México, siglo XVI," *Faculty Publications* 51, Louisiana State University, LSU Digital Commons, 2001, https://digitalcommons.lsu.edu/cgi/viewcontent.cgi?article=1050&context=geoanth_pubs.
50. Sluyter, "Ecological Origins and Consequences," 171.
51. Sluyter, 174. On the spread northward, see Miguel Aguilar-Robledo, "Ganadería, tenencia de la tierra e impacto ambiental en la Huasteca Potosina: Los años de la Colonia," in L. Hernández, *Historia ambiental de la ganadería*, 9–24.
52. Carl Christian Sartorius, *Mexico About 1850* (Stuttgart: F. A. Brockhaus, 1961), 9.
53. Elena Lazos Chavero, "Ciclos y rupturas: Dinámica ecológica de la ganadería en el sur de Veracruz," in L. Hernández, *Historia ambiental de la ganadería*, 133–53.
54. Quoted in Lazos Chavero, "Ciclos y rupturas," 133.
55. Bradley Skopyk, *Colonial Cataclysms: Climate, Landscape, and Memory in Mexico's Little Ice Age* (Tucson: University of Arizona Press, 2020), 20.
56. My recounting follows Melville, *Plague of Sheep*; quoted phrases from page xx.
57. See Dusenberry, *Mexican Mesta*.
58. Melville, *Plague of Sheep*, quotes Motolinía on the ten plagues, vi.
59. Lane Simonian, *Defending the Land of the Jaguar: A History of Conservation in Mexico* (Austin: University of Texas Press, 1995), 38–39.
60. Mendoza's ban and edicts from Simonian, *Defending*, 36–39. Dusenberry, *Mexican Mesta*, 111.
61. The most judicious survey in English I've found is Kathleen Ann Myers, *Fernández de Oviedo's Chronicle of America: A New History for a New World* (Austin: University of Texas Press, 2007).
62. Gonzalo Fernández de Oviedo y Valdés, *Historia General y Natural de las Indias, Islas y Terra Firme del Mar Oceano* (Madrid: Academia de la Historia, 1818–78), bk. 5, ch. 5, p. 142; bk. 4, ch. 13, p. 50; bk. 6, ch. 4, p. 70; bk. 6, ch. 5, p. 172; bk. 12, ch. 29, p. 447. Available through Internet Archive: https://archive.org/details/generalynatural01fernrich.
63. José de Acosta, *Natural and Moral History of the Indies*, ed. Jane E. Mangan, trans. Frances M. López-Morillas (Durham: Duke University Press, 2002), 100.
64. Acosta, *Natural and Moral History*, 100.
65. Ignacio Bernal, "Introduction," in *The Aztecs: The History of the Indies of New Spain*, by Fray Diego Durán, trans. Doris Heyden and Fernando Horcasitas (New York: Orion Press, 1964), xxi, xxx.
66. Fray Diego Durán, *Book of the Gods and Rites and the Ancient Calendar*, trans. and ed. Fernando Horcasitas and Doris Heyden (Norman: University of Oklahoma Press, 1971), 51. Durán, *Aztecs*, xxxi.

67. De Landa, *Yucatan*, 112, 111.
68. Information on Hernández from Rick A. López, "Nature as Subject and Citizen in the Mexican Botanical Garden, 1787–1829," in *A Land Between Waters: Environmental Histories of Modern Mexico*, ed. Christopher R. Boyer (Tucson: University of Arizona Press, 2012), 75–76.
69. The best summary of the enterprise is Iris H. W. Engstrand, *Spanish Scientists in the New World: The Eighteenth-Century Expeditions* (Seattle: University of Washington Press, 1981). On the transit project, see also Iris Higbie Wilson, "The 1769 Transit of Venus Observed by Velazquez from Lower California," *Astronomical Society of the Pacific*, leaflet 419 (May 1964). For original accounts, see Jean Chappe d'Auteroche, *A Voyage to California, to Observe the Transit of Venus* (London: Edward and Charles Dilly, 1778); José Longinos Martínez, *Journal: Notes and Observations of the Naturalist of the Botanical Expedition in Old and New California and the South Coast, 1791–1792*, ed. and trans. Leslie Byrd Simpson (San Francisco: J. Howell-Books, 1961).

    On the second age generally, see William H. Goetzmann, *New Lands, New Men: America and the Second Great Age of Discovery* (New York: Viking, 1986); and Stephen J. Pyne, *The Great Ages of Discovery: How Western Civilization Learned About a Wider World* (Tucson: University of Arizona Press, 2021).
70. On the Linnaeus episode, see Gunhild Weimarck, *Ulfshult: Investigations Concerning the Use of Soil and Forest in Ulfshult, Parish of Örkened, During the Last 250 Years*, Acta Universitatis Lundensis section 2, no. 6 (Lund: Gleerup, 1968), 45–48.
71. Crespí and Longinos Martínez quotes from Lowell John Bean and Harry W. Lawton, "Some Explanations for the Rise of Cultural Complexity in Native California with Comments on Proto-Agriculture and Agriculture," in *Patterns of Indian Burning in California: Ecology and Ethnohistory*, by Henry T. Lewis, Ballena Press Anthropological Papers 1 (Ramona, Calif.: Ballena Press, 1973), xix–xx. Pineda quoted in Engstrand, *Spanish Scientists*, 81.
72. Alexander von Humboldt, *Personal Narrative*, abridged and trans. Jason Wilson (London: Penguin Books, 1995), 38, 82, 95, 135, 178, 185, 199, 229.
73. Alexander von Humboldt, *Political Essay on the Kingdom of New Spain*, trans. John Black, ed. Mary Maples Dunn, abridged ed. (Norman: University of Oklahoma Press, 1972), 134. On the agricultural uses (e.g., sugar), see Humboldt, *Political Essay*, unabridged, vol. 3, pp. 11, 32, 45.
74. Sartorius, *Mexico About 1850*, 179–80. A. K. Shepard, *Land of the Aztecs; or, Two Years in Mexico* (Albany: Weed, Parsons, and Company, 1859), 673, 677. Frederick A. Ober, *Travels in Mexico and Life Among the Mexicans* (Boston: Estes and Lauriat, 1887), 82. Carl Lumholtz, *New Trails in Mexico: An Account of One Year's Exploration in North-Western Sonora, Mexico, and South-Western Arizona 1909–1910* (New York: Charles Scribner's Sons, 1912), 253. Edward B. Tylor, *Anahuac; or Mexico and the Mexicans, Ancient and Modern* (London: Longmans, Green, Reader, and Dyer, 1861), 208–9. Julius Froebel, *Seven Years' Travel in Central America, Northern Mexico, and the Far West of the United*

States (London: R. Bentley, 1859), p. 119, bk. 3, p. 456. A good example is John Russell Bartlett, *Personal Narrative of Explorations and Incidents in Texas, New Mexico, California, Sonora, and Chihuahua* (New York: D. Appleton, 1854), 27–28, 67, 88–89, 295, 296, 525.

75. Mrs. Alec Tweedie, *Mexico as I Saw It* (London: Thomas Nelson & Sons, 1911), 45, 49. J. W. Audubon, *Illustrated Notes of an Expedition Through Mexico and California* (New York: J. W. Audubon, 1852), 32.
76. Hans Gadow, *Through Southern Mexico, Being an Account of the Travels of a Naturalist* (London: Witherby & Co., 1908), 49.
77. See, for example, Toribio de Benavente [Motolinía], *Historia de los Indios de la Nueva España* (Barcelona: Herederos de Juan Gili, 1914), especially ch. 7. Humboldt quote from *Personal Narrative*, 221.
78. Humboldt, *Personal Narrative*, 150–51.
79. Humboldt, *Political Essay*, abridged ed., 162.

FOURTH SUN

1. My principal primary source is Juan Palacios, "Memoria sobre el incendio del pozo de petróleo de 'Dos Bocas,'" *Boletín de la Sociedad Mexicana de Geografía y Estadística*, 5th ser., 3, no. 1 (1908). Quote from Myrna I. Santiago, *The Ecology of Oil: Environment, Labor, and the Mexican Revolution, 1900–1938* (Cambridge: Cambridge University Press, 2006), 135, a book I found invaluable in explaining Mexico's relationship to oil.
2. Santiago, *Ecology of Oil*, 134–35.
3. Santiago, 122.
4. Quoted in Eyler N. Simpson, *The Ejido: Mexico's Way Out* (Chapel Hill: University of North Carolina Press, 1937), vii.
5. Carl Sauer, "The Road to Cibola," in *Land and Life: A Selection from the Writings of Carl Ortwin Sauer*, ed. John Leighly (Berkeley: University of California Press, 1963), 53. On industrial transport, see John H. Coatsworth, *Growth Against Development: The Economic Impact of Railroads in Porfirian Mexico* (DeKalb: Northern Illinois University Press, 1981), 17–18.
6. See Stephen J. Pyne, *Fire: A Brief History*, 2nd ed. (Seattle: University of Washington Press, 2019), for details about this pyric transition.
7. I detail the process of converting from second- to third-fire in Pyne, *Brief History*, ch. 7; and in even briefer compass in *The Pyrocene: How We Created an Age of Fire, and What Happens Next* (Berkeley: University of California Press, 2021), ch. 4.
8. Coatsworth, *Growth Against Development*, 36–37 (tables). A fascinating, person-centered case study of railroad building in the Porfiriato, specifically the Monterey and Mexican Gulf Railway, is Sonja Pyne, *Leon and the Colonel: Leon Noel Stuart, Colonel Joseph A. Robertson, and Their Quest for Citrus and a Railroad* (self-pub., 2013).

9. John Bigelow, "The Railway Invasion of Mexico," *Harper's New Monthly Magazine*, October 1882, 745–48.
10. An interesting commentary on wood and coal is available in Matías Romero, *Railways in Mexico: In Answer to an Article of the Hon. John Bigelow Entitled "The Railway Invasion of Mexico"* (Washington, D.C.: W. H. Moore, 1882), 15–16.
11. Santiago, *Ecology of Oil*, 65–67.
12. Germán Vergara, *Fueling Mexico: Energy and Environment, 1850–1950* (Cambridge: Cambridge University Press, 2021), 44.
13. For a wonderful review, see Germán Vergara, "How Coal Kept My Valley Green: Forest Conservation, State Intervention, and the Transition to Fossil Fuels in Mexico," *Environmental History* 23 (2018): 82–105. Vergara opens his essay with Río de la Loza's observations, p. 83. Christopher R. Boyer, *Political Landscapes: Forests, Conservation, and Community in Mexico* (Durham: Duke University Press, 2016), 35, 37.
14. Don Pedro L. Monroy, "Observaciones sobre algunos combustibles minerales de México," *La Naturaleza* 1, no. 1 (1869): 87–94, quote at 87.
15. For an overview of the Ministerio de Fomento, see Casey Marina Lurtz, "Developing the Mexican Countryside: The Department of Fomento's Social Project of Modernization," *Business History Review* 90 (Autumn 2016): 431–55. On coal and forests, and for the quote from Pacheco, see Vergara, "How Coal Kept My Valley Green," 88. "Fuel famine" quote from Romero, *Railways in Mexico*, 16.
16. Once again, I'm relying for dates on Vergara, "How Coal Kept My Valley Green," 93–94.
17. I follow the interesting study by Matthew Vitz, "'To Save the Forests': Power, Narrative, and Environment in Mexico City's Cooking Fuel Transition," *Mexican Studies / Estudios Mexicanos* 31 (Winter 2015): 125–55; quotes and figures from pp. 129–30. Railroad car estimate from Boyer, *Political Landscapes*, 89.

    Whetten quote from Nathan L. Whetten, *Rural Mexico* (Chicago: University of Chicago Press, 1948), 276.
18. Quote from Simpson, *Ejido*, 260n17.
19. Quotes from Vergara, "How Coal Kept My Valley Green," 96. On the source of firewood, see Vitz, "To Save the Forests," 135.
20. Santiago, *Ecology of Oil*, 49.
21. Frank Tannenbaum, *Mexican Agrarian Revolution* (Washington, D.C.: Brookings Institution, 1929), 89–90.
22. Quotes from Coatsworth, *Growth Against Development*, 181. On Chihuahua railroads, see Boyer, *Political Landscapes*, 44.
23. For statistics on the nineteenth century, see International Bureau of the American Republics, *Mexico* (Washington, D.C.: GPO, 1904), 363.
24. Gamio quoted in George McCutchen McBride, *The Land Systems of Mexico*, American Geographical Society Research Series 12 (New York: American Geographical Society, 1923), ix.

25. Quotes from David Barton Bray, *Mexico's Community Forest Enterprises: Success on the Commons and the Seeds of a Good Anthropocene* (Tucson: University of Arizona Press, 2020), 7. For the background source for the figures, see Héctor M. Robles Berlanga, "Ejidos y comunidades en México: Problemas y perspectivas," in *Memorias del seminario: "Propiedad social y servicios ambientales," 8 de noviembre de 2011, Ciudad de México*, ed. Juan Antonio Reyes González and Susana D'Acosta Lafon (n.p.: Proyecto de Cooperación Registro Agrario Nacional e Instituto Interamericano de Cooperación para la Agricultura, 2012), 30–36.

The literature on the ejido—its origins, its application—is vast. I've relied on a small body of histories, mostly by U.S. authors, to guide my understanding since my focus is not on the details or ideologies of the program but on its consequences for landscape fire. Those basic works include: Emily Wakild, *Revolutionary Parks: Conservation, Social Justice, and Mexico's National Parks, 1910–1940* (Tucson: University of Arizona Press, 2011); McBride, *Land Systems*; Christopher R. Boyer and Emily Wakild, "Social Landscaping in the Forests of Mexico: An Environmental Interpretation of Cardenismo, 1934–1940," *Hispanic American Historical Review* 92 (2012): 73–106; Diana Hadley, "Landholding Systems and Resource Management in the Sky Island Borderlands," in *Connecting Mountain Islands and Desert Seas: Biodiversity and Management of the Madrean Archipelago II*, ed. Gerald J. Gottfried et al., Proceedings RMRS-P-36 (Fort Collins, Colo.: USDA Forest Service, 2005), 15–26; Susan R. Walsh Sanderson, *Land Reform in Mexico: 1910–1980* (Orlando: Academic Press, 1984); Simpson, *Ejido*; and, from Mexican authors, Sergio Reyes Osorio, *Estructura agraria y desarrollo agrícola en México: Estudio sobre las relaciones entre la tenencia y uso de la tierra y el desarrollo agrícola de México* (Mexico City: Fondo de Cultura Económica, 1974).

26. Information on the Cárdenas regime from Wakild, *Revolutionary Parks*, 38; on the 1960s–1970s from Sanderson, *Land Reform in Mexico*, 83, 138, with quote at 71.

27. Tannenbaum, *Mexican Agrarian Revolution*, 79, 87. I also found useful, as a profile of the mid-twentieth century, Gonzalo Blanco, *Agriculture in Mexico* (Washington, D.C.: Pan American Union, 1950).

28. Starker Leopold quotes from an oral history: Sierra Club History Committee, "A. Starker Leopold," *Sierra Club Nationwide* 2 (1984): 12; and A. Starker Leopold, "Adios, Gavilán," *Pacific Discovery*, January–February 1949, 5–6. On the association between ejidos and broken fire regimes, see Emily K. Heyerdahl and Ernesto Alvarado, "Influence of Climate and Land Use on Historical Surface Fires in Pine-Oak Forests, Sierra Madre Occidental, Mexico," in *Fire and Climatic Change in Temperate Ecosystems of the Western Americas*, ed. Thomas T. Veblen et al., Ecological Studies 160 (New York: Springer-Verlag, 2003), 196–217; see especially graph on p. 207.

29. My guide is Myrna I. Santiago's wonderfully informative *Ecology of Oil* (see note 1 above); quote from p. 2.
30. On the local version, see Santiago, *Ecology of Oil*, 83. For the official version, see Palacios, "Memoria," 18.
31. Santiago, *Ecology of Oil*, 137–38; supplemented by Palacios, "Memoria," 18–29 (which includes fascinating photos); and Juan D. Villarello, "El pozo de petróleo de Dos Bocas," *Parergones del Instituto Geológico de México* 3, no. 1 (1909).
32. Santiago, *Ecology of Oil*, 156–62, gives a good account told from the perspective of the workers, along with some striking photos. A technical account, again with fascinating photos, is available in Arthur B. Clifford, "Extinguishing an Oil-Well Fire in Mexico, and the Part Played Therein by Self-Contained Breathing-Apparatus," *Transactions, Institution of Mining Engineers* 63 (1921–22): 2–12, quote from p. 7.
33. Clifford, "Extinguishing an Oil-Well Fire," 6–7.
34. Clifford, 8–9. On the airplane experiment, see *New York Times*, May 13, 1932.
35. Santiago, *Ecology of Oil*, 133, 122–24, 126. Mexico City episode reported in the *New York Times*, February 28, 1932.
36. Silverware story from Santiago, *Ecology of Oil*, 83.
37. Carl Christian Sartorius, *Mexico About 1850* (Stuttgart: F. A. Brockhaus, 1961), 179–80.
38. See, for example, Santiago, *Ecology of Oil*, 42–47, 49.
39. My survey on fire in Mexico City derives from Anna Rose Alexander, *City on Fire: Technology, Social Change, and the Hazards of Progress in Mexico City, 1860–1910* (Pittsburgh: University of Pittsburgh Press, 2016); quote from p. 1.
40. Quote from Mauricio Tenorio-Trillo, *I Speak of the City: Mexico City at the Turn of the Twentieth Century* (Chicago: University of Chicago Press, 2012), xxii. Background from Alexander, *City on Fire*; the introduction offers a distilled summary, 1–14.
41. Quotes from Alexander, *City on Fire*, 33. The examples also come from this book, scattered throughout its close analysis.
42. For a good overview, see Matthew Vitz, "La ciudad y sus bosques: La conservación forestal y los campesinos en el valle de México, 1900–1950," *Estudios de Historia Moderna y Contemporánea de México* 43 (January–June 2012): 135–72.
43. A. Starker Leopold, *Wildlife of Mexico: The Game Birds and Mammals* (Berkeley: University of California Press, 1972), 57. Interestingly, William Vogt raised the money for the project and directed it during "its formative years."
44. For an excellent survey in English of Quevedo's career, see Lane Simonian, *Defending the Land of the Jaguar: A History of Conservation in Mexico* (Austin: University of Texas Press, 1995), 67–84. See note 59 below for biographical sources on Quevedo's life and achievements.

    For his urban park work, see Emily Wakild, "Naturalizing Modernity: Urban Parks, Public Gardens and Drainage Projects in Porfirian Mexico City," *Mexican Studies / Estudios Mexicanos* 23 (Winter 2007): 101–23, esp. 108–10; and

Christopher R. Boyer, "Revolución y paternalismo ecológico: Miguel Ángel de Quevedo y la política forestal en México, 1926–1940," *Historia Mexicana* 57 (July–September 2007): 91–138. Statistics from Simonian, *Defending*, 72–73; Vitz, "La ciudad y sus bosques," 143–44.

45. Boyer, "Revolución y paternalismo ecológico," 91–138. Vitz, "La ciudad y sus bosques," 154–55.
46. Alexander, *City on Fire*, 14.
47. My discussion of Mexico's national parks relies heavily on Wakild, *Revolutionary Parks*; and Emily Wakild, "Border Chasm: International Boundary Parks and Mexican Conservation, 1935–1945," *Environmental History* 14, no. 3 (July 2009): 453–75; as well as Boyer and Wakild, "Social Landscaping in the Forests of Mexico."
48. Phrases, and an excellent summary of her idea, can be found in Wakild, *Revolutionary Parks*, 1–5, 42.
49. Dante Arturo Rodríguez Trejo, "Fire Regimes, Fire Ecology, and Fire Management in Mexico," *Ambio* 37, no. 7–8 (December 2008): 550. See also W. Lauer, "Timberline Studies in Central Mexico," *Arctic and Alpine Research* 10, no. 2 (1978): 393. On Orizaba, see Larissa L. Yocom and Peter Z. Fulé, "Human and Climatic Influences on Frequent Fire in a High Elevation Tropical Forest," *Journal of Applied Ecology* 49, no. 6 (2012): 1359.
50. Hector Ortiz-Contla, "Análisis del manejo tradicional de pastizales con fuego en el Ajusco" (master's thesis, Universidad Autónoma Chapingo, 2006), quoted phrase at 24. On the basis behind traditional burning ("un conocimiento empírico y heredado de generaciones previas"), see p. 1.
51. For a survey of traditional practices, see Wakild, *Revolutionary Parks*, 72–73, 112; quote from Quevedo on p. 112.
52. Miguel Ángel de Quevedo, "La cuestión del carbón vegetal y su explotación en los bosques del Distrito Federal, que importa poner en veda," *México Forestal* 10, no. 11–12 (November–December 1932): 133–36.
53. Quevedo, "La cuestión del carbón vegetal," 134.
54. The role of firewood gathering in reducing fire hazard around settlements should seem obvious, but it seems we need to be reminded from time to time. For a recent description, based on archaeological evidence and fire-scarred trees, see Christopher Roos et al., "Native American Fire Management at an Ancient Wildland-Urban Interface in the Southwest United States," *PNAS* 118, no. 4 (2021): e2018733118, https://doi.org/10.1073/pnas.2018733118.
55. Wakild, "Border Chasm," 456.
56. "Meteor Starts Forest Fire," *New York Times*, April 26, 1910. Cigarette and field fire: "Airplanes Fight Fire in Mexican Oil Field," *New York Times*, May 14, 1932. San Luis: "Mexican Oil Zone Fears Forest Fires," *New York Times*, May 4, 1928. Hurricane: "2,000 Mexicans Homeless as Fire Destroys Village," *New York Times*, April 15, 1934. Fanatic fire: "Forest Fires in Mexico," *New York Times*, May 14, 1886. Earthquake: "Thousands Made Homeless in Mex. 'Quake,'" *Indi-*

ana Gazette (Pennsylvania), April 16, 1941, 1. *Excélsior* reports: "Mexican Fires Drive Panthers to Villages," *New York Times*, April 30, 1928. Prospectors: "Forest Fires in Northern Mexico," *The Sentinel* (Carlisle, Penn.), May 21, 1898. Army: "Grazing Lands Burned," *New York Times*, April 20, 1916. Laredo: *New York Times*, April 26, 1920. On locomotive fires in Chihuahua, see Boyer, *Political Landscapes*, 43–44.

57. Such notices could reach the fraying ends of U.S. news, as Earhart's did by making the *Vidette-Messenger of Porter County* in Valparaiso, Indiana, on May 5, 1935.
58. On the 1861 law, see Simonian, *Defending*, 54. Simonian's book and Boyer, *Political Landscapes*, which chronicles developments from the Díaz era to the present, are the fountainheads for the history of the subject in English. Río de la Loza's writings were finally published as *Escritos de Leopoldo Río de la Loza* (Mexico City: Imprenta I. Escalante, S.A., 1911); on fire, see pp. 238–39. On the Díaz administration's forestry measures, see Simonian, *Defending*, 64–66; and Boyer, *Political Landscapes*, 25–59, with details of 1894 legislation on p. 52.
59. Unsurprisingly, there is an impressive literature on Quevedo and his influence. The source for most biographical information is his memoir *Relato de mi vida* (Mexico City: n.p., 1943). A modern biography in Spanish is available in J. Humberto Urquiza García, *Miguel Ángel de Quevedo: El proyecto conservacionista y la disputa por la nación 1840–1940* (Mexico City: UNAM, 2018). In English the best summaries are Simonian, *Defending*, 67–84; and Emily Wakild, "'It Is to Preserve Life, to Work for the Trees': The Steward of Mexico's Forests, Miguel Ángel de Quevedo, 1862–1946," *Forest History Today*, Spring–Fall 2006, 4–14. For a valuable interpretation of Quevedo and *paternalismo científico*, see Boyer, "Revolución y paternalismo ecológico."
60. Quevedo, *Relato de mi vida*, 11.
61. Quote from Tom Gill, "Miguel de Quevedo (1862–1946)," *Journal of Forestry* 45 (February 1947): 126.
62. Numbers on urban parks from Simonian, *Defending*, 72–73. On the French influence, see Antonio H. Sosa, "La influencia de Francia en el desarrollo forestal de México," *México Forestal* 14, no. 9–10 (September–October 1936): 73–75; and for a good summary, Boyer, *Political Landscapes*.
63. "Torres-observatorios para descubrir incendios de bosques," *México Forestal* 1, no. 11 (1928): 11.
64. Miguel A. de Quevedo, "Los incendios de bosques y quemas de pastos en la actual temporada de secas," *México Forestal* 7, no. 3 (1929): 51–52.
65. Quevedo, "Los incendios de bosques," 52.
66. Quevedo, "La cuestión del carbón vegetal," 133–36.
67. Antonio H. Sosa, "El problema forestal del estado de Tlaxcala," *México Forestal* 10, no. 11–12 (1932): 144.
68. Ricardo de la Vega, "El fuego y la expansión agraria hacia el bosque como principales motivos de la deforestación de territorio patrio," *México Forestal* 11, no. 11–12 (November–December 1933): 205–9, quotes from p. 205.

69. See Stephen J. Pyne, "Fire in the Mind: Changing Understandings of Fire in Western Civilization," *Philosophical Transactions of the Royal Society B* 371, no. 1696 (2016): http://dx.doi.org/10.1098/rstb.2015.0166.
70. The best survey of European fire remains Stephen J. Pyne, *Vestal Fire: An Environmental History, Told Through Fire, of Europe and Europe's Encounter with the World* (Seattle: University of Washington Press, 1997). Other useful sources are Axel Steensberg, *Fire-Clearance Husbandry: Traditional Techniques Throughout the World* (Henning: Poul Kristensen, 1993); and François Sigaut, *L'Agriculture et le feu: Rôle et place du feu dans les techniques de préparation du champ de l'ancienne agriculture européenne* (Paris: Mouton & Co., 1975).
71. Stephen J. Pyne, "Nataraja: India's Cycle of Fire," *Environmental History Review* 18, no. 3 (1994): 1–20.
72. On Madagascar, see Christian A. Kull, *Isle of Fire: The Political Ecology of Landscape Burning in Madagascar* (Chicago: University of Chicago Press, 2004). For the global survey, see Georges Kuhnholtz-Lordat, *La Terre incendiée* (Nimes: Éditions de la Maison Carrée, 1938).
73. Gilberto Fabila, "Organización del Servicio de Protección Forestal y especialmente contra incendios en los Estados Unidos de Norte América, por el Ing. Agrónomo Gilberto Fabila, agregado a la legación mexicana, en Washington," *México Forestal* 1, no. 9–10 (1928): 94–95.
74. Boyer, "Revolución y paternalismo ecológico," 92–93, for quotes from Quevedo, here loosely paraphrased. For a general background, see Boyer and Wakild, "Social Landscaping in the Forests of Mexico."
75. Miguel A. de Quevedo, "Proyecto de plan sexenal 1934–40 en lo relativo a la protección forestal del territorio," *México Forestal* 11, no. 11–12 (November–December 1933): 187.
76. See *Boletín del Departamento Forestal y de Caza y Pesca* 3, no. 8 (September–November 1937): the presidential instructions are on pp. 1–4, Quevedo's reply on pp. 4–19.
77. For an example of the publicity side of the project, see "Chihuahua y Durango se unen en la campaña contra incendios," *Boletín Forestal: Órgano de la Unión de Madereros de Chihuahua* 1, no. 7 (April–May 1949): 1–2.
78. Miguel Ángel de Quevedo, "Anteproyecto del plan sexenal para el periodo 1941–1946 en los ramos forestal y de caza y pesca," *Boletín del Departamento Forestal y de Caza y Pesca* 4, no. 4 (March–May 1939): 1–2.
79. Antonio H. Sosa, "Exploración forestal en la Alta Sierra Tarahumara del estado de Chihuahua," *Boletín del Departamento Forestal y de Caza y Pesca* 3, no. 9 (December 1937–February 1938): 187–235, at 187–90.
80. Sosa, "Exploración forestal," 228, 224.
81. Sosa, 212, 222, 220.
82. Lázaro Cárdenas, *El problema indígena de México* (Mexico City: Departamento de Asuntos Indígenas, 1940).

83. On his understanding of the pressures for revolution, see Miguel Ángel de Quevedo, *Algunas consideraciones sobre nuestro problema agrario* (Mexico City: Imprenta Victoria, 1916).
84. Cárdenas, *El problema indígena*, especially p. 11. For sample complaints, see Wakild, *Revolutionary Parks*, 149.
85. Aldo Leopold, "The Conservationist in Mexico," in *The River of the Mother of God and Other Essays by Aldo Leopold*, ed. Susan L. Flader and J. Baird Callicott (Madison: University of Wisconsin Press, 1991), 240. Quote on *trincheras* (ancient rock terraces) from Robert A. McCabe, "The Prehistoric Engineer-Farmers of Chihuahua," *Wisconsin Academy of Sciences, Arts and Letters* 44 (1955): 76.
86. Aldo Leopold, "Conservationist in Mexico," 240.
87. Starker Leopold quotes come from Sierra Club History Committee, "A. Starker Leopold," 11–12. On the witches'-broom, see A. Starker Leopold, "Adiós, Gavilán," 4. I'm indebted to Jordan Fisher Smith for alerting me to the oral history from which the quotes come.
88. Sierra Club History Committee, "A. Starker Leopold," 12. Aldo Leopold, "Piute Forestry vs Forest Fire Prevention," *Southwestern Magazine* 2 (1920): 12–13.
89. Aldo Leopold, "Conservationist in Mexico," 244, 243.
90. A. Starker Leopold, "Adiós, Gavilán," 6.
91. A great introduction to the complex character of Vogt is available in Charles C. Mann, *The Wizard and the Prophet* (New York: Vintage Books, 2018).
92. Vogt's memo quoted from Mann, *Wizard and the Prophet*, 72. Guillermo Vogt [William Vogt], *El hombre y la tierra*, Biblioteca Enciclopédica Popular 32 (Mexico City: Secretaría de Educación Pública, 1944).
93. Vogt, *El hombre y la tierra*, 46, 76, 94.
94. William Vogt, *Road to Survival* (New York: William Sloan Associates, 1948), 14, 5. William Vogt, "A Continent Slides to Ruin," *Harper's Magazine* 196, no. 1177 (June 1948): 484. Mann, *Wizard and the Prophet*, 368.
95. Dementia quote from Mann, *Wizard and the Prophet*, 396. "Biological bankruptcy" quote from Vogt, "Continent Slides to Ruin," 481. Other quotes from Vogt, *Road*, 28, 33, 172.
96. Vogt, *Road*, 28, 33, 172.
97. Tom Gill, *Land Hunger in Mexico* (Washington, D.C.: Charles Lathrop Pack Forestry Foundation, 1951), 1–4, 52, 54.
98. Gill, *Land Hunger*, 74.
99. Enrique Beltrán, Gonzalo Blanco Macías, and Roberto Villaseñor, *Homenaje al Dr. Tom Gill 1891–1972* (Mexico City: Ediciones del Instituto Mexicano de Recursos Naturales Renovables, 1972), 28. Original wording: "sincero amigo de México, y de ello existen pruebas incontables."
100. "Mexicans to Light Aztec Sacred Fire," *New York Times*, March 15, 1925.
101. "Toltec Sun Legend Enacted in Mexico," *New York Times*, October 12, 1953.

102. Fernando Ortiz Monasterio et al., *Tierra profanada: Historia ambiental de México* (Mexico City: Instituto Nacional de Antropología e Historia, 1987), 232.
103. Guillermo Gaitán M., "Control de incendios," in *Memoria de la Primera Convención Nacional Forestal: Agosto de 1941* (Mexico City: Secretaría de Agricultura y Fomento, Dirección General Forestal y de Caza, 1942), 299–300. I am grateful to Dante Rodríguez and Ing. Miguel Gallegos for tracking down the elusive proceedings.
104. A good distillation of fire-related programs in these years is available in Dante Arturo Rodríguez Trejo, *Incendios de vegetación: Su ecología, manejo e historia* (Guadalajara: Colegio de Postgraduados, 2014–15), 2:1436–38. My observation on the 5 to 12 percent commitment devoted to fire reflects the proportion of papers devoted to the subject in Mexican forest congresses.
105. Boyer, *Political Landscapes*, 4.
106. On forest loss, see Boyer, *Political Landscapes*, 16.
107. Gill, *Land Hunger*, 46, 45, 36.
108. Abel López Caballero, "Control y combate de los incendios forestales" (bachelor's thesis, Escuela Nacional de Agricultura, Chapingo, Mexico, 1955). Guillermo Moreno Arellano, "Organización para el combate de incendios forestales" (bachelor's thesis, Escuela Nacional de Agricultura, Chapingo, 1957).
109. Comité de Asesoría Técnica Forestal, *La situación forestal de México es grave: Contribución al conocimiento de las causas que la engendran y a la posible solución de sus problemas* (n.p.: Cámara Nacional de las Industrias del Papel, 1958), quotes from 45, 51. On the cynical arguments from the timber sector, see Boyer, *Political Landscapes*, 148–49. See, for example, Antonio Sierra Pineda and Guillermo Gaitán Medina, "Incendios forestales," in *II Convención Nacional Forestal* (Mexico City: SRF, 1959), 325–26.
110. Quotations from Boyer, *Political Landscapes*, 85, 155.
111. Land reform statistics from Sanderson, *Land Reform in Mexico*, 90; and Bray, *Mexico's Community Forest Enterprises*, esp. 71. On land distribution, see Dan Klooster, "Campesinos and Mexican Forest Policy During the Twentieth Century," *Latin American Research Review* 38, no. 2 (2003): 103; and Sanderson, *Land Reform in Mexico*, 138.
112. Enrique Beltrán, "Man Versus Land," *Americas* 2 (April 1950): 24, 46.
113. For a survey of Beltrán's extraordinary career, see the reflections offered in Academia Nacional de Ciencias Forestales, *Homenaje al Dr. Enrique Beltrán* (Mexico City: n.p., 1980).

Beltrán became the face of Mexican conservation and the director of the Instituto Mexicano de Recursos Naturales Renovables. Four publications, each printed a decade after the other—1939, 1949, 1959, and 1969—track the trajectory of his thinking: "Los recursos naturales de México: Lineamientos para una política de conservacíon," *Revista de la Sociedad Mexicana de Historia Natural* 1 (1939): 33–43; *La protección de la naturaleza: Principios y problemas,*

Biblioteca Enciclopédica Popular Tercera Época 206 (Mexico City: Secretaría de Educación Pública, 1949); *Los recursos naturales del Sureste y su aprovechamiento*, 3 vols. (Mexico City: n.p., 1959), with Beltrán as editor; and "Los recursos naturales y la reforma agraria," in *Los recursos naturales y la reforma agraria (tres ensayos)* (Mexico City: Ediciones Productividad, 1969), 3–26. For a U.S. audience, he wrote a nice explanation of how Mexican forestry aligns with the Mexican Revolution: "Forestry and the Public Domain: A Mexican Point of View," parts 1 and 2, *American Forests*, December 1969, 36, 57–59; January 1970, 37, 46–47.

114. Enrique Beltrán, "Los incendios forestales: Prevención y combate," in *Temas forestales 1946–1960* (Mexico City: n.p., 1961), 243–60; quotes at 243–45, 251.
115. Beltrán, "Los incendios forestales," 246, 251.
116. Beltrán, 253–54, 258, 255, 259.
117. "Plan para combatir los incendios forestales en México," *México y Sus Bosques* 2, no. 4–5 (October 1962–March 1963): 21–25, 38; quote from p. 22.
118. A detailed description of the fire brigades is given in Hermenegildo Cuenca Díaz, "La creación de las brigadas contra incendio," in *Memoria de la III Convención Nacional Forestal* (Mexico City: [Secretaría de Agricultura?], 1966), 240–56.
119. "Plan para combatir," 21–25, 38. A good summary of the era is in Rodríguez Trejo, *Incendios de vegetación*, 2:1440–45.
120. Enrique Beltrán, *La batalla forestal: Lo hecho, lo no hecho, lo por hacer* (Mexico City: n.p., 1964), 1. For the official report on the sexenio, see Departamento de Prevención y Combate de Incendios Forestales, *Dirección General De Protección y Repoblación Forestales: Seis años de actividades forestales y de la fauna, 1959–1964* (Mexico City: SFF México, 1964), 103–26.
121. Beltrán, *La batalla forestal*, 35, 116.
122. Beltrán, 116. Gill, *Land Hunger*, 51.
123. Miguel A. de Quevedo, "Discurso de inauguración del Primer Congreso Forestal Mexicano," *México Forestal* 8, no. 3 (1930): 31. Coert duBois, *Systematic Fire Protection in the California Forests* (Washington, D.C.: Government Printing Office, 1914), 3. DuBois's model became the basis for the national forest system. On the earliest texts, see A. D. Folweiler and A. A. Brown, *Fire in the Forests of the United States* (St. Louis: John S. Swift, 1946), revised as Brown and Folweiler, *Fire in the Forests of the United States*, in 1953.
124. Brown and Folweiler, *Fire in the Forests*, 3.
125. On the background to the NAFC and the fire group, see James Sorenson, *Seventeen Years of Progress Through International Cooperation 1962–1978* (Atlanta: U.S. Forest Service, Southeastern Area State and Private Forestry, 1979); and Vicente Arriaga Martínez, Bill de Groot, and Tom Harbour, "Fire Management Working Group of the North American Forest Commission: History Report, Revised in 2008," available online at https://studylib.net/doc/11244815/fire-management-working-group-history-report-of-the-north.

126. Quote from Sorenson, *Seventeen Years*, 13. On the translation, see Forest Resources Development Branch, *Wildland Fire Management Terminology*, FAO Forestry Paper 70 (Rome: FAO, 1986).
127. "Eighth Reunion of the Forest Fire Management Group," in "Fire Management Study Group: Organization, History, Accomplishment, Opportunities" (unpublished report, ca. 1974), personal copy, p. 8.
128. French is also an official language of Canada, so it was likely also an official language of NAFC. But I have found no documents for the fire working group in French.
129. On forestry's role in European expansion and fire control, see Pyne, *Vestal Fire*, 465–99.
130. India's story is particularly significant. For a snapshot history, see Pyne, "Nataraja."
131. On the reformation in U.S. fire policy, see Stephen J. Pyne, *Between Two Fires: A Fire History of Contemporary America* (Tucson: University of Arizona Press, 2015).
132. On the industrial brigades, see *Reestructuración de corporaciones contra incendios forestales* (Mexico City: Subsecretaría Forestal y de la Fauna, 1969); and an interesting thesis by Antonio Quiñónez Silva, "Corporaciones de defensa contra incendios forestales" (professional thesis, Escuela Nacional de Agricultura, Chapingo, 1969).

    José Verduzco Gutiérrez and Guillermo Gaitán Medina, *Combate de incendios forestales*, rev. ed. (Mexico City: Secretaría de Agricultura y Ganadería, Dirección General de Protección y Repoblación Forestales, 1968); Alfonso Gutiérrez Palacio and José Verduzco Gutiérrez, *Texto guía, para la enseñanza popular de los principios de conservación forestal* (Mexico City: Departamento de Divulgación Forestal y de la Fauna, 1970); José Verduzco Gutiérrez, *Protección forestal* (Chapingo: Editorial de México, 1976).
133. Sosa, "Exploración forestal."
134. Antonio H. Sosa, "A través de la Tarahumara," parts 1 and 2, *México Forestal* 29, no. 4 (July–May 1965): 1–4; 29, no. 5 (September October 1965): 1–4. The incident is wonderfully described, as an interaction with the Rarámuri, in Boyer, *Political Landscapes*, 1–3.
135. Quoted in William Forbes, "Revisiting Aldo Leopold's 'Perfect' Land Health: Conservation and Development in Mexico's Rio Gavilan," (PhD diss., University of North Texas, 2004), 76. The dissertation includes photos from the trek.
136. A. Starker Leopold, *Wildlife of Mexico*, 55.
137. A. S. Leopold et al., "Wildlife Management in the National Parks," Advisory Board on Wildlife Management Appointed by Secretary of the Interior Udall (March 4, 1963); reprinted in *Administrative Policies for the National Parks and National Monuments of Scientific Significance (Natural Area Category)* (Washington, D.C.: Government Printing Office, 1970), appendix D.
138. See A. Starker Leopold, "Adiós, Gavilán," 4–13; and Forbes, "Leopold's 'Perfect' Land Health," 71–76.

139. Edward Abbey, *Abbey's Road* (New York: E. P. Dutton, 1979), 81–82, 96. Abbey published his original essay (later included in the anthology) in 1975; see James M. Cahalan, *Edward Abbey: A Life* (Tucson: University of Arizona Press, 2001), 321.
140. Unless otherwise noted, my account of Michoacán relies on Boyer, *Political Landscapes*, 187–93; quote from p. 189.
141. A. Starker Leopold, *Wildlife in Mexico*, 55.
142. See Griselda Benítez Badillo, "Efectos del fuego en la vegetación herbácea de un bosque de *Pinus hartwegii* Lindl. de la Sierra del Ajusco," in *Aportes a la ecología urbana de la Ciudad de México*, ed. Eduardo H. Rapoport and Ismael R. López-Moreno (Mexico City: Editorial Limusa, 1987), 111–52; and Hector Ortiz-Contla et al., "Manejo tradicional de pastizales en el sur del Valle de México" (unpublished contribution to Proyecto Ajusco at the Universidad Autónoma Chapingo).
143. Sierra Pineda and Gaitán Medina, "Incendios forestales," 325–26. Rodríguez Trejo, *Incendios de vegetación*, 2:1438–40, gives a good summary of developments.
144. "Departamento de Prevención y Combate de Incendios," in *Subsecretaría Forestal y de la Fauna: Seis años de actividades forestales y de la fauna, 1959–1964* (Mexico City: SFF, 1964), 103–36. Rodríguez Trejo, *Incendios de vegetación*, 2:1440, also includes some personal communications from Garduño and Sierra Pineda.
145. Roberto Garduño García, "Protección forestal," in *Memoria de la III Convención Nacional Forestal* (Mexico City: [Secretaría de Agricultura?], 1966), 277, 288.
146. On Cofre de Perote: Julián Cerano-Paredes et al., "Effects of Climate on Historical Fire Regimes (1451–2013) in *Pinus hartwegii* Forests of Cofre de Perote National Park, Veracruz, Mexico," *Dendrochronologia* 65 (2021): 125784. The likely cause for the sharp drop-off in burning was a local manifestation of the national Programa de Desarrollo Forestal. On Pico de Orizaba: Wilhelm Lauer and Dieter Klaus, "Geoecological Investigations on the Timberline of Pico de Orizaba, Mexico," *Arctic and Alpine Research* 7, no. 7 (1975): 328–29; Yocom and Fulé, "Human and Climatic Influences," 1356–64. On Chignahuapan: Julián Cerano-Paredes et al., "Régimen histórico de incendios y su relación con el clima en un bosque de *Pinus hartwegii* al norte del estado de Puebla, México," *Bosque* 37, no. 2 (2016): 389–99.
147. Jesús Vásquez Soto, "The Significance of Forest Fires in Mexico," in *Fire in the Environment: Symposium Proceedings, May 1–5, 1972, Denver, Colorado*, FS-276 (n.p.: U.S. Forest Service, 1972), 28.
148. Boyer, *Political Landscapes*, 153–55. Quotes from Elisa García Villafan, "Chihuahua y los incendios forestales," *Ciencia Forestal* 10, no. 54 (1985): 49.
149. Peter Z. Fulé, José Villanueva-Díaz, and Mauro Ramos-Gómez, "Fire Regime in a Conservation Reserve in Chihuahua, Mexico," *Canadian Journal of Forest Research* 35 (2005): 320–30.
150. An easily accessible summary of the fire history studies is available in Peter Z. Fulé and W. Wallace Covington, "Changing Fire Regimes in Mexican Pine

Forests," *Journal of Forestry* 94, no. 10 (October 1996): 33–38. For a more technical summary, which includes the cited rebounding fires, see Fulé, Villanueva-Díaz, and Ramos-Gómez, "Fire Regime in a Conservation Reserve," especially p. 327; and Heyerdahl and Alvarado, "Influence of Climate and Land Use."

151. Sierra de los Ajos: John H. Dieterich, "Cronología de los incendios forestales de la Sierra de los Ajos, Sonora, Méx.," *Revista Dasonomía Mexicana* 3 (1985): 20–24; and Thomas W. Swetnam and Christopher H. Baisan, "Fire Histories of Montane Forests in the Madrean Borderlands," in *Effects of Fire on Madrean Province Ecosystems: A Symposium Proceedings*, coord. Peter F. Ffolliott et al., USDA Forest Service Gen. Tech. Report RM-GTR-289 (Fort Collins, Colo.: U.S. Forest Service, 1996), 15–36. Tutuaca: Fulé, Villanueva-Díaz, and Ramos-Gómez, "Fire Regime in a Conservation Reserve." Charcos: Iván M. Molina-Pérez et al., "Historical Fire Frequency (1779–2013) in Pine-Oak Forests in the Community of Charcos, Mezquital, Durango," *Revista Chapingo Serie Ciencias Forestales y del Ambiente* 23, no. 1 (2016): 91–104. Other sites: Citlali Cortés Montaño et al., "Linking Old-Growth Forest Composition, Structure, Fire History, Climate and Land-Use in the Mountains of Northern Mexico," *Ecosphere* 3, no. 11 (November 2012): 1–16; Peter Z. Fulé et al., "Testing a Pyroclimatic Hypothesis on the Mexico–United States Border," *Ecology* 93, no. 8 (2012): 1830–40; Peter Z. Fulé and W. Wallace Covington, "Spatial Patterns of Mexican Pine-Oak Forests Under Different Recent Fire Regimes," *Plant Ecology* 134, no. 2 (1998): 197–209.

152. Peter Z. Fulé and W. Wallace Covington, "Fire Regime Changes in La Michilía Biosphere Reserve, Durango, Mexico," *Conservation Biology* 13, no. 3 (June 1999): 640–52.

153. S. A. Drury and T. T. Veblen, "Spatial and Temporal Variability in Fire Occurrence Within the Las Bayas Forestry Reserve, Durango, Mexico," *Plant Ecology* 197 (2008): 299–316.

154. Peter Z. Fulé et al., "Fire Regime in a Mexican Forest Under Indigenous Resource Management," *Ecological Applications* 21, no. 3 (April 2011): 764–75

155. See Larissa L. Yocom Kent et al., "Climate Drives Fire Synchrony but Local Factors Control Fire Regime Change in Northern Mexico," *Ecosphere* 8, no. 3 (March 2017), for a successful summary of northern Mexican fire histories; p. 6 distills the distillation.

156. Larissa L. Yocom et al., "El Niño-Southern Oscillation Effect on a Fire Regime in Northeastern Mexico Has Changed over Time," *Ecology* 91, no. 6 (June 2010): 1660–71; quote from p. 1666. The authors identify as an "unusual feature" the "almost complete dominance of dormant-season fires" (1667), but this is wholly consistent with the scar records from elsewhere in Mexico, so I do not understand why this set should seem anomalous.

157. Yocom et al., "El Niño-Southern Oscillation Effect," 1667.

158. Larissa L. Yocom et al., "Fine-Scale Factors Influence Fire Regimes in Mixed-Conifer Forests on Three High Mountains in Mexico," *International Journal of*

Wildland Fire 23, no. 7 (2014): 8, http://dx.doi.org/10.1071/WF13214. Another study in the nearby Parque Ecológico Chipinque exists: Marco Aurelio González Tagle et al., "Reconstrucción del historial de incendios y estructura forestal en bosques mixtos de pino-encino en la Sierra Madre Oriental," *Madera y Bosques* 13, no. 2 (2007): 51–63. Its purpose was to correlate regeneration episodes with fires, of which it identified five events. It thus stands somewhat outside the other, more comprehensive surveys.

159. On the tierra caliente, see Elena Lazos Chavero, "Ciclos y rupturas: Dinámica ecológica de la ganadería en el sur de Veracruz," in *Historia ambiental de la ganadería en México*, ed. Lucina Hernández (Xalapa: Instituto de Ecología, 2001), 133–53.
160. Regarding modern inquiries into fire's ancient purposes in pasture, see José Francisco Villanueva Avalos et al., "Respuesta de seis gramíneas tropicales a la quema prescrita en la costa oeste de México," *Téc Pecu México* 46, no. 40 (2008): 397–411.
161. Rodríguez Trejo, *Incendios de vegetación*, vol. 1, offers the best digest of fire ecology: ch. 8 ("Pastizales") reviews the literature on grasslands, including those in the north; and ch. 9 ("Matorrales") discusses shrubs. On the use of cacti as emergency fodder, see pp. 529–31.
162. Donald D. Brand, "The Early History of the Range Cattle Industry in Northern Mexico," *Agricultural History* 35, no. 3 (1961): 132–39, numbers from 136.
163. Brand, "Early History of the Range Cattle Industry," 137.
164. Brand, 137.
165. H. M. Poulos et al., "Human Influences on Fire Regimes and Forest Structure in the Chihuahuan Desert Borderlands," *Forest Ecology and Management* 2908 (2013): 1–11.
166. John Sakulich and Alan H. Taylor, "Fire Regimes and Forest Structure in a Sky Island Mixed Conifer Forest, Guadalupe Mountains National Park, Texas, USA," *Forest Ecology and Management* 241 (2007): 62–73.
167. Jed Meunier, William H. Romme, and Peter M. Brown, "Climate and Land-Use Effects on Wildfire in Northern Mexico, 1650–2010," *Forest Ecology and Management* 325 (2014): 49–59; quote about "an active and unresolved area of study" on p. 56.
168. Fulé et al., "Testing a Pyroclimatic Hypothesis."
169. See Rodríguez Trejo, *Incendios de vegetación*, vol. 1, pp. 448, 456; Fernando Ibarra-F., M. Martin-R., J. R. Cox, and H. Miranda-Z., "The Effects of Prescribed Burning to Control Brush Species on Buffelgrass Pastures in Sonora, Mexico," in Ffolliott et al., *Effects of Fire on Madrean Province Ecosystems*, 195–204.
170. For the backstory on ranching in the region, see R. Aguirre et al., eds., *Sonora: 400 años de ganadería*, 2nd ed. (Hermosillo: Sociedad Sonorense de Historia / Unión Ganadera Regional de Sonora, 2007).

    On fire ecology, see Rodríguez Trejo, *Incendios de vegetación*, vol. 1, pp. 457–537 (a survey of matorral) and 626–28 (on invasives); p. 466 includes a map of various groupings of matorral. The particular variety of vetch is *Vicia villosa*;

see James C. Sorenson, "A Look at Fire Prevention in Mexico," *Fire Management Today* 48, no. 2 (1987): 26. On fire and buffelgrass, see Ibarra-F. et al., "Effects of Prescribed Burning"; but especially C. M. D'Antonio, "Fire, Plant Invasions, and Global Changes," in *Invasive Species in a Changing World*, ed. H. A. Mooney and R. J. Hobbs (Washington, D.C.: Island Press, 2000), 65–93.

171. For the history of the region, I rely mostly on Richard A. Minnich and Ernesto Franco Vizcaíno, *Land of Chamise and Pines: Historical Accounts and Current Status of Northern Baja California's Vegetation* (Berkeley: University of California Press, 1980); for an article-length précis of the book, see Richard A. Minnich and Ernesto Franco-Vizcaíno, "Protecting Vegetation and Fire Regimes in the Sierra San Pedro Mártir of Baja California," *Fremontia* 25, no. 3 (July 1997): 13–21. Minnich's research has been particularly pertinent in contemporary times and his comparisons with fire history in the United States sufficiently controversial to spark studies by others.

Another good distillation is Michael G. Barbour et al., "A Land Above: Protecting Baja California's Sierra San Pedro Martir Within a Biosphere Reserve," *Journal of the Southwest* 39, no. 3–4 (Autumn–Winter 1997): 613–95; quote on the Kiliwas at 614–16.

172. Minnich and Franco Vizcaíno, *Land of Chamise and Pines*, 105–9, and esp. 648–50. "Substantially" quote from Scott L. Stephens, Carl N. Skinner, and Samantha J. Gill, "Dendrochronology-Based Fire History of Jeffrey Pine-Mixed Conifer Forests in the Sierra San Pedro Martir, Mexico," *Canadian Journal of Forest Research* 33 (2003): 1090–1101, quote from p. 1096. Sources disagree on the persistence of sheep after 1964, when national park officials closed the site to sheep. Certainly most of the twentieth century had flocks of transhumant sheep in the park.

173. Carl Sauer and Peveril Meigs, "Lower Californian Studies, I: Site and Culture at San Fernando de Velicatá," *University of California Publications in Geography* 2, no. 9 (September 30, 1927): 271–302, quote at 295.

174. R. A. Minnich et al., "Californian Mixed-Conifer Forests Under Unmanaged Fire Regimes in the Sierra San Pedro Mártir, Baja California, Mexico," *Journal of Biogeography* 27 (2000): 105–29, with lightning and fire regimes discussed mostly on pp. 120–21.

175. For a discussion of the pivot around 1800, see Stephens, Skinner, and Gill, "Dendrochronology-Based Fire History," 1095–97.

On Indigenous burning, see studies of Alta California, especially J. Timbrook, J. R. Johnson, and D. D. Earle, "Vegetation Burning by the Chumash," *Journal of California and Great Basin Anthropology* 4 (1982): 163–86; and Henry T. Lewis, *Patterns of Indian Burning in California: Ecology and Ethnohistory*, Ballena Press Anthropological Papers 1 (Ramona, Calif.: Ballena Press, 1973).

176. The most closely argued analyses are in Minnich and Franco Vizcaíno, *Land of Chamise and Pines*, 115–21; and Stephens, Skinner, and Gill, "Dendrochronology-Based Fire History."

177. The dominant interpretation remains Minnich's.

The prevailing fire regimes around the Sierra San Pedro Mártir, as documented in fire scars, suggest a regime with a mean return interval of 20–30 years in the lowlands, 50–70 years in the matorral-clothed foothills, and 70–80 years in the mountains. But fire scars, both scars on trees and scarred land, tend to underestimate frequency because they miss small fires and grassy sites don't leave records. See Minnich and Franco Vizcaíno, *Land of Chamise and Pines*, 115–21.

178. For an example of an alternative view written at the same time, see Jon E. Keeley and C. J. Fotheringham, "Historic Fire Regime in Southern California Shrublands," *Conservation Biology* 15, no. 6 (December 2001): 1536–48.
179. My overview of this era is informed by Herman W. Konrad, "Tropical Forest Policy and Practice During the Mexican Porfiriato, 1876–1910," in *Changing Tropical Forests: Historical Perspectives on Today's Challenges in Central and South America*, ed. Harold K. Steen and Richard P. Tucker (Durham, N.C.: Forest History Society, 1992), 123–43.
180. Konrad, "Tropical Forest Policy and Practice," 132.
181. A. Starker Leopold, *Wildlife of Mexico*, 46–47.
182. Still a foundational document for the formal study of the Mayan milpa and, indirectly, for landscape fire is E. Hernández-Xolocotzi, "La agricultura en la península de Yucatán," in Beltrán, *Los recursos naturales del Sureste*, 3–57; see especially his conclusions on p. 53. As a summary (and tribute), see Efraím Hernández Xolocotzi, Eduardo Bello Baltazar, and Samuel Levy Tacher, eds., *La milpa en Yucatán: Un sistema de producción agrícola tradicional*, 2 vols. (Montecillo, Mex.: Colegio de Postgraduados, 1995).
183. Hernández-Xolocotzi, "La agricultura en la península de Yucatán," quotes on pp. 4, 37. A more comprehensive survey that situates his contribution is Hernández Xolocotzi, Baltazar, and Tacher, *La milpa en Yucatán*.
184. On the intercultivation, see Hernández-Xolocotzi, "La agricultura en la península de Yucatán," with a photo on p. 43; for his criticisms, see pp. 52–53. For the perspective of a U.S. journalist on developments, see Joel Simon, *Endangered Mexico: An Environment on the Edge* (San Francisco: Sierra Club Books, 1997), 110–20.
185. Vergara, *Fueling Mexico*, makes this point with repeated examples—e.g., p. 163 on roads.
186. Yocom Kent et al., "Climate Drives Fire Synchrony."

FIFTH SUN

1. A simple box score of the fires, written shortly after the season ended, is Dante A. Rodríguez Trejo and Stephen J. Pyne, "Mexican Fires of 1998," *International Forest Fire News* 20 (March 1999): 61–63; updated by Rodríguez Trejo as "Update of the 1998 Fire-Season in Mexico" (unpublished manuscript, courtesy of its author). A brisk account from someone on the front lines is available in

Roberto Martínez Domínguez, "Roberto Martínez Domínguez," in *Seis vidas en fuego: Seis cuentos de México*, ed. Stephen J. Pyne and Said Infante Gil (Texcoco: Editorial del Colegio de Postgraduados, 2023), 189–92.

An official summary was authored by Oscar Cedeño Sánchez, "Incendios forestales en México en 1998: Magnitud, extensión, combate y control," in *Incendios forestales y agropecuarios: Prevención e impacto y restauración de los ecosistemas*, ed. Hortensia Santiago Fragoso et al. (Mexico City: UNAM, 1999), 1–18; and for the FAO an unpublished report: Oscar Cedeño, "Resultados de la campaña antiincendios de México 1998," https://www.fao.org/3/x2095s/x2095s10.htm. For a photo gallery from the southern states, see "Photo Archive: Mexico; Some Visual Impressions of Fire in Chiapas, Mexico," *International Forest Fire News* 17 (July 1997): https://gfmc.online/iffn/country/mx/mx_1.html. On ENSO effects at Peña Nevada, see Larissa L. Yocom et al., "El Niño-Southern Oscillation Effect on a Fire Regime in Northeastern Mexico Has Changed over Time," *Ecology* 91, no. 6 (June 2010): 1660–1771. For a useful overview, see also Mark A. Cochrane, *Tropical Fire Ecology: Climate Change, Land Use, and Ecosystem Dynamics* (Chichester, UK: Springer-Praxis, 2009), 10–14. Probably the most comprehensive attempt to understand the fires is Santiago Fragoso et al., *Incendios forestales y agropecuarios.*

For a lively account of how several Americans detailed to the fires saw them, see Kelly Andersson, "Wildfire in Mexico!," *Wildland Firefighter*, July 1998.

2. Rodríguez Trejo and Pyne, "Mexican Fires of 1998"; and Rodríguez Trejo, "Update." For an interesting survey of the fires in Los Chimalapas, see Salvador Anta Fonseca and Antonio Plancarte Barrera, "Los incendios forestales," in *Chimalapas: La última oportunidad* (n.p.: World Wildlife Fund Programa México / Secretaría de Medio Ambiente, Recursos Naturales y Pesca, Taller El Castor, 2001), 123–38. On the final area burned, see CONAFOR, "Programa de Manejo del Fuego, Centro Nacional de Manejo del Fuego (Cierre 2019)," https://www.gob.mx/cms/uploads/attachment/file/522446/Cierre_de_la_Temporada_2019.pdf.
3. The literature on the fires is large, but a concise, systematic survey is available in Cedeño Sánchez, "Incendios forestales en México en 1998." Quote from Associated Press, "Forest Fire Deaths Mar Mexican Anniversary," *Democrat and Chronicle*, May 6, 1998, 12.
4. Rodríguez Trejo and Pyne, "Mexican Fires of 1998," 3. Dante A. Rodríguez Trejo, "A Brief History of Forest Fires in Mexico," *International Forest Fire News* 19 (September 1998): 3–4. Glickman quote from Susan Ferriss, "Glickman Calls Blazes in Mexico a Transnational Disaster," *Austin American-Statesman*, June 6, 1998, 15. "Unwanted enemy" quote from Anta Fonseca and Plancarte Barrera, "Los incendios forestales," 135.
5. Associated Press, "Forest Fire Deaths."
6. David A. Sonnenfeld, "Mexico's 'Green Revolution,' 1940–1980: Toward an Environmental History," *Environmental History Review* 16, no. 4 (Winter 1992): 28–52, data on p. 28.

7. Statistic on financial outflow from Colin M. MacLachlan and William H. Beezley, *El Gran Pueblo: A History of Greater Mexico*, 3rd ed. (Upper Saddle River, N.J.: Prentice Hall, 2004), 416.
8. Statistics on emissions from Our World in Data, https://ourworldindata.org/co2/country/mexico.
9. A. Jernelöv and O. Lindén, "Ixtoc I: A Case Study of the World's Largest Oil Spill," *Ambio* 10, no. 299 (1981).
10. For a very accessible source about the outcomes of Mexico's agricultural modernization project, see Salvador Mena Munguía y Mario Ramírez Martínez, *Panorama de la agricultura en México* (Guadalajara: Universidad de Guadalajara, 2014).
11. For an interesting discussion of the impact the oil industry was having on agriculture, at a time when petroleum's future seemed boundless, see P. Lamartine Yates, *Mexico's Agricultural Dilemma* (Tucson: University of Arizona Press, 1981), 254–57.
12. For a comprehensive survey of biodiversity and cultural diversity, see Eckart Boege, *El patrimonio biocultural de los pueblos indígenas de México: Hacia la conservación* in situ *de la biodiversidad y agrodiversidad en los territorios indígenas* (Córdoba, Mex.: Instituto Nacional de Antropología e Historia / Comisión Nacional para el Desarrollo de los Pueblos Indígenas, 2008). Curiously, the book succeeds in avoiding fire as a critical process.
13. Beltrán quote from Martínez Domínguez, "Roberto Martínez Domínguez," 163.
14. Figures and assessment of Verduzco from Martínez Domínguez, 164.
15. Martínez Domínguez, 168; Dante Arturo Rodríguez Trejo, "Dante Arturo Rodríguez Trejo," in Pyne and Gil, *Seis vidas*, 314–21.With a tinge of exasperation, Martínez blamed some of the mismanagement on the six-year purges that accompanied elections ("el eterno tránsito de cambio de nombre por el que siempre han pasado las dependencias de gobierno mexicanas"). He helped guide a thesis at UACh on aircraft in fire control, which offers good insight into thinking at the time, just before the 1988 and 1989 seasons: Gilberto Galeote Rivera, "Evaluación del uso de equipo aéreo en el combate de incendios forestales en México" (bachelor's thesis, Universidad Autónoma Chapingo, 1987).

    On EXTINGE, see Dante Arturo Rodríguez Trejo and Antonio Sierra Pineda, "Evaluación de los combustibles forestales en los bosques del Distrito Federal," *Revista de Ciencia Forestal en México* 20, no. 77 (January–June 1995).
16. Martínez Dominguez, "Roberto Martínez Domínguez," 168–69.
17. James H. Perkins and George A. Roby, "Fire Management Training in International Forestry," *Fire Management Notes* 48, no. 1 (1987): 18–20. Interview with Alexandra Zamenski, USFS, 2019.
18. Martínez Domínguez, "Roberto Martínez Domínguez," 168–69, 176–77. The lexicon included English, Spanish, Italian, German, and French: Forest Resources Development Branch, Forest Resources Division Forestry Depart-

ment, FAO, *Wildland Fire Management Terminology*, FAO Forestry Paper 70 (Rome: UN Food and Agriculture Organization, 1986).

19. Martínez Domínguez, "Roberto Martínez Domínguez," 170–72. On hand tools, see Dante Rodríguez Trejo, *Incendios de vegetación: Su ecología, manejo e historia* (Guadalajara: Colegio de Postgraduados, 2014–15), 2:1452. See Armando González-Cabán and David V. Sandberg, "Fire Management and Research Needs in Mexico," *Journal of Forestry* 87, no. 8 (1989): 20–26.
20. Rodríguez Trejo, "Dante Arturo Rodríguez Trejo," 313. For a longer but thumbnail history, see the official history page on the UACh website: https://chapingo.mx/historia/.
21. José Verduzco Gutiérrez, *Protección forestal* (Mexico City: Patena, 1976).
22. Dispatch center and EXTINGE information from Rodríguez Trejo, *Incendios de vegetación*, 2:1449–50.
23. For the United States' "fire revolution," see Stephen Pyne, *Between Two Fires: A Fire History of Contemporary America* (Tucson: University of Arizona Press, 2015).
24. Jesús Sánchez-Córdova and J. H. Dieterich, "Efecto de quemas controladas en *Pinus durangensis* en Madera, Chihuahua," in *Nota Técnica* 9 (n.p.: Secretaría de Agricultura y Recursos Hidráulicos, Instituto Nacional de Investigaciones Forestales, Centro de Investigaciones Forestales del Norte, 1983); and in English, Héctor E. Alanis-Morales, "Prescribed Fire in the Pine Forests of Northwestern Chihuahua," in *Effects of Fire on Madrean Province Ecosystems: A Symposium Proceedings*, coord. Peter F. Ffolliott et al., USDA Forest Service Gen. Tech. Report RM-GTR-289 (Fort Collins, Colo.: U.S. Forest Service, 1996), 193–94. See also Héctor Eligio Alanis Morales and Jesus Sánchez Córdova, *Quemas prescritas en una masa jóven de* Pinus durangensis *Mtz. en Madera, Chih.*, Folleto Técnico 5 (n.p.: Secretaría de Agricultura y Recursos Hidráulicos, Instituto Nacional de Investigaciones Forestales, 1994).
25. González-Cabán and Sandberg, "Fire Management and Research Needs," 26.
26. On Hurricane Gilbert, see Edward N. Rappaport and Colin J. McAdie, preparers, *Hurricane Gilbert (1988) in Review and Perspective*, NOAA Technical Memorandum NWS NHC 45 (Coral Gables, Fl.: National Hurricane Center, 1991); Dennis F. Whigham et al., "The Impact of Hurricane Gilbert on Trees, Litterfall, and Woody Debris in a Dry Tropical Forest in the Northeastern Yucatán Peninsula," *Biotropica* 23, no. 4a (1991): 434–41.

    In 2007 Hurricane Dean, another Category 5 storm, with maximum winds at 315 km/h, repeated the scenario. It was estimated that the winds affected 5,054,539 ha, 25% at high severity, 20% at medium severity, and 55% at low severity. Jackstrawed fuels ranged from 40 to 137 tons/ha. See Rodríguez Trejo, *Incendios de vegetación*, 1:52–54; and Dante Rodríguez Trejo et al., "Modelaje del peligro de incendio forestal en las zonas afectadas por el Huracán Dean," *Agrociencia* 45, no. 5 (2011): 593–608.

27. Martínez Domínguez, "Roberto Martínez Domínguez," 173–76. Quotes from Alfredo Nolasco Morales, "Alfredo Nolasco Morales," in Pyne and Gil, *Seis vidas*, 269.
28. Xavier García Cuevas, Bartolo Rodríguez Santiago, and Javier Chavelas Pólito, "Regeneración natural en sitios afectados por el Huracán Gilberto e incendios forestales en Quintana Roo," *Revista Ciencia Forestal en México* 17, no. 72 (July–December 1992): 75–99, with burned area from p. 75. Xavier García Cuevas, Bartolo Rodríguez Santiago, and Javier Chavelas Pólito, "Regeneración natural en sitios afectados por el huracán Gilberto e incendios forestales en Quintana Roo," *Revista Ciencia Forestal en México* 17, no. 72 (July–December 1992): 75–99.

    Alfredo Nolasco alliterated the impact by referring to "el derribo, desramado, descopado, y defoliación de árboles" in Pyne and Gil, *Seis vidas*; quotes from p. 269. Rodríguez Trejo, *Incendios de vegetación*, 2:1449. On the shifting perception of the hurricane-conflagration cycle, see Jorge López-Portillo et al., "Los incendios de Quintana Roo: ¿Catástrofe ecológica o evento periódico?," *Ciencia y Desarrollo* 16, no. 91 (March–April 1990): 43–57.
29. The 1998 and 2011 fires acquired generally accepted labels. The 1989 fires did not. I queried the three men about a suitable name, and the paragraph records the answers they sent me by email: Roberto, on August 15, 2022; Alfredo, on August 14, 2022; and Dante, on August 12, 2022.
30. Figures from Dirección General de Protección Forestal, Secretaría de Agricultura y Recursos Hidráulicos, *Resultados del Programa Nacional de Prevención y Combate de Incendios Forestales (1989–1994)* (Mexico City: SARH, 1994).
31. For a recent assessment, which includes video of Hansen's presentation, see Peter Sinclair, "Judgment on Hansen's '88 Climate Testimony: 'He Was Right,'" *Yale Climate Connections*, June 20, 2018, https://yaleclimateconnections.org/2018/06/judgment-on-hansens-88-climate-testimony-he-was-right/.
32. See Whigham et al., "Impact of Hurricane Gilbert," 434–42; and Dennis F. Whigham et al., "Impacts of Hurricanes on the Forests of Quintana Roo, Yucatán Peninsula, Mexico," in *The Lowland Maya Area: Three Millennia at the Human-Wildland Interface*, ed. Scott Fedick et al. (Binghamton, N.Y.: CRC Press, 2003), 193–212, with quotes from p. 209. An interesting booklet, published in 1989 and sponsored by the Fondo de Cultura Económica, made an ecological argument for fire's presence: see Carlos Vázquez Yanes and Alma Orozco Segovia, *La destrucción de la naturaleza*, La Ciencia desde México 83 (Mexico City: Secretaría de Educación Pública, 1989).
33. Oscar Cedeño Sánchez, "Las quemas controladas: Posible solución a la reforestación natural en bosques del Ajusco," in *Memoria Congreso Mexicano Forestal, Toluca, Edo. de Méx., 19–22 Jul. 1989*, vol. 2 (Mexico City: Protinbos/SARH/ANCF/CIAM, 1989), 717–18. On the global scene, see Johann Georg Goldammer, "Thirty Years International Wildland Fire Conferences: Review and Achievements of a Circumglobal Journey from Boston to Campo Grande,"

Biodiversidade Brasileira 11, no. 2 (2021): 6–52, https://www.doi.org/10.37002/biobrasil.v11i2.1743.

34. World Bank, "Personal Remittances Received (Current US$)—Mexico," https://data.worldbank.org/indicator/BX.TRF.PWKR.CD.DT?locations=MX; and World Bank, "Foreign Direct Investment, Net Inflows (BoP, Current US$)," https://data.worldbank.org/indicator/BX.KLT.DINV.CD.WD?locations=MX. Ana Gonzalez-Barrera and Mark Hugo Lopez, "A Demographic Portrait of Mexican-Origin Hispanics in the United States," Pew Research Center, May 1, 2013, https://www.pewresearch.org/hispanic/2013/05/01/a-demographic-portrait-of-mexican-origin-hispanics-in-the-united-states/.
35. The novelty of Los Diablos, from time to time, grabs media attention. For a brief overview, see https://wildfiretoday.com/tag/los-diablos; and for a video produced by *The Atlantic*, https://www.theatlantic.com/video/index/480354/los-diablos/.
36. In sorting through the neoliberal reforms, I found helpful Judith A. Teichman, *Privatization and Political Changes in Mexico* (Pittsburgh: University of Pittsburgh Press, 1995).
37. Quote from MacLachlan and Beezley, *El Gran Pueblo*, 470.

    On these topics, I found especially helpful Wayne A. Cornelius and David Myhre, eds., *The Transformation of Rural Mexico: Reforming the Ejido Sector* (La Jolla: Center for U.S.-Mexican Studies, UC San Diego, 1998); David Barton Bray, Leticia Merino-Pérez, and Deborah Barry, eds., *The Community Forests of Mexico: Managing for Sustainable Landscapes* (Austin: University of Texas Press, 2005); and David Barton Bray, *Mexico's Community Forest Enterprises: Success on the Commons and the Seeds of a Good Anthropocene* (Tucson: University of Arizona Press, 2020).
38. Quote from Bray, *Mexico's Community Forest Enterprises*, 74.
39. David Hutchins, *A Discussion of Australian Forestry* (Perth: n.p., 1916), 142.
40. Ramón Silva-Flores, José Ciro Hernández-Díaz, and Christian Wehenkel, "Does Community-Based Forest Ownership Favour Conservation of Tree Species Diversity? A Comparison of Forest Ownership Regimes in the Sierra Madre Occidental, Mexico," *Forest Ecology and Management* 363 (2016): 218.
41. I found three books particularly helpful in parsing what happened (and didn't happen): Bray, *Mexico's Community Forest Enterprises*; Christopher R. Boyer, *Political Landscapes: Forests, Conservation, and Community in Mexico* (Durham: Duke University Press, 2015); and Bray, Merino-Pérez, and Barry, *Community Forests of Mexico*.
42. The literature on community-based forestry in Mexico is large, but most of it does not consider fire in any depth. Andrew S. Mathews does, both in *Instituting Nature: Authority, Expertise, and Power in Mexican Forests* (Cambridge, Mass.: MIT Press, 2011) and in "Building the Town in the Country: Official Understandings of Fire, Logging and Biodiversity in Oaxaca, Mexico, 1926–2004," *Social Anthropology* 14, no. 3 (2006): 335–59. See also Peter Z. Fulé et al.,

"Fire Regime in a Mexican Forest Under Indigenous Resource Management," *Ecological Applications* 21, no. 3 (April 2011).

43. My thoughts on community-based conservation and Mexico are particularly influenced by Silva-Flores, Hernández-Díaz, and Wehenkel, "Does Community-Based Forest Ownership Favour"; Barbara Pazos-Almada and David Barton Bray, "Community-Based Land Sparing: Territorial Land-Use Zoning and Forest Management in the Sierra Norte of Oaxaca, Mexico," *Land Use Policy* 78 (2018): 219–26; and Eric Van Vleet, David Barton Bray, and Elvira Durán, "Knowing but Not Knowing: Systematic Conservation Planning and Community Conservation in the Sierra Norte of Oaxaca, Mexico," *Land Use Policy* 59 (2016): 504–15.
44. See https://www.gob.mx/conanp. For an interesting overview, see Roberto Lindig-Cisneros, "Ecological Restoration in Mexico: The Challenges of a Multicultural Megadiverse Country," in "Restoration in Mexico," special issue, *Ecological Restoration* 28, no. 3 (August 2010): 232–33.
45. The charter for biosphere reserves explained that their overarching objective was "to develop the basis within the natural and social sciences for the rational use and conservation of the resources of the biosphere and for the improvement of the global relationship between man and the environment, to predict the consequences of today's actions on tomorrow's world and thereby to increase man's ability to manage efficiently the natural resources of the biosphere." See https://en.unesco.org/mab/50years/history.

    Statistics from UNESCO: https://en.unesco.org/biosphere.
46. Figures from UNESCO: https://en.unesco.org/biosphere.
47. Bray, *Mexico's Community Forest Enterprises*, 74. The forest programs were Protinbos (Protectora e Industrializadora de Bosques), which morphed into Probosque (Promotora de Bosques). Figures for fire equipment upgrades from Rodríguez Trejo, *Incendios de vegetación*, 2:1452–53.
48. Martínez Domínguez, "Roberto Martínez Domínguez," 186.
49. Martínez Domínguez, 176–77.
50. Martínez Domínguez, 177–78.
51. Herbert Kaufman, *The Forest Ranger: A Study in Administrative Behavior* (Baltimore: Johns Hopkins Press, 1960); and Francis Fukuyama, "America in Decay: Sources of Political Dysfunction," *Foreign Affairs*, September–October 2014, https://www.foreignaffairs.com/united-states/america-decay.
52. Martínez Domínguez, "Roberto Martínez Domínguez," 186–87.
53. The Madagascan experience makes an instructive complement to Mexico. See Christian A. Kull, *Isle of Fire: The Political Ecology of Landscape Burning in Madagascar* (Chicago: University of Chicago Press, 2004).
54. The best narrative survey of the U.S. fire scene during this period is Pyne, *Between Two Fires*.
55. See what is perhaps the classic expression: T. W. Swetnam and J. L. Betancourt, "Fire-Southern Oscillation Relations in the Southwestern United States," *Sci-*

ence 249, no. 4972 (1990): 1017–20. For an excellent summary of the techniques involved, with Mexican examples, see Julián Cerano-Paredes et al., "Using Tree-Rings to Reconstruct Fire History Information from Forested Areas," *Journal of Visualized Experiments* 164 (2020): e61968, https://doi.org/10.3791/61698.

The controversy in the United States was formally ignited by James Rodney Hastings and Raymond M. Turner, *The Changing Mile: An Ecological Study of Vegetation Change with Time in the Lower Mile of an Arid and Semiarid Region* (Tucson: University of Arizona Press, 1965), followed by numerous publications, including books. The original consensus was that the change reflected climate, but as more studies amassed, consensus shifted to culture and land use as explanations.

56. Mexico was part of a continental project. See Ellis Q. Margolis et al., "The North American Tree-Ring Fire-Scar Network," *Ecosphere* 13, no. 7 (March 7, 2022): https://doi.org/10.1002/ecs2.4159.
57. Dante Arturo Rodríguez Trejo, *Incendios forestales* (Mexico City: Universidad Autónoma Chapingo / Mundi-Prensa México, 1996), 522–23. Note: at this time, INIFAP was INIF, but this does not change the state of fire research and seemed an unnecessary complication for most readers.
58. Rodríguez Trejo, *Incendios forestales*, "Agradecimientos" and "Acerca del autor."
59. Rodríguez Trejo, *Incendios forestales*, frontispiece and 1–3.
60. The source for this profile is Ernesto Alvarado Celestino, "Ernesto Alvarado Celestino," in Pyne and Gil, *Seis vidas*. Particular sources and quotes from pp. 8–9, 48.
61. Alvarado Celestino, "Ernesto Alvarado Celestino," 10.
62. Alvarado Celestino, 19.
63. Rodríguez Trejo, *Incendios de vegetación*, 2:1472. Ernesto Alvarado Celestino, "Comportamiento del fuego en rodales poco perturbados de *Pinus montezumae* Lam." (master's thesis, Colegio de Postgraduados, Montecillo, 1986).
64. Quote from email to author and co-authors of *Seis vidas en fuego*, November 25, 2019.
65. The source for this profile is José Germán Flores Garnica, "José Germán Flores Garnica," in Pyne and Gil, *Seis vidas*. Quote from p. 70.
66. Flores Garnica, "José Germán Flores Garnica," 72.
67. José Germán Flores Garnica and Dante Arturo Rodríguez Trejo, eds., *Incendios forestales: Definiendo el problema, ecología social, fortalecimiento de capacidades, educación y divulgación* (Mexico City: Mundi-Prensa México, 2006); and José Germán Flores Garnica, ed., *Impacto ambiental de incendios forestales* (Mexico City: Mundi-Prensa México / Instituto Nacional de Investigaciones Forestales, Agrícolas y Pecuarias / Colegio de Postgraduados, 2011).
68. Quote from Hutchins, *Discussion of Australian Forestry*, 24.
69. The source for this profile is Enrique J. Jardel Peláez, "Enrique J. Jardel Peláez," in Pyne and Gil, *Seis vidas*. Quote from p. 112.
70. Jardel Peláez, "Enrique J. Jardel Peláez," 112.

71. Jardel Peláez, 113–14.
72. Jardel Peláez, 134.
73. Jardel Peláez, 138–41, 147–48.
74. Probably the most comprehensive attempt to understand the fires is Santiago Fragoso et al., *Incendios forestales y agropecuarios.*
75. SEMARNAP, *Foro regional sobre prevención y combate a los incendios forestales y regeneración de áreas afectadas, 26 a 28 de agosto de 1998* (Mexico City: SEMARNAP, 1998). Santiago Fragoso et al., *Incendios forestales y agropecuarios.*
76. Quote from Eréndira Salgado, adjunct secretary for the Supreme Court: Eréndira Salgado Ledesma, "Normatividad en materia de quemas agropecuarias," in Santiago Fragoso et al., *Incendios forestales y agropecuarios*, 143–58.
77. Presentation by Citlali Cortés Montaña at the Association for Fire Ecology's 6th International Fire Ecology and Management Congress, San Antonio, Texas, 2015.
78. Quotes from Ferriss, "Glickman Calls Blazes," 15. North American Fire Management Team Final Report (unpublished report, March 20, 2001), 2.
79. Roberto Martínez Domínguez and Paul Weeden, "Mexico National Fires Prevention and Restoration Program: Final Report October 1998 to August 2000" (unpublished consultant report). I'm grateful to Paul for an interview at an early stage in my research. A subsequent report that reviews the program through 2003 is Thomas M. Catterson, Oscar Cedeño Sánchez, Stephen Lenzo, and Margaret Haines, "Assessment and Diagnostic Review of Fire Training and Management Activities Funded by USAID/Mexico from 1998–2003" (unpublished report, USAID/Mexico, 2004). North American Fire Management Team Final Report (unpublished report, March 20, 2001), 2–3, 5.
80. Martínez Domínguez and Weeden, "Mexico National Fires Prevention," 1–2, 4 ("Executive Summary").
81. Martínez Domínguez and Weeden, 2 ("Executive Summary"). On the border agreement, see https://www.fs.usda.gov/inside-fs/delivering-mission/apply/international-agreements-help-fire-management. On the U.S. National Fire Plan, see Pyne, *Between Two Fires*, 346–47, and scattered throughout.
82. Alfred Nolasco Morales, "Reporte final de la asistencia técnica y entrenamiento en el manejo de helicópteros en emergencias por incendios forestales, Quintana Roo, 1999" (unpublished report to SEMARNAP and USFS).
83. Associated Press, "Forest Fire Deaths," 12.
84. Nigel Dudley and Jean-Paul Jeanrenaud, *The Year the World Caught Fire* (Gland, Switzerland: WWF, 1997). On the hockey-stick graph, see M. E. Mann, R. S. Bradley, and M. K. Hughes, "Global-Scale Temperature Patterns and Climate Forcing over the Past Six Centuries," *Nature* 392 (1998): 779–87.
85. A good introduction to the changes that flowed from the 1998 disaster is available in Rodríguez Trejo, *Incendios de vegetación*, 2:1460–62.
86. The responses to the 1998 season rippled throughout the Mexican fire community. A particularly helpful chronology is the career of Roberto Martínez

Domínguez, who was involved with nearly all aspects at an operational level; see Martínez Domínguez, "Roberto Martínez Domínguez," especially 189–227.

87. A clear explanation in English of the NOM's provisions is available in Catterson et al., "Assessment and Diagnostic Review," 25. For quote, amid a critical legal commentary, see Salgado Ledesma, "Normatividad en materia de quemas agropecuarias," 156.

88. A general account of the revisions is available in Rodríguez Trejo, *Incendios de vegetación*, 2:1492.

89. A thorough summary of developments is available in English in Catterson et al., "Assessment and Diagnostic Review." On the Big Black Book of Fire, see Paul Weeden and Roberto Martínez Domínguez, *Mexico National Fires Prevention and Restoration Program—Final Report October 1998 to August 2000* (n.p.: SEMARNAP / USAID / USDA Forest Service / BLM, 2001); quote from p. 8.

90. CONAFOR has sponsored a brief, popular history of its first decade: Gustavo A. Del Ángel-Mobarak, ed., *La Comisión Nacional Forestal: En la historia y el futuro de la política forestal de México* (Mexico City: Centro de Investigación y Docencia Económicas, 2012).

91. See Catterson et al., "Assessment and Diagnostic Review," 43–44, and appendix A, p. 2.

92. Catterson et al., 45–47. On sponsored brigades, see Martínez Domínguez, "Roberto Martínez Domínguez," 195.

93. Catterson et al., "Assessment and Diagnostic Review," 46. On differing perceptions of NOM-015, see Rodríguez Trejo, *Incendios de vegetación*, 2:1490.

94. Land ownership figures from Silva-Flores, Hernández-Díaz, and Wehenkel, "Does Community-Based Forest Ownership Favour," 218.

95. Catterson et al., "Assessment and Diagnostic Review," 35; and Martínez Domínguez, "Roberto Martínez Domínguez," 205. The relocation was apparently prompted by the new director of CONAFOR, Alberto Cárdenas, a former governor of Jalisco.

96. A good summary is in Catterson et al., "Assessment and Diagnostic Review," 35, 43.

97. See Catterson et al., "Assessment and Diagnostic Review." For abbreviated summaries, see also Rodríguez Trejo, *Incendios de vegetación*, 2:1469–71; and Martínez Domínguez, "Roberto Martínez Domínguez," 206.

98. Catterson et al., "Assessment and Diagnostic Review," appendix D, p. 22.

99. See Catterson et al., 9.

100. On the Big Book, see Catterson et al., 39, and for a roster of its recommendations, p. 40. On state fire and forestry programs, see Hubert Tchikoué Maga, Dante Arturo Rodríguez Trejo, Javier Santillán Pérez, and Hugo Ramírez Maldonado, "Informe técnico: Evaluación del Programa Nacional de Prevención y Combate de Incendios Forestales; Ejercicio fiscal 2004" (report to CONAFOR, November 2005), which includes a state-by-state assessment. And see Catterson et al., 31, for a highlighting of Michoacán.

101. Catterson et al., "Assessment and Diagnostic Review," appendix A, p. 2.
102. Tchikoué Maga et al., "Informe técnico" (see note 100 above); Hubert Tchikoué, Dante Arturo Rodríguez Trejo, and Javier Santillán Pérez, "Informe técnico: Evaluación del Programa Nacional de Prevención y Combate de Incendios Forestales; Ejercicio fiscal 2006" (report to CONAFOR, November 2007); Dante Arturo Rodríguez Trejo, Hubert Tchikoué, Javier Santillán Pérez, and Hugo Ramírez Maldonado, "Informe Final: Evaluación complementaria del ejercicio del Programa E013 Prevención y Combate de Incendios Forestales; Ejercicios fiscales 2008–2010" (report to SEMARNAT and CONAFOR, November 2001).
103. Boege, *El patrimonio biocultural*. The book does include a photo of a milpa, recently burned, but makes no mention of the burning in the text.
104. An interesting survey that compares Australia and the United States is Daniel Joshua May, "Taking Fire: The Historical and Contemporary Politics of Indigenous Burning in Australia and the Western United States" (PhD thesis, Australian National University, Canberra, 2020).
105. Enrique J. Jardel Peláez, "Necesidades y oportunidades de investigación aplicada en ecología y manejo del fuego y de desarrollo de estrategias y prácticas de manejo del fuego en México," in Catterson et al., "Assessment and Diagnostic Review," appendix F, pp. 27–39.
106. On the fire that burned into Los Alamos (Cerro Grande fire), see Pyne, *Between Two Fires*, 345, 350–56.
107. Catterson et al., "Assessment and Diagnostic Review," 28–29 (on the 1999 law), 57–58 (on the need for suppression).
108. Enrique J. Jardel Peláez, *Planificación del manejo del fuego* (Autlán, Jalisco: Universidad de Guadalajara / Fundación Manantlán para la Biodiversidad de Occidente A.C. / Consejo Civil Mexicano para la Silvicultura Sostenible A.C. / Fondo Mexicano para la Conservación de la Naturaleza A.C., 2010).
109. See Tchikoué Maga et al., "Informe técnico" (November 2005), 130–32.
110. Sadly, a biography of Myers does not exist. A brief profile, "One Foot in the Black," is available in Stephen J. Pyne, *Florida: A Fire Survey* (Tucson: University of Arizona Press, 2016), 109–19.
111. Jeff Hardesty, Ron Myers, and Wendy Fulks, "Fire, Ecosystems and People: A Preliminary Assessment of Fire as a Global Conservation Issue," *George Wright Forum* 22, no. 2 (2005): 78–86. Rodríguez Trejo, *Incendios de vegetación*, vol. 2, gives a participant's view. For the two summary publications, see Ronald L. Myers, *Living with Fire—Sustaining Ecosystems and Livelihoods Through Integrated Fire Management* (n.p.: The Nature Conservancy, 2006); and A. Shlisky et al., *Fire, Ecosystems and People: Threats and Strategies for Global Biodiversity Conservation* (n.p.: The Nature Conservancy, 2007).
112. TNC Global Fire Initiative, FY06 annual report (2006); Catterson et al., "Assessment and Diagnostic Review," 20. Mary Huffman, "Community-Based Fire Management at La Sepultura Biosphere Reserve" (presentation to Warner College of Natural Resources, Colorado State University, November 19, 2008).

See also Rodríguez Trejo, "Dante Arturo Rodríguez Trejo," 245–47, for a participant's perspective.

113. Information on the end of the program from Blane Heumann, email of August 13, 2018; and interview with Ron Myers, September 19, 2018. Rodríguez Trejo, *Incendios de vegetación*, 2:1471. On Nolasco, see resignation letter from Alfredo Nolasco to TNC Mexico Program, February 23, 2009 (personal copy, courtesy of TNC).
114. Rodríguez Trejo, *Incendios de vegetación*, 2:1471–89, provides a detailed, even loving profile of fire research, rich with photo galleries of major figures and UACh students. For a historical review from the perspective of the major governmental institution, see José Germán Flores-Garnica, "Antecedentes y perspectivas de la investigación en incendios forestales en el INIFAP," *Revista Mexicana de Ciencias Forestales*, 12, no. Especial-1 (2021): https://doi.org/10.29298/rmcf.v12iEspecial-1.981.
115. Data from Fire Research Institute master bibliography on fire, courtesy Jason Greenlee. Jesús Sánchez Córdova, "Los incendios forestales y las prioridades de investigación en México" (presentation to Congreso Forestal Mexicano, Toluca, July 1989), 722. Rodríguez Trejo, *Incendios forestales*.
116. Information on UACh students from Rodríguez Trejo, *Incendios de vegetación*, 2:1479. An email from Dante on October 13, 2022, updated the numbers slightly.
117. The 2003 symposium: Lourdes Villers Ruiz and Jorge López Blanco, eds., *Incendios forestales en México: Métodos de evaluación* (Mexico City: UNAM, 2004). Flores Garnica and Rodríguez Trejo, *Incendios forestales*.
118. J. G. Flores Garnica and D. A. González Moreno, "Análisis de la investigación en incendios forestales en México" (presentation to Congreso Forestal de Cuba, Havana, April 17–20, 2007). Flores Garnica subsequently wrote a concept paper that elaborated on the idea of a research coordination center: "Centro nacional de investigación sobre incendios forestales" (unpublished grey literature, Mexico, November 2007). See also Flores-Garnica, "Antecedentes y perspectivas."
119. Enrique J. Jardel Peláez et al., *Prioridades de investigación en manejo del fuego en México: Memorias del taller realizado en el Campus Morelia de la Universidad Nacional Autónoma de México, 23–24 de octubre de 2008* (Mexico City: Fondo Mexicano para la Conservación de la Naturaleza, 2010).
120. Flores Garnica, *Impacto ambiental de incendios forestales*. Enrique J. Jardel Peláez, Diego Pérez Salicrup, Ernesto Alvarado, and Jorge E. Morfín Ríos, *Principios y criterios para el manejo del fuego en ecosistemas forestales: Guía de campo* (Zapopan: CONAFOR, 2014). Also mentioned: Rodríguez Trejo, *Incendios de vegetación*, vols. 1 and 2.
121. Martínez Domínguez, "Roberto Martínez Domínguez," 166.
122. Martínez Domínguez, 169.
123. Martínez Domínguez, 173.
124. Martínez Domínguez, 171–72.
125. Martínez Domínguez, 217.

126. Dante Arturo Rodríguez Trejo, "Fire Regimes, Fire Ecology, and Fire Management in Mexico," *Ambio* 37, no. 7 (2008): 548–56; and Dante Arturo Rodríguez Trejo et al., "The Present Status of Fire Ecology, Traditional Use of Fire, and Fire Management in Mexico and Central America," *Fire Ecology* 7, no. 1 (2011): 40–56.
127. Distillation with abundant figures in Rodríguez Trejo, *Incendios de vegetación*, 2:1462–68. Official report is CONAFOR, *Informe de incendios forestales en el estado de Coahuila, temporada 2011* (n.p.: [CONAFOR], 2011).
128. CONAFOR, *Factores relacionados con los incendios forestales en Coahuila*, Informe de incendios forestales 4 (n.p.: [CONAFOR], 2011). Summarized in Rodríguez Trejo, *Incendios de vegetación*, 2:1463.
129. See Adin H. Velázquez and Jose M. Iniguez, "Manejo del fuego a escala de paisaje en el estado de Coahuila, México" (unpublished report).
130. CONAFOR, *Informe de incendios forestales*, has a full accounting. A précis is available in Rodríguez Trejo, *Incendios de vegetación*, 2:1464–67. On his service in the fires, email from Dante Rodríguez Trejo, February 8, 2012; and his preliminary report "Evaluación preliminar de incendios en La Sabina y El Bonito, municipios de Múzquiz, Ocampa, Zaragoza y Acuña, del estado de Coahuila: Informe ejecutivo" (unpublished report submitted to CONAFOR, June 2011).
131. *Estrategia y lineamientos de manejo del fuego en áreas naturales protegidas*, 2nd ed. (Mexico City: CONANP, 2011); CONANP et al., *Guía para la elaboración de programas de manejo del fuego en áreas naturales protegidas y sitios de interés (contenido del guión modelo)* (Mexico City: SEMARNAT, 2012).
132. Rodríguez Trejo, *Incendios de vegetación*, 2:1490.
133. Source for this profile comes from Nolasco Morales, "Alfredo Nolasco Morales," 233–306. Quotes from pp. 235, 237.
134. Nolasco Morales, 233.
135. Nolasco Morales, 238–39, 241.
136. Nolasco Morales, 241–42.
137. Nolasco Morales, 243–44, 246–48.
138. Nolasco Morales, 251–52.
139. Nolasco Morales, 263.
140. Nolasco Morales, 275–76. A. Nolasco, "La protección contra incendios forestales en el estado de Quintana Roo: Memoria de experiencia profesional" (thesis, Universidad Autónomo Chapingo, Forest Sciences Division, 1993).
141. Nolasco Morales, "Alfredo Nolasco Morales," 290.
142. Nolasco Morales, 290, 292.
143. Email memo from Alfredo Nolasco Morales to Mexico Program, subject: Salida de TNC, February 23, 2009 (courtesy of Alfredo Nolasco).
144. Nolasco Morales, "Alfredo Nolasco Morales," 296.
145. See Nolasco Morales, 299–303.
146. The butterflies have attracted considerable attention. I found the best sources for my needs to be Diego Rafael Pérez-Salicrup et al., *Plan integral del manejo*

del fuego: Reserva de la Biósfera Mariposa Monarca; Acción e investigación participativa y adaptable (Mexico City: FMCN/CONANP/Alternare/IIES-UNAM, 2020); Diego R. Pérez-Salicrup et al., "Restauración de un proceso: El fuego en la Reserva de la Biosfera Mariposa Monarca en los estados de México y Michoacán," in *Experiencias mexicanas en la restauración de los ecosistemas*, ed. Eliane Ceccon and Cristina Martínez Garza (Cuernavaca: Universidad Nacional Autónoma de México / Universidad Autónoma del Estado de Morelos / Comisión Nacional para el Conocimiento y Uso de la Biodiversidad, 2016), 215–34; Héctor Martínez-Torres et al., "Fires and Fire Management in the Monarch Butterfly Biosphere Reserve," in *Monarchs in a Changing World: Biology and Conservation of an Iconic Butterfly*, ed. Karen S. Oberhauser, Kelly R. Nail, and Sonia Altizer (Ithaca: Cornell University Press, 2015), 179–80; and Jesús E. Sáenz-Ceja and Diego R. Pérez-Salicrup, "Dendrochronological Reconstruction of Fire History in Coniferous Forests in the Monarch Butterfly Biosphere Reserve, Mexico," *Fire Ecology* 15 (2019): 18, https://doi.org/10.1186/s42408-019-0034-z.

147. Pérez-Salicrup et al., *Plan integral del manejo del fuego*, 29–31; and Sáenz-Ceja and Pérez-Salicrup, "Dendrochronological Reconstruction of Fire History," 4.
148. *Programa de manejo de la Reserva de la Biosfera Mariposa Monarca, México* (n.p.: CONANP, 2001), 44–45.
149. Martínez-Torres et al., "Fires and Fire Management," 188–89. See also "Incendios," Zotero, http://www.zotero.org/fuego_mex.
150. Information from the remarkable study by Sáenz-Ceja and Pérez-Salicrup, "Dendrochronological Reconstruction of Fire History," especially pp. 13–14.
151. On the Codex Magliabechiano, see the Foundation for the Advancement of Mesoamerican Studies website: http://www.famsi.org/research/graz/magliabechiano/img_page173.html.
152. Quotes from Rosa María Román Cuesta, "Forest Fire Situation in the State of Chiapas, Mexico," *International Forest Fire News* 27 (July 2002): 11–12.

    For the narrative of two towns I rely on Laura Patricia Ponce Calderón et al., "El nacimiento de los 'Sin fuego': Caso Parque Nacional Lagunas de Montebello, Chiapas, México," *Nova Scientia* 12, no. 25 (2020), which distills the doctoral thesis of Laura Patricia Ponce Calderón, "Percepciones ambientales sobre el uso y manejo del fuego en el Parque Nacional Lagunas de Montebello" (PhD thesis, Colegio de la Frontera Sur, 2020).
153. Ponce Calderón et al., "El nacimiento de los 'Sin fuego,'" 31, 35. For an updated, English-language analysis, see Laura Patricia Ponce Calderón et al., "Fire Management in Pyrobiocultural Landscapes, Chiapas, Mexico" (unpublished paper, 2022).
154. Laura Ponce Calderón, email to author, April 19, 2022. Her two papers (and thesis) cited in notes 152 and 153 coin the term "Sin Fuego."
155. The La Sepultura story attracted wide attention. An excellent distillation of the project is Ronald L. Myers, "CBFiM in Mexico: La Sepultura Biosphere

Reserve in Chiapas," in *Community-Based Fire Management: A Review*, FAO Forestry Paper 166 (Rome: FAO, 2011), annex 3, https://www.fao.org/3/i2495e/i2495e00.pdf. For a Mexican perspective, see V. Pantoja-Campa, J. D. Cruz-López, V. Negrete-Paz, and A. Vázquez-Vázquez, "Sistematización del manejo del fuego en la Reserva de la Biosfera La Sepultura, Chiapas, México" (conference paper, V Simposio Internacional sobre Manejo Sostenible de Recursos Forestales y II Taller Internacional sobre Manejo del Fuego, Universidad de Pinar de Río, Cuba, 2008). On the plan, see CONANP-REBISE, FMCN, and TNC, *Programa de manejo integral del fuego de la Reserva de la Biosfera La Sepultura* (Mexico City: CONANP, 2004).

For the actual manual produced, see V. Negrete-Paz, A. Vázquez-Vázquez, and J. D. Cruz-López, *Guía para la construcción de un plan comunitario de manejo integral del fuego* (Tuxtla Gutiérrez: CONANP, 2008).

156. Myers, "CBFiM in Mexico," 60; Pantoja-Campa et al., "Sistematización del manejo del fuego," 1, 4–5.
157. Myers, "CBFiM in Mexico," 58.
158. Myers, 62–63. Quotes from Huffman, "Community-Based Fire Management" (see note 112 above). For the full write-up, see Mary R. Huffman, "Community-Based Management at La Sepultura Biosphere Reserve, Chiapas, Mexico" (PhD diss., Colorado State University, 2010); on the 40 factors, ii. For an expanded model, see Mary Huffman, "The Many Elements of Traditional Fire Knowledge: Synthesis, Classification, and Aids to Cross-cultural Problem Solving in Fire-Dependent Systems Around the World," *Ecology and Society* 18, no. 4 (December 2013).
159. Huffman, "Community-Based Fire Management"; Myers, "CBFiM in Mexico," 60. On drip torches, I relied on a personal interview with Huffman in 2019.
160. Myers quotes from Myers, "CBFiM in Mexico," 58, 61; on the value of the La Sepultura experiment for Mesoamerica, 64–65. On the manual, see Dante Arturo Rodríguez Trejo et al., *Instructivo de quemas prescritas para el manejo integral del fuego en el municipio de Villaflores, Chiapas* (Chapingo: Fondo Mexicano para la Conservación de la Naturaleza / USDA Forest Service / USAID / Biomasa, A.C. / Universidad Autónoma Chapingo, 2018); on its significance, see pp. 12–13. See also Dante Arturo Rodríguez Trejo, Pedro Martínez Muñoz, and Pedro Jerónimo Martínez Lara, "Efectos del fuego en el arbolado de un bosque tropical de pino y en el de una selva baja caducifolia en Villaflores, Chiapas," *Ciencia Forestal* 29, no. 3 (July–September 2019): 1–15. Other information, especially on the role of Pedro Martínez Muñoz, comes from an email to me from Dante Rodríguez Trejo on October 4, 2022.
161. Dante Arturo Rodríguez Trejo et al., *Instructivo de quemas prescritas.*
162. My treatment of Sierra Juárez fire history relies almost exclusively on Mathews, *Instituting Nature*, and several articles drawing on that original research (based on his dissertation), including "Power/Knowledge, Power/Ignorance: Forest Fires and the State in Mexico," *Human Ecology* 33, no. 6 (2005): 795–820; and

especially "Suppressing Fire and Memory: Environmental Degradation and Political Restoration in the Sierra Juárez of Oaxaca 1887–2001," *Environmental History* 8, no. 1 (2002): 77–108.

163. Mathews, "Suppressing Fire and Memory," 87–89, for fire scar analysis. In *Instituting Nature* (p. 86) Mathews proposes a six-year return interval, but the sample size was small, and the conclusion falls easily within the norm for most of Mexico's pine-oak forests.
164. Cassiano Conzatti, "La repoblación arbórea del Valle de Oaxaca," *Boletín de la Estación Agrícola Experimental de Oaxaca* 1 (1914): 1–13; quotes from pp. 3–4.
165. Mathews, "Suppressing Fire and Memory," 92, 94.
166. Memo quoted in Mathews, "Suppressing Fire and Memory," 93.
167. Claudia María Monzón Alvarado, "Governance of Fire Management in Calakmul, Mexico: Institutions, Local Practice and Human-Environmental Outcomes" (PhD diss., University of Florida, 2014), 13.
168. Mathews, *Instituting Nature*, 138. Mathews, "Suppressing Fire and Memory," 100.
169. The region's fire history is deftly documented in a series of conferences: see Leonard DeBano et al., tech. coords., *Biodiversity and Management of the Madrean Archipelago: The Sky Islands of Southwestern United States and Northwestern Mexico; September 19–23, 1994, Tucson, AZ*, General Technical Report RM-GTR-264 (Fort Collins, Colo.: U.S. Forest Service, 1995); Gerald J. Gottfried et al., eds., *Connecting Mountain Islands and Desert Seas: Biodiversity and Management of the Madrean Archipelago II, Proceedings*, RMRS-P-36 (Fort Collins, Colo.: U.S. Forest Service, 2005); Gerald J. Gottfried et al., eds., *Merging Science and Management in a Rapidly Changing World: Biodiversity and Management of the Madrean Archipelago III and 7th Conference on Research and Resource Management in the Southwestern Deserts; May 1–5, 2012, Tucson, Arizona, Proceedings*, RMRS-P-67 (Fort Collins, Colo.: U.S. Forest Service, 2013).
170. The travails of modern ranching can be hard to sort through. Since I was mostly concerned with Sonora, I found an excellent introduction in Jacob C. Brenner, "Pasture Conversion, Private Ranchers, and the Invasive Exotic Buffelgrass (*Pennisetum ciliare*) in Mexico's Sonora Desert," *Annals of the Association of American Geographers* 101, no. 1 (2011): 84–106; supplemented by Eric P. Perramond, *Political Ecologies of Cattle Ranching in Northern Mexico: Private Revolutions* (Tucson: University of Arizona Press, 2010). On buffelgrass, see pp. 110–13 of the latter.
171. Like cheatgrass, buffelgrass is both applauded and condemned, and like many modernizing projects, it begins with enthusiasm and concludes with messes. Two treatments for Sonora are A. Búrquez-Montijo and A. Martínez-Yrizar, "Mexican Grasslands, Thornscrub, and the Transformation of the Sonoran Desert by Invasive Exotic Buffelgrass (*Pennisetum ciliare*)," in *Invasive Exotic Species of the Sonoran Region*, ed. B. Tellman (Tucson: University of Arizona

Press / Arizona-Sonora Desert Museum, 2002), 126–46; and Kim A. Franklin et al., "Buffelgrass (*Pennisetum ciliare*) Land Conversion and Productivity in the Plains of Sonora, Mexico," *Biological Conservation* 127 (2006): 62–71; cited figures from p. 62.

172. Quotes from Brenner, "Pasture Conversion," 102.

173. See, for example, the project sponsored by the USFS International Programs, The Nature Conservancy Programa México, and Biodiversidad y Desarrollo Armónico: Baruk G. Maldonado Leal, "Diagnóstico sobre la situación actual y potencial del papel del fuego en Sierra San Luis y cuenca del Río San Pedro, en Sonora, México" (unpublished grey literature, Hermosillo, Sonora BIDA Biodiversidad y Desarrollo Armónico, December 2006); and the one sponsored by USFS International Programs, The Nature Conservancy Programa México, and CONANP: Baruk Giovani Maldonado Leal, consultant, *Plan integrado de manejo del fuego para la fracción IV de la Reserva Forestal Nacional y Refugio de Fauna Silvestre Sierra de los Ajos-Bavispe, Sonora, México* (Hermosillo: CONANP, 2006); for fire history, pp. 30–32.

The sites were topics of presentations by Mexicans at international conferences. See, for example, J. Mario Cirett-Galán, "Conservation Management of the Ajos Bavispe National Forest Preserve and Wildlife Refuge [abstract]," in Gottfried et al., *Merging Science and Management*, 18; and Jose A. Sarakhan-Kermez, "Pine-Oak Woodlands in Mexico: Challenges and Opportunities in Conserving Sky Islands," in Gottfried et al., *Merging Science and Management*, xi.

174. See, for example, Naturalia Comité para la Conservación de Especies Silvestres, "Plan de quema prescrita para la unidad de quema Los Fresnos I del Rancho Los Fresnos, municipio de Santa Cruz, estado de Sonora, México" (working plan, October 21, 2005); and Biodiversidad y Desarrollo Armónico, TNC Programa México, CONAFOR, and CONANP, "Plan de quema prescrita para Academia de Quemas Prescritas: Fase II, ejecución del plan de quema Rancho 'Los Fresnos,' municipio de Santa Cruz, Sonora, México" (working plan, October 23, 2009). I'm grateful to Baruk Giovani Maldonado Leal of CONAFOR (Sonora) for his help in obtaining documents of these events.

175. See https://afefirecongress.org/wp-content/uploads/2019/11/Fire-Congress-2019-Program-Final.pdf.

176. An overview of BLP's governance history is available in Arturo Curiel Ballesteros, "Bosque La Primavera a 29 años del decreto: Balance y prospectiva," in *II Foro de Investigación y Conservación del 'Bosque La Primavera,' 5–6 marzo de 2009, Zapopan, Jalisco* (Guadalajara: Universidad de Guadalajara, 2009), 1–7, https://www.organicaeditores.mx/biblioteca/ii_foro_blp/contenido.html.

Agustín de Castillo, *La Primavera en llamas* ([Guadalajara?]: CONAFOR / Siémbrate en La Primavera / Público Milano, 2006), 24–25.

177. For a thumbnail introduction to BLP's fire scene, see Enrique J. Jardel Peláez, "Para entender los incendios en el Bosque La Primavera" (unpublished paper, April 18, 2021).

178. Information on origins from de Castillo, *La Primavera en llamas*, 32–33, 40–41, 93; quote from p. 33.

179. De Castillo, *La Primavera en llamas*, 76, 40.

180. De Castillo, *La Primavera en llamas*, 109.

181. Centro Universitario de Ciencias Biológicas y Agropecuarias de la Universidad de Guadalajara, *I Foro de Investigación y Conservación del Bosque La Primavera: Resumen Ejecutivo; Zapopan, Jalisco, 22 y 23 de septiembre de 2005* (Zapopan: Universidad de Guadalajara, 2006), https://www.organicaeditores.mx/biblioteca/i_foro_blp/index.html. Germán Flores was deeply involved with INIFAP's contribution to the forum, with papers on nearly all aspects of the BLP firescape.

On the the second forum, see https://www.organicaeditores.mx/biblioteca/ii_foro_blp/contenido.html; on the third, https://udg.mx/es/evento/iii-foro-de-conocimiento-y-conservacion-del-bosque-la-primavera.

182. Quotes from an email from Enrique Jardel Peláez to the author, April 21, 2022.

183. Enrique José Jardel Peláez, "Programa de manejo del fuego del Bosque La Primavera" (unpublished report, final draft v. 1.4, Departamento de Ecología y Recursos Naturales / Centro Universitario de la Costa Sur / Universidad de Guadalajara, 2020). While Jardel Peláez was the principal author, the document was a collaborative project, and it was reviewed by the Scientific Committee of BLP.

Jardel Peláez created a suite of complementary presentations and essays that provide, for the uninitiated, easier access to the issues. See Enrique J. Jardel Peláez, "Para entender los incendios en el Bosque La Primavera" (unpublished paper, courtesy of author); and Enrique J. Jardel Peláez, "Programa de manejo del fuego del Bosque La Primavera" (presentation to the Junta de Gobierno del OPD-Bosque La Primavera, November 18, 2021). For a collective presentation, see E. J. Jardel Peláez et al., "Evaluación de los incendios forestales de 2021 en el Bosque La Primavera" (presentation, Universidad de Guadalajara, November 2021).

184. Population statistics from World Population Review, https://worldpopulationreview.com/world-cities/mexico-city-population. Area growth from Germán Vergara, *Fueling Mexico: Energy and Environment, 1850–1950* (Cambridge: Cambridge University Press, 2021), 194. Comparative stats from S. A. Edgerton et al., "Particulate Air Pollution in Mexico City: A Collaborative Research Project," *Journal of the Air and Waste Management Association* 49 (1999): 1222s.

185. Erick D. Gamas et al., "Exhaust Emissions from Gasoline- and LPG-Powered Vehicles Operating at the Altitude of Mexico City," *Journal of the Air and Waste Management Association* 49 (1999): 1180.

186. J. D. Crouse et al., "Biomass Burning and Urban Air Pollution over the Central Mexican Plateau," *Atmospheric Chemistry and Physics* 9 (2009): 4936–37.

187. Crouse et al., "Biomass Burning and Urban Air Pollution," 4936–37.

188. For a review of legislation and protection issues as of 1988, see Oscar Cedeño Sánchez, "¿Están los bosques de la Serranía del Ajusco en peligro de extinción?," *Dasonomía Mexicana* 12, no. 7 (July–December 1989): 11–16.
189. For a good summary of Proyecto Ajusco, see Dante Arturo Rodríguez-Trejo et al., "Integrating Fire Management: Contrasting Cases of Central and Southern Mexico" (presentation, 5th International Wildland Fire Conference, Sun City, South Africa, May 9–13, 2011), https://www.researchgate.net/publication/322518512_Integrating_Fire_Management_Contrasting_Cases_of_Central_and_Southern_Mexico. A sequel focused on Mesoamerica is available in Dante Arturo Rodríguez Trejo and Arturo Cruz Reyes, "Avances en manejo integral del fuego en el centro de México," in *Memorias del Cuarto Simposio Internacional sobre Políticas, Planificación y Economía de los Incendios Forestales: Cambio climático e incendios forestales*, tech. coord. Armando González Cabán, General Technical Report PSW-GTR-245 (Albany, Calif.: U.S. Forest Service, 2013), 363–81. Dante offers a more personal perspective in Rodríguez Trejo, "Dante Arturo Rodríguez Trejo," 333–40. Information on Ajusco's origins as Dante conceived it comes from "Dante Arturo Rodríguez Trejo," 333–36; and from an email to me on October 4, 2022, which clarified motives not available in the published literature.
190. Rodríguez-Trejo et al., "Integrating Fire Management."
191. José Luis Romo Lozano et al., "Public Preferences and the Economic Value of Forests Affected by Fires in Mexico City," *Forest Ecology and Management* 234S (2006): S11, https://doi.org/10.1016/j.foreco.2006.08.026.
192. Dante Arturo Rodríguez Trejo et al., "Ecología del fuego en bosques de *Pinus hartwegii*," in *Incendios forestales en México*, ed. Lourdes Villers Ruiz and Jorge López Blanco (Mexico City: Universidad Nacional Autónoma de México, 2004), 103–20. On his comparison to North American fire exemplars, see Rodríguez Trejo, "Dante Arturo Rodríguez Trejo," 324–26.
193. Unless otherwise indicated, my biographical profile of Dante relies on his "Dante Arturo Rodríguez Trejo," in Pyne and Gil, *Seis vidas*; quote on p. 309.
194. Rodríguez Trejo, "Dante Arturo Rodríguez Trejo," 314, 316, 318, 316.
195. Rodríguez Trejo, 320.
196. Rodríguez Trejo, 320.
197. Rodríguez Trejo, 323–24. See also the biographical notes in Rodríguez Trejo, *Incendios forestales* (see note 57 above). Dante Arturo Rodríguez Trejo, *La lucha contra el fuego* (Chapingo: University Autónoma Chapingo, División de Ciencias Forestales / Secretaría de Agricultura y Recursos Hidráulicos, 1994).
198. Rodríguez Trejo, "Dante Arturo Rodríguez Trejo," 326–27.
199. D. A. Rodríguez-Trejo and P. Z. Fulé, "Fire Ecology of Mexican Pines and a Fire Management Proposal," *International Journal of Wildland Fire* 12 (2003): 23–37. Rodríguez Trejo, "Fire Regimes, Fire Ecology." Flores Garnica and Rodríguez Trejo, *Incendios forestales*. D. A. Rodríguez T. and L. L. Myers, "Using Characteristics of Oaks as Guides to Restoring Fire Regimes in Mexican Pine-Oak and

Oak Forest," *Ecological Restoration* 28 (2010): 304–23. Quote from Rodríguez Trejo, "Dante Arturo Rodríguez Trejo," 343. On expressing Mexico's quest for "new horizons" in integrated fire management, see Dante Arturo Rodríguez Trejo, "Nuevos horizontes en el manejo del fuego," *México Forestal* 19 (October 5, 2006).

200. Quote from Dante Rodríguez Trejo, "Fire Management Conflict Among Urban and Rural Populations and Fire Related Ecosystems in the Mexico City Forests" (unpublished report on Proyecto Ajusco).

201. Rodríguez Trejo, "Dante Arturo Rodríguez Trejo," 364.

EPILOGUE

1. See Ioan Grillo, "Mexico's Fuel Thieves Must Be Stopped," *New York Times*, March 26, 2019, https://www.nytimes.com/2019/03/26/opinion/mexico-oil.html. Quote from Germán Vergara, *Fueling Mexico: Energy and Environment, 1850–1950* (Cambridge: Cambridge University Press, 2021), 226.

   On pollution, see "Pemex Among Top 10 Global Polluters," *El Universal*, October 10, 2019; and for a graphic rendition, https://ourworldindata.org/co2/country/mexico.

2. At the time of my visit Dante was halfway through his doctoral studies at the University of Florida. We had struck up an email correspondence, and he invited me to give a talk at UACh and tour with him some of the sites of the recent burns. We were also able to visit the Cerro de la Estrella, site of the Aztec New Fire Ceremony, still bearing some scars from the 1998 fire that had burned over it.

CREDITS

Map 1. Cartography by Matt Toro.

Map 2. Cartography by Matt Toro.

Figure 1. Photo by the author.

Figure 2. FAMSI, public domain.

Figure 3. National Museum of Anthropology and History, Mexico. GNU Free Documentation License.

Figure 4. From Paul Kirchoff, Lina Odena Guemes, and Luis Reyes García, *Historia Tolteca-Chichimeca* (Mexico City: Instituto Nacional de Antropología e Historia, 1976). Internet Archive: https://archive.org/details/historia-tolteca-chichimeca/page/n1/mode/2up.

Figure 5. Graph created by the author. Data from L. B. Simpson, *Exploitation of Land in Central Mexico in the XVI Century* (Berkeley: University of California Press, 1952); and Rodolfo Acuna-Soto et al., "Megadrought and Megadeath in 16th Century Mexico," *Emerging Infectious Diseases* 8, no. 4 (2002): 360–62.

Figure 6. Graph created by the author. Data from Violeta R. Núñez Rodríguez, "A cien años del inicio de La Reforma Agraria en México," *REDPOL*, no. 9: http://redpol.azc.uam.mx/index.php/articulos-publicados/redpol-9; Emily Wakild, *Revolutionary Parks: Conservation, Social Justice, and Mexico's National Parks, 1910–1940* (Tucson: University of Arizona Press, 2011); and "UNESCO Biosphere Reserves in Latin America and the Caribbean," https://en.unesco.org/biosphere/lac.

Figure 7. From https://ourworldindata.org/co2/country/Mexico. CC BY 4.0.

Figure 8. Graph created by the author. Data from Fire Research Institute.

Figure 9. Graph created by the author. Data from CONAFOR.

Figure 10. Photo by the author.

INDEX

Note: An italic *f* following a page number indicates a figure.

1988 fires: globe, 185, 242–44; Mexico, 116*f*, 260, 270
1989 fires, 116*f*, 260, 296–97, 300–302, 305, 307, 310
1998 fires, 16, 115*f*, 116*f*, 203–4, 255, 271, 296–97, 299–303, 305, 308, 314–15, 325–26, 329, 336, 340, 345, 347; after-effects of, 272–95; global setting, 276; narrative of, 225–28; U.S. responses to, 273–75
2011 fires, 16, 210, 292, 300–303, 305, 337–38; comparison to U.S., 302

Abbey, Ed, 191, 193–94
aboriginal fire economies, 28–32; in Mexico, 29–32, 76
Acosta, Juan de, 30, 94–95, 97
Agricola, 71, 92
Ajusco, 17, 145, 147, 244, 326, 330–32, 334, 336–38. *See also* Proyecto Ajusco
Alanís, Héctor, 264
Alec-Tweedie, Ethel, 103
Alexander, Anna Rose, 145
Alta California, 69, 81–83, 100
Alvarado Celestino, Ernesto, 239, 259, 267, 294–95; profile of, 260–64
Andagoya, Pascal de, 29
año critico, 241–44
Antelá, 313
Anthropocene, 277
anthropogenic fire, arrival in Mexico, 17–18, 25–27
Apache, 29–30, 80, 140, 167, 206, 208
Arrillaga, José Joaquín, 82, 210
article 17 (1917 constitution), 244–45
article 27 (1917 constitution), 124, 244–46
Association for Fire Ecology, 291, 300, 323
Atwood, Brian, 274
Audubon, John, 104

Babbitt, Bruce, 275
Baja California, 10, 99–100, 164, 221, 225, 257, 264, 351; fire history of, 210–14; weather, 12–14. *See also* Sierra de San Pedro Mártir
Barca, Frances Calderón de la, 72, 79
Beltrán Castillo, Enrique, 173, 179–84, 192, 197, 234, 254

Big Black Book of Fire, 280, 284
biosphere reserves, 249–50. *See also individual reserves by name*
Boege, Eckart, 286–87
Bosque La Primavera (BLP), 265, 324–28
Boyer, Christopher, 144, 176, 348
Brand, Donald, 206
Bray, David Barton, 133, 246
British Columbia, 252–53, 307
Brotherhood of Prometheus, 245, 297
Budowski, Gerardo, 269
buffelgrass, 321
Butzer, Karl, 64

Calderón, Felipe, 301, 309
California, 27, 226, 237, 253, 304, 328, 340; aboriginal economy, 28–30, 82–83, 100–101; contrast to Mexico, 213–14; contrast with Baja California, 213–14; failure to control fire, 304, 340; geology, 10–11; USFS in, 183–85. *See also* Alta California; Baja California
Canada, 149, 152, 239, 259, 264, 266, 272, 275, 281, 293, 310–13, 323; assistance to Mexico, 227, 251, 296; comparison to Mexico, 19, 186–87, 286, 295, 297; conservation policy, 143, 187; in NAFC (Fire Management Working Group), 182, 184, 235, 275; national parks, 146, 149. *See also* British Columbia
Carabias, Julia, 251, 273–74, 277, 284
Cárdenas, Lázaro, 124, 133, 146, 148, 150, 162–66, 195, 197, 246
Cedeño Sánchez, Oscar, 244, 252, 274, 283, 298
Central Mexico: early fire protection, 196–99. *See also* Mexico City; Trans-Mexican Volcanic Belt; Valley of Mexico
Cerro de la Estrella, 2–4, 18, 228
charcoal, 103, 142, 196, 317, 324, 341; Chicxulub, 10–11; criticism of, 89, 143–44, 153, 158, 170, 341; as fuel, 127–29, 147–49; relationship to mines, 70–72, 94; relationship to wildfire, 152, 169, 175, 179, 181
Charles Lathrop Pack Forestry Foundation, 171, 180
Chihuahua, 16, 18, 37, 39–40, 72, 122, 132, 146, 151, 157, 164, 178, 185, 191, 193, 199–200, 206–7, 225, 237, 240, 264, 266, 322, 332
Chiapas, 11–12, 92, 215, 237, 334; 1998 fires in, 16, 225–26, 281; ejido program, 133; Highlands, 17, 78; modern fire management in, 313–17; Nature Conservancy in, 281, 291, 315–17; wildifres in, 78. *See also* Chimalpas; Mayan milpa; Reserva Biosfera de La Sepultura
Chicxulub, 9, 119
Chignahuapan, 198
Chimalapas, Los, 18, 225–26, 287
chinampa, 35, 40, 46, 48
Chiricahua Mountains, 208–9
Cicero, 94, 107
Ciudad Real, Antonio de, 77
coal, 125–26, 128. *See also* pyric transition; third-fire
COCODA, 235, 330, 334
COCODER, 235, 330, 334
Codex Magliabechiano, 313
Coe, Michael, 25, 49–50, 67
Cofre de Perote National Park, 198, 268–69
Colegio de la Santa Cruz de Tlatelolco, 92–93, 343
Colegio de Postgraduados (Chapingo), 216, 238, 262, 335, 337
Colorado State University, 266, 316
Comisión Nacional para el Conocimiento y Uso de la Biodiversidad, 227, 238, 273, 294, 302
Common Border Wildfire Protection Agreement, 275
community-based conservation, 247–49, 286
community forestry, 247–49, 251, 280–81, 317, 319

comunidad, 133, 249, 282, 287, 324
CONABIO. *See* Comisión Nacional para el Conocimiento y Uso de la Biodiversidad
CONAFOR. *See* National Forestry Commission
Conquest: ecological significance of, 57–65; fire omens of, 57; intellectual consequences, 90–98; Spanish fire practices in Mexico, 76–79
Conrad, Mike, 274
conservation: decolonization, 248; and fire, 186–88
Cook, Sherburne, 29
Cordero, Lt. Col. Don Antonio, 30
Cortés, Hernán, 48, 58, 67, 77, 83, 99, 214
Cortés Montaño, Citlali, 323, 348
Covarrubias, José, 121
Covington, Wallace, 257
Crespí, Fray Juan, 100
Crosby, Alfred, 58
Cuenca Los Ojos, 322
cultural burning, 286, 292

d'Auteroche, Jean Chappe, 99
DC (despues de Coahuila), 303
decolonization, 248
Denver symposium, 199
Departmento contra Incendios Forestales, 176
desagüe (Great Drainage Project), 155–56
Desierto de los Leones, 116*f*, 142, 146, 157, 190, 226, 326, 344–45
Díaz, Bernal, 46
Díaz, Porfirio, 106, 124–26, 130, 133, 143. *See also* Porfiriato
Dos Bocas, 119–20, 136–38, 231, 293. *See also* oil
duBois, Coert, 183
Durán, Diego, 23, 95–97
Durango, 72, 121, 133, 200–202, 206, 225–26, 235, 262

earthquakes, 11, 14, 17, 151, 154, 165, 235, 237, 352
Ecological Research Institute, 257
Ejido Pino Gordo, 39, 202–3
ejido, 62, 114*f*, 124, 144, 146, 158, 162–65, 168, 177, 192, 195–96, 210, 221, 244, 246–47, 249–50, 261–63, 282, 289, 306, 311, 316, 324–25; colonial era, 68; contemporary era, 246–47; impact on fire history, 199–208, 217; post-Revolutionary era, 132–34
El Bonito fire, 300–301
El Niño-Southern Oscillation (ENSO), 13–14, 203, 225, 257, 276, 300, 312
encomienda, 66
Enlightenment, and Mexican fire, 98–105
Enríquez, Molina, 68
Escuela Nacional de Agricultura (National Agricultural School), 178, 238, 261. *See also* Universidad Autónoma Chapingo (UACh)
EXTINGE, 235, 239, 331

Fabila, Gilberto, 161
Fábricas Papeleras de Tuixtepec, 319–20
Fernow, Bernhard, 160
fields of fire, 28, 47, 126
fire: adaptations to, 16–17; biological characteristics of, 14–15; domestication, 33; gods of, 5–6; impact of humans on, 18; Mexican characteristics, 15–19
fire ceremonies, pre-Hispanic, 3–5, 25. *See also* New Fire Ceremony
fire education, 237–38, 259–60
Fire Learning Network, 281, 291
Fire Management Working Group, 266, 296; fire research, 237, 294; meetings, 199, 253; origin, 182, 184, 188; projects, 184–85, 235, 237, 245, 275; post-Gilbert review, 241
fire policy, U.S. reforms, 188, 192, 240, 253, 271, 285, 290
fire research (Mexico), 238–40, 292–95; U.S., 238
fire science, absence of, 160–62. *See also* fire research

firewood: criticism of, 44, 71, 89, 127, 142–43, 148; fire prevention measure, 142, 152; as fuel, 72, 74, 127–28; mines, 72–74; regulations on, 47, 131, 165; traditional gathering of, 25, 49, 129, 142, 144, 147–48, 151, 158; types of wood, 38, 129, 158. *See also* charcoal
fire-based agriculture, 32–36; Mayan milpa, 40–45; Sierra Tarahumara, 37–40; Valley of Mexico, 45–47
first-fire, 232, 344
Fish and Wildlife Service (U.S.), 253, 322
Florentine Codex, 56–57, 93
Flores Garnica, Germán, 239, 293–95, 336; profile of, 264–67
Florida, 12, 226, 272, 276, 290, 336
Fondo Mexicano para la Conservación de la Naturaleza (FMCN), 238, 249, 270, 273–74, 283, 289, 303, 315–16, 337; response to 1998 fires, 279–82
Food and Agriculture Organization (FAO), 171, 182; 1998 fires, 272–73; Global Fire Monitoring Center, 185; manual on fire control, 183; Quintana Roo, 307; translation project, 236. *See also* North American Forestry Commission
forest congresses (Mexico), 158, 172, 175, 178, 197–98, 244
Forest Fire News, 184–85
forest laws: 1894 law, 153; 1926 law, 144, 157–59, 175; 1942–1949 laws, 175–76; 1960 law, 180–81, 192, 197; 1992 law, 246, 248–49; pre-Hispanic, 47
forestry: European origins, 159–62; French influence, 155–56, 161–62, 190; Mexico, 154–66. *See also* French forestry; Quevedo
Foshag, William F., 24
Fox Quesada, Vicente, 277
French forestry, influence on Mexico, 155–56, 161–62, 190
Froebel, Julius, 103
fuel. *See* charcoal; coal; firewood; oil
fuel famine, 125, 128
Fulé, Pete, 258, 336, 348

Gadow, Hans, 104
Gaitán Medina, Guillermo, 175, 189, 197
Galeote Rivera, Gilberto, 297–98, 307
Galicia, Daniel, 150
Gamio, Manuel, 65–67, 133
García Gil, Manuel, 241
Garcidueñas, Apolo R., 259
Garduño, Roberto, 197–98
geology, of Mexico, 10–12, 19
Gila National Forest, 264
Gleeson, Matt, 348
Glickman, Daniel, 274
Global Fire Initiative, 289–92, 308, 317. *See also* Nature Conservancy
Gómez, Marte R., 238
Gómez Pompa, Arturo, 268
González, D. A., 294
González Cabán, Armando, 237, 262
González Roa, Fernando, 121
Gran Chichemeca, 11, 32, 37, 70, 80
grandes incendios. See 2011 fires
Green Revolution: Alvarado, Ernesto, 261, 263; and education, 261, 263; Flores, German, 264; compared to Mayan milpa, 216–17; Mexican Miracle, 174, 191, 229; origins, 238; ranching, 209, 321, 342; relationship to fire, 135, 179, 273, 313; relationship to oil, 218, 231–32, 342; relationship to Red Revolution, 250, 266, 291; as third nature, 218, 231–32
Guadalajara, 11, 127, 154, 165, 266, 283, 324–28. *See also* Bosque la Primavera
Guadarrama, Ramón, 348
Guevara, Sergio, 62
Gutiérrez Palacio, Alfonso, 189, 259

haciendas, colonial era, 67–68
Hansen, James, 243

Hartman, Carl, 32, 36
Hermosillo, 291, 322
Hernández, Francisco, 98–99, 101
Hernández Xolocotzi, Efraím, 216–17, 262
Herrera, Gabriel Alonso de, 75, 92, 107
Huasteca, 34, 119, 136–37, 140
Huehuetéotl, 3, 5–6, 23, 25, 55, 110*f*, 259
Huffman, Mary, 316
Huixachtlan. *See* Cerro de la Estrella
Humboldt, Alexander von, 70–71, 101–2, 104–5, 127, 130
Hurricane Gilbert, 17, 242–43, 252, 260, 307. *See also* 1989 fires
Hutchins, D. E., 248

ignitions, other than lightning, 16. *See also* lightning
Incendios de vegetación, 295, 337, 339
Incendios Forestales, 259–60, 293, 295, 335
incident command system, 237, 241, 275, 296, 302, 307–8
industrial fire. *See* lithic landscapes; pyric transition; third-fire
INIF. *See* Instituto Nacional de Investigaciones Forestales
INIFAP. *See* Instituto Nacional de Investigaciones Forestales, Agrícolas y Pecuarias
Instituto Nacional de Investigaciones Forestales (INIF), 252, 264
Instituto Nacional de Investigaciones Forestales, Agrícolas y Pecuarias (INIFAP), 238, 240, 259, 262, 264, 266, 282, 293–94, 327
integrated fire management: Bosque la Primavera, 328; CONAFOR, 302; CONANP, 312, 322; Dante Rodríguez Trejo, 331–32, 336; different meanings of, 254, 281; international context, 245, 299; La Sepultura, 315, 338; within Mexico, 278, 284, 309; Proyecto Ajusco, 331–32, 336; TNC contribution, 291, 294, 308, 317, 323, 337
International Wildland Fire Conferences, 244, 253, 275
Ixtacamaxtitlán, 226–27

Jalisco, 133, 252, 261, 264–65, 281, 324–25
Jardel Paláez, Enrique, 239, 283, 287, 289, 294–95, 326–27; profile of, 267–72
Johnson, Floyd, 167
Juárez, Benito, 68

Konrad, Herman, 214–15
Koontz, Rex, 25, 49–50
Kuhnholtz-Lordat, George, 161

Laboratorio Natural Las Joyas, 240, 269–70, 272, 289, 327. *See also* Sierra de Manantlán Biosphere Reserve
La Farge, Oliver, 78
Lagunas de Zempoala, 226
Landa, Fray Diego de, 77, 97
land reforms, post-independence, 123
land use, colonial, 67–74
Las Bayas forest reserve, 202
Las Casas, Bartolomé, 29, 56, 92–95, 338
Las Marismas, 84
Leander, Birgitta, 76
Leopold, Aldo, 134, 167–68, 171, 173, 201, 257
Leopold, Starker, 142, 173, 192, 195–96, 215–16
light burning controversy, 164, 256
lightning: and 2011 fires, 300, 309; as fire, 9; as percent of fires, 16, 175, 187; dry lightning, 14, 160, 300, 309; as ignition source in Baja California, 21; as ignition source in cities, 141; as ignition source in Desierto de Los Leones, 345; as ignition source in Madrean archipelago, 322; as ignition source in Mexico, 14, 17, 138, 141, 147, 194, 198, 200; as ignition source in oil well, 136; as ignition source in Reserva de la Biosfera Mariposa Monarca, 311;

compared to people, 18, 39, 142, 202–4, 208
lines of fire, 28, 47, 122
lithic landscapes, 35, 107, 115*f*, 122, 134, 185, 199, 218, 221, 230, 303, 339, 342, 344. *See also* fossil fuel; pyric transition
livestock: colonial era, 79–81; from Conquest, 61–64; fire histories of, 204–10; Veracruz example, 83–84. *See also* ranching
Longinos Martínez, José, 100, 210
López Caballero, Abel, 178
López Mateos, Aldolfo, 180–81, 197
Los Diablos, 245
lost decade, 228–29
Luis Terrazas hacienda, 206
Lumholtz, Carl, 38–39, 103

Maderas del Carmen, 206–7
maize, origin of, 33–34
Malaspina, Alessandro, 100
Maldonado, Baruk, 348
Marana courses: 1989 fires, 241; impact on Mexicans, 239, 259, 295–97, 305, 307, 333–34; origin of, 236; propagation of, 236–37, 244, 252–53, 275, 296; syllabus, 254
Martínez Muñoz, Pedro, 317
Martínez Rodríguez, Víctor, 234
Martínez, Roberto, 235, 239, 242, 262, 274, 292, 303, 307, 336–37, 347; profile of, 295–98
Mathews, Andrew, 318, 320
matorral, 18, 205–6, 212–13, 300
Mayan milpa, 40–45, 216; supplementary practices to, 43
Meigs, Peveril, 211
Melville, Elinor, 86
Mendoza, Antonio de, 76, 89
Mesa de las Guacamayas, 208
Mesta, 62; Mexican Mesta, 63, 81, 87
mestizaje ecológico, 76
metepantli system, 86
Mexican Miracle, 174–75; impact on fire, 174–94; meltdown, 229–30; sample pyrogeographies, 194–218
Mexico City, 89, 92, 98, 100–101, 121, 127–29, 138, 171–73, 197, 211, 226, 231, 235, 237, 235, 239, 335, 343–44; air quality, 328–30; as model, 140, 143, 145, 197–99
fire issues, 140–46; surrounding forests, 156–58, 235, 330–32. *See also* Valley of Mexico
Mexico: climate, 12–14, 19; contrast with Canada, 19; encounter with Spain, 57–65; fire characteristics, 15–19; fire triangle, 27, 344; geology of, 10–12, 19; global context of, 220–21; impact of humans on fire, 18; recent status in fire, 338–42; weather systems, 12–14, 19
Michilía Biosphere Reserve, La, 202, 226
Michoacán, 127, 133, 151, 162, 170, 194–95, 251, 284, 311; fires in, 71; mines, colonial era, 70–73
Michoacana de Occidente, La, 194–95
Ministerio de Fomento, Colonización, Industria y Comercio, 127–28, 130, 132, 214
missions, colonial era, 68–70
Moctezuma, 23, 46, 50
Monterrey (California), 81, 83
Monterrey (Mexico), 12, 125, 128, 146, 152, 204, 235
Monzón, Claudia, 319
Moreno Arellano, Guillermo, 178
Morfín, Jorge, 295
Mota y Escobar, Alfonso de la, 89
Motolinía, Fray Toribio de Benevente, 76
Myers, J. G., 44
Myers, Ron, 290–92, 308, 317, 322, 336–37, 348

NAFTA, for fire, 235–37
Nanahuatzin, 1, 93
National Advanced Resources and Technology Center, 236

National Astronomical Observatory, 211, 225
National Center for Forest Fire Control, 252
National Fire Plan (U.S.), 275, 286, 291
National Forest Fires Program (Mexico), 282, 285–86, 292
National Forestry Commission (CONAFOR): 2011 fires, 301; Bosque La Primavera, 326–27; budget, 283, 323, 338; conference participant, 294; fire center (Zapopan), 282, 326–27, 339, 347; fire responsibility, 286; Martínez, Roberto, 292, 298; Martínez Muñoz, Pedro, 317; national evaluation of, 285, 290, 336; Nolasco, Alfredo, 292, 305, 308–9; origin of, 280, 282–83, 297; prescribed fire training facility, 302; relations to states, 284–85; Rodríguez Trejo, Dante, 336–37; sponsored research, 266, 295, 303; U.S. influence, 283–84. See also *servicio forestal*
National Institute for Ecology, 292
National Park Service (U.S.), 150, 192–94, 236, 240, 243, 253, 288, 290
national parks, 324; contrast to U.S., 149–50; Mexico, 114*f*, 146–51, 157, 162, 197, 211, 250, 286, 289
National Prescribed Fire Training Center, 308
National Program for Forest Fire Protection, 252
National System for Civil Protection, 237
Nature Conservancy, origins, 290; Fire Learning Community, 281; Global Fire Initiative, 289, 291, 308, 317; Myers, Ron, 290–91; Nolasco, Alfredo, 308; prescribed fire workshops, 297–98, 322; Rodríguez Trejo, Dante, 336; role, 284, 279, 299; Sepultura, 315
New Fire Ceremony, Aztec, 1–2, 111*f*, 173–74
New Laws, 66
Nezahualcoyotl, 47
Nolasco Morales, Alfredo, 239, 242, 292, 322, 337; profile of, 304–10
NOM-015 (Norma Oficial Mexicana-015-SEMARNA/SAGAR-1997), 279, 282, 292, 298, 309–10, 315, 317
Nopaltzin, 47
North American Conference on the Conservation of Natural Resources, 156
North American Fire Management Team, 274–75
North American Forestry Commission (NAFC), 182, 184, 186, 188, 199, 235, 237, 241, 252, 253, 266, 275, 294, 296. *See also* Fire Management Working Group
Nuevo León, 146, 203, 225–26, 294

Ober, Frederick, 102
Odum, Howard, 36
oil: competition with fire, 120, 129, 219, 227, 277; Dos Bocas, 119; as fuel, 147, 342; impact on society, 229, 231–32; as lithic landscape, 107, 277, 303; Mexican oil industry, 121, 126, 128, 135–38, 219; politics of, 121, 125, 246; Potrero del Llano No. 4, 136–37. *See also* pyric transition; third-fire
Old God, 3, 5–6, 19, 25, 27, 36, 48–49, 51, 110*f*, 257. *See also* Huehuetéotl; Xiuhtecuhtli-Huehuetéotl
Otomís, 29, 76, 86–87
Oviedo y Valdés, Fernández de, 29, 93–94, 99, 116, 338

Pack, Charles, 171
Parícutin (volcano), 11, 24–25
Parkinson, Northcote, 260
Parque Chipinque, 226, 235, 296
Parque Nacional Cumbres del Ajusco, 147, 331
parteaguas, 226, 303
Partido Revolucionario Institucional (PRI), 124, 230

paternalismo ecológico, 144
Pemex, 163, 218, 232, 244, 246
Peña Nevada, 203, 226
People Without Fire (Sin Fuego), 314
Pérez Salicrup, Diego, 294–95, 323, 348
Pico de Orizaba, 147, 198
Plaza de las Tres Culturas, 343
Pleistocene, 13–14, 19, 26, 55
Ponce Calderón, Laura, 314, 348
Porfiriato: economics of, 124, 128, 153, 229; view of fire, 255; forest policy of, 152–53; political challenge to, 106; role of *cientificos*, 153
Potrero del Llano No. 4, 137–37
prescribed fire: differing views of, 254, 316, 338; legal status, 279, 299, 317, 339; manuals for, 272; Nature Conservancy, 292, 322; as progressive, 310; relation to traditional knowledge, 341; research on, 240, 244, 266, 331; struggle to achieve, 199, 254, 256, 288; training facility for, 302; U.S. experience with, 271, 288, 308, 315, 336
Programa para el Aprovechamiento Integral de Recursos Naturales, 251
Program for Fire Protection, 234
Program for the Prevention and Combat of Fires, 197
Proyecto Ajusco, 330–33
Puebla, 127, 157, 159, 197–98, 211, 226, 237
Pulido, Dionisio, 24
pyric transition, 218–21, 234, 341
Pyrocene, 14, 277, 285, 337. *See also* pyric transition
pyrodendrochronology, 257

quemas controladas, 244. *See also* prescribed fire
quemas prescritas, 265. *See also* prescribed fire
Quevedo, Miguel Ángel de: Cárdenas administration, 163–66; conflicts with traditional use, 148, 162, 158, 165–66; early years, 154–55; forest law (1926), 144; French influence, 154–56, 162; role in Mexican forestry, 143–44, 153–58, 163–66; Sociedad Forestal Mexicana, 148; view of fire, 143, 148
Quintana Roo, 13–14, 17, 164, 197, 215, 226, 241, 243–44, 251, 275, 305–8. *See also* 1989 fires; Hurrican Gilbert; Yucatán

railroads, 124–27
ranching: economics of, 209; fire history, 204–10; Green Revolution in, 209; northern Mexico, 320–24
Rancho Los Fresnos, 322
Rarámuri, 37–40, 165, 191, 200, 202, 208, 258, 322
Raygoza Martínez, Arturo, 297
Red Revolution, 238, 266–67, 277–300, 310, 328, 333, 338, 344
Renaissance, and Mexican fire, 91–98
repartimiento, 66
Reserva de la Biosfera La Sepultuira, 292, 315–17, 331
Reserva de la Biosfera Mariposa Monarca, 311–13
resin-tapping, 147, 158, 165, 195
Revilla, Juan, 85
Rio de la Loza, Leopoldo, 127, 153
Rivera y Moncada, Captain Fernando, 81
Rodríguez Trejo, Dante, 242, 259, 266, 292–94, 301, 303, 308, 330–31, 345, 347–48; profile of, 333–38
Romahn de la Vega, Carlos Francisco, 259
roza-tumba-quema, 34–35, 40–41, 92, 94, 216–17, 242, 286, 313. *See also* Mayan milpa
Rzedowski, Jerzy, 239

Sabina fire, 300–301
Sahagún, Bernardino de, 2, 57, 92–93, 96, 99, 161
Salgado, Eréndira, 273
Salinas de Gortari, Carlos, 245–46, 249

San Bernardino National Forest, 274
Sanchez, Jésus, 264, 293
Sandberg, David, 237, 262
Sanderson, Susan, 134
San Juan Capistrano (mission), 69
Santiago, Myrna, 120, 130, 135, 137
Santillan Pérez, Javier, 259
SARH. *See* Secretaria de Agricultura y Recursos Hidráulicos
Sartorius, Carl, 102, 139
Sauer, Carl, 11, 31, 33, 56, 106–7, 121, 193, 211
Schlich, William, 247
Schmieder, Oscar, 77
second-fire, 232–33, 314, 323, 340, 342, 344
second nature, 24–25, 35, 107, 122, 340
Secretaria de Agricultura y Recursos Hidráulicos (SAHR), 237, 251, 264, 298, 324
Secretaría de Defensa Nacional (SEDENA), 275, 282, 301
Secretaría de Medio Ambiente y Resursos Naturales (SEMARNAT), 251, 282–83, 298, 301, 307–8
Secretaría de Medio Ambiente, Recursos Naturales, y Pesca (SEMARNAP), 226–27, 251, 253, 264, 272, 274–75, 277, 279–83, 297
servicio forestal, 158, 178, 215; 1988–1989 fires, 242–44; 1998 fires, 226–27; 2011 fires, 300–303; under Beltrán, 179–83; Comisión Nacional Forestal (CONAFOR), 282–86; comparison internationally, 159; conflicts with agrarian reform, 162, 165–66, 176–77, 318; early years, 153–57; focus on central Mexico, 156–57, 196–99; under Fox regime, 282–86; French influence on, 155–56, 161–62, 190; Junta Central de Bosques, 143, 156–57; memorandum of understanding with Mexico, 184, 237; origin, 143, 156; parks, 151; post-Beltrán era, 242, 244, 250–51; post-war era, 175–79; purpose, 163, 181–82, 198; under Quevedo, 154–66; relationship to army, 181; relationship to industry, 157–59, 190–91; Sierra Juárez experience, 318–20; U.S. connections, 156–57, 161, 184, 237, 141, 252–56, 274, 283, 301; standing in U.S., 184, 190, 238, 252–53, 340; USAID connection, 241, 274, 301; under Zedillo regime, 250–52, 274–75. *See also* Marana courses; North American Forestry Commission
Shepard, Ashbel, 102
Sierra de los Ajos, 18, 201, 240, 257, 322
Sierra de Manantlán Biosphere Reserve, 270–72, 287–89, 327
Sierra de Manantlán, 17, 235, 266, 269–71, 287–89, 308, 326
Sierra de Santa Marta, 85, 205
Sierra Juárez (Oaxaca), recent fire history, 317–20
Sierra Madre Occidental, 12, 14, 32, 34, 37, 135, 193, 200–201, 203, 257; fire histories, 199–203. *See also* Sierra Tarahumara
Sierra Madre Oriental, 12, 14, 202–4, 206–7, 257; fire histories, 203–4
Sierra Pineda, Antonio, 197, 235, 330, 334–35, 345
Sierra San Luis, 18, 71–72, 207–8, 322–24
Sierra San Pedro Mártir, 17, 331; fire history of, 210–14
Sierra Tarahumara, 37–40, 164, 191, 322
Skopyk, Bradley, 86
Smokey Bear, 184–85
solar calendar (Aztec), 3, 111*f*
Solís, Isidoro, 253
Sonora, 13, 18, 37, 72, 193, 200, 207, 230, 240, 292, 308, 322–23
Sosa, Antonio H., 159, 164–65, 190–91
Stephens, John L., 42, 78

Tall Timbers Research Station, 290–91
Tamaulipas, 12–13, 17, 221, 226, 232

Tannenbaum, Frank, 131, 134
Taube, Karl, 23
Templo Mayor, 2–4, 25, 49
Tenochtitlan, 4, 25, 46, 49, 57–58, 64, 67, 75, 91, 99, 230
Tenorio-Trillo, Mauricio, 141
Teotihuacan, 3–4, 49, 86
Texas, 12, 147, 206–7, 226–27, 238, 261, 302
third-fire, 121, 130, 230, 232–34, 314, 340–42, 344; and modernity, 218–19. *See also* lithic landscapes; pyric transition
third nature, 107, 121–22, 230, 250, 340–41, 344
Thompson, J. E. S., 43
tierra caliente, 12, 60, 102, 204
tierra fría, 12
tierra templada, 12, 61, 204
Tlaloc, 25, 113*f*
Tlatelolco Massacre, 343
Toltecs, 4, 49
Toluca, 23–24, 157
Tom Gill, 171–73, 177, 180, 182, 193
Tonatiuh, 3
Torquemada, Fray Juan de, 34, 76
towns: colonial era, 73–74; and fire, 138–41; ties to countryside, 141–45
Townsend, Camilla, 50
transhumance, 60, 62–63, 80, 83–84, 87–88, 207, 213
Trans-Mexican Volcanic Belt, 10–11, 13, 25, 196
tree-ring research, 201
Turner II, B. L., 60
Tutuaca, 200–201
Tylor, E.B., 103
Tziscao, 313

UNESCO, 250
Universidad Autonoma Chapingo (UACh), 189, 216, 238, 262, 264, 308, 330
Universidad Nacional Autónoma de México (UNAM), 238, 251, 262, 273, 293–94, 332
University of Florida, 290, 308, 335–36
University of Guadalajara, 238, 240, 262, 265, 269, 289, 294, 327, 332
University of Washington, 261–63, 281, 294
U.S. Agency for International Development (USAID), 262, 307, 337; 1998 fires, 271, 274, 280–81; 2011 fires, 301–3; agreements with SEMARNAP, 280; assessment (2004), 287; CONAFOR, 282–83; Environment Program, 283; fire suppression as foundation, 287–88; Fondo Mexicano para la Conservacion de la Naturaleza, 273, 279; Forest Fire Protection and Rehabilitation of Burned Areas, 280; impact of, 299; Marana courses, 236; Nature Conservancy, as counterweight, 291; North American Forestry Commission, 185; Wildfire Prevention and Restoration Program, 279
U.S. Forest Service (USFS), 167, 171, 248, 290, 308; 1989 fires, 241; 1998 fires, 227, 276, 280; 2011 fires, 301–3; fire policy, 193, 207, 240, 253; international forestry program, 262; in North American Forestry Commission, 184; Marana courses, 236, 307; memorandum of understanding with Mexico, 184, 237; Mexican connections, 161, 178, 184, 237, 239, 253, 283, 308, 337; standing in U.S., 184, 190, 238, 252–53, 340; USAID connection, 241, 274, 301. *See also* Marana courses
U.S.-Mexico, memoranda of understanding, 184, 236–37

Vaca, Cabeza de, 29
Valle Del Mezquital, 86–88
Valley of Mexico, 24, 87, 94, 204, 210, 258; 1998 fires, 225–26; air quality, 233, 328–30; as model, 197–99; fire protection, 141–45, 157–59, 176, 181, 196–99,

234–35, 244; geography of, 10–11; national parks, 146–47; New Fire Ceremony, 2–3; pre-Conquest history, 26, 45–47, 49; pyrodendrology, 257; urban fire in, 139–42; water problems, 154–56. *See also* Central Mexico; Mexico City; Trans-Mexican Volcanic Belt
Vásquez, Rigoberto, 172
Vazques Soto, Jesús, 199
Vega, Ricardo de la, 159
Velasco, Luis de, 65–66
Velázques de León, Joaquín, 99, 210
Vélez Muñoz, Ricardo, 236
Veracruz, 11–12, 56, 79, 81, 83–85, 119, 121, 126–35, 151, 155–57, 205, 217, 232, 304. *See also* Huasteca
Verduzco Gutiérrez, José, 169, 189, 239, 259
Verduzco Tapia, Manuel, 234, 298
Vergara, Germán, 129, 344
Villaflores (Chiapas), 317, 337
Villalobos, Víctor Manuel, 273
Viveros de Coyoacán, 143, 252
Vogt, William, 168–71, 193

Wakild, Emily, 146, 159
Weeden, Paul, 274, 384
Whitmore, Thomas, 60
Wildfire Prevention and Restoration Program (FMCN), 279
World Conservation Union, 291
World Wide Fund for Nature, 291

Xiuhcoatl, 3, 5, 23
Xiuhtecuhtli-Huehuetéotl, 3, 5, 23, 25, 55, 113*f*

Yellowstone National Park, 185, 243–44, 252, 270
Yucatán Peninsula, 9–10, 12–13, 15, 17, 34, 40–42, 44, 49, 55, 75, 97, 100, 106, 165, 252, 300, 305, 307; año critico, 241–243; Mayan agriculture in, 214–17

Zedillo Ponce de León, Ernesto, 227, 251–52, 259, 270, 277, 284, 297
Zerecero, Guillermo, 264

ABOUT THE AUTHOR

Stephen J. Pyne is currently a fire historian, urban farmer, and emeritus professor at Arizona State University. Most of his books narrate the long saga of humanity and fire. Others include histories based on exploration, a meditation on Antarctica, a cultural history of the Grand Canyon, and guides for writing nonfiction. His most recent book is *Pyrocene Park: A Journey Through the Fire History of Yosemite* (UAP, 2023).